Biological chemistry

Cambridge Texts in Chemistry and Biochemistry

GENERAL EDITORS

D. T. Elmore
Professor of Biochemistry
The Queen's University, Belfast

J. Lewis
Professor of Inorganic Chemistry
University of Cambridge

K. Schofield
Professor of Organic Chemistry
University of Exeter

J. M. Thomas
Professor of Physical Chemistry
University of Cambridge

Biological chemistry

THE MOLECULAR APPROACH TO
BIOLOGICAL SYSTEMS

K. E. SUCKLING

Department of Biochemistry
University of Edinburgh Medical School

C. J. SUCKLING

Department of Pure and Applied Chemistry
University of Strathclyde

CAMBRIDGE UNIVERSITY PRESS
Cambridge
London   New York   New Rochelle
Melbourne   Sydney

Published by the Press Syndicate of the University of Cambridge
The Pitt Building, Trumpington Street, Cambridge CB2 1RP
32 East 57th Street, New York, NY 10022, USA
296 Beaconsfield Parade, Middle Park, Melbourne 3206, Australia

First published 1980

Typeset by H. Charlesworth & Co Ltd, Huddersfield

Printed in Great Britain by
Redwood Burn Limited, Trowbridge & Esher

British Library Cataloguing in Publication Data

Suckling, Keith E
Biological chemistry. – (Cambridge texts in chemistry and biochemistry)

1. Biological chemistry
I. Title II. Suckling, Colin J III. Series
574.1'92   QH345   79-41468
ISBN 0 521 22852 2 hard covers
ISBN 0 521 29678 1 paperback

CONTENTS

FOREWORD

Living systems display an enormous and fascinating range of chemical processes; indeed, almost the entire development of organic chemistry over its first 100 years or so flowed from studies of natural materials. During this early period, there was particular emphasis on determination of the structures of isolated pure substances and gradually it was recognised by some of the pioneers that relationships existed among the structures which had been elucidated. So Nature did not produce a haphazard set of substances. There was order and it was correctly deduced that this must be a reflection of there being patterns in the biosynthetic processes. Though there was intense speculation at that time about possible biosynthetic pathways, the means (isotopes of carbon and hydrogen) which were used by organic chemists some 40 years later to discover biosynthetic routes were not then available.

However, vigorous development was occurring in other studies of living systems and the area of science we now know as biochemistry grew apace. Fundamentally though, these and other branches were growing from the same tree. Titles for the branches such as bio-, organic, bio-organic, bio-inorganic and so on tend to obscure the fact that scientists in all these areas have a basic common interest, namely, the chemistry of living systems. The danger is that one branch can stay rather isolated from another and if students reading biochemistry fail to grasp the basic principles of reaction mechanism and reactivity, they are hamstrung in their approach, for example, to enzymes and coenzymes. Similarly organic chemists who regard biochemistry as 'messy and non-crystalline' are missing exciting opportunities to enrich their studies.

This book aims, on the one hand, to help chemists see the beauty and order in the chemistry of living systems and to recognise that the processes are related to familiar reactions. On the other hand, it aims to help students of biochemistry to think in molecular and mechanistic terms. Such a weaving of these various threads is not an easy task and this book, written by two authors whose experience and interests span the various branches, is therefore particularly welcome. There will be few readers whether from chemical or biochemical backgrounds who do not benefit from it.

University Chemical Laboratory
Cambridge

A. R. Battersby

PREFACE

Many modern scientific and technological problems require contributions from scientists trained in several different disciplines. Nowhere is this more apparent than in the interaction of the life sciences with chemistry, where the chemist must understand the nature of the biological system that is under study and his biologist colleague must be conversant with the language and logic of chemistry. Biochemistry courses can easily underestimate the value of the molecular approach: similarly there is little room in chemistry courses to give students a feeling for the nature of the problems in the life sciences which require a contribution from them.

In this book we attempt to guide students of both chemistry and biochemistry through the common ground of their subject in what we believe is a novel way, so that they can appreciate fully the potential of the molecular approach in the life sciences. To achieve this we have concentrated more on communicating patterns of thought rather than on presenting a comprehensive account of modern biological chemistry. We have found this approach to be successful in teaching both chemistry students in Strathclyde and biochemistry students in Edinburgh and it also seems to be popular with students.

Our discussion develops from a principally chemical beginning to consideration of biological systems of increasing complexity, and we attempt to show the power of molecular thinking in all these systems. Thinking in molecular terms requires practice, and we have included problems designed to help the student acquire this desirable ability. Many have been tested in classes in Edinburgh and Strathclyde. Answers will be found at the end of the book or in the references cited in the problem.

We have included an outline of some central metabolic pathways in Appendix 1 to help the student not already familiar with these. Since mechanistic ideas are frequently expressed using the curly arrow formalism we have also included a note on their use (Appendix 2). Mrs Barbara Stewart and Miss Helen Scott made a valuable contribution to the typing of the manuscript and Professor D. T. Elmore and Mr Andrew Scott are to be thanked for their critical reading of the earlier drafts. We would also like to thank the editorial staff of Cambridge University Press for their help in producing the volume which you now have before you.

Keith Suckling
Colin Suckling

1   Chemistry and nature

1.1.   Introduction: our aim

Chemistry is the fundamental science that deals with the properties
and behaviour of molecules. It is a mature scientific discipline in that its
axioms have been substantiated by experiment and can also be derived mathematically by rigorous theoretical methods. In the past, the chemist has practised
his art upon relatively small molecules, and both theory and experiment with
these systems have led to the establishment of a number of general chemical
principles. For instance, up to 1950 molecules of molecular weights greater
than several hundred were of little interest to the chemist; synthetic polymer
chemistry was in its infancy and natural-product chemists were only concerned
with molecules smaller than chlorophyll. Nevertheless, within this range
chemistry developed into a more or less self-consistent body of experimental
fact and related principles. If these principles are any good, they should also
allow the chemistry of much larger molecules and molecular aggregates, such
as are common in biochemistry, to be described.

We want to show you how a thorough understanding of the principles of
chemistry derived from and applied to biochemistry allows us to say a good
deal about what goes on in nature and to produce socially useful end results.
If you are trained as a chemist, we shall try to demonstrate how to extend
experience of simple reactions to understand complex biochemicals with
many interactions both within and without the molecule. If you are a biochemist by background, we hope to convince you that disciplined chemical
thought can help you in a more penetrating study of your subject.

We begin by tidying up our ideas about the elementary principles of organic
chemistry and see in general terms how they relate to biologically important
molecules.

1.2.   Systematic organic chemistry

One of the greatest difficulties for students of organic chemistry has
always been the vast quantity of experimental information pertaining to the
subject. To assimilate such material and to apply the knowledge, systematisation

of organic chemistry has been essential. Traditionally, since the early nineteenth century systematisation has been built around the recognition that the same **functional group** in one compound undergoes more or less the identical reactions to the same functional group in another compound; conversely, different functional groups behave differently under the same conditions. Thus hydroxylamine, for instance, forms an oxime with both the simple aldehyde acetaldehyde, and with the complex aldehyde retinal, but the carboxylic acids, acetic (simple) and glucaric (complex) form salts (Figure 1.1). We usually take it that a hydrocarbon chain is inert to most reactions (a paraffin derivative) and concentrate upon the behaviour of the functional group. What is 'functional' depends to some extent upon the reaction in question.

Having classified our compounds according to functional groups, we can say a good deal about the sort of chemistry that we would expect, provided that we know a few simple facts about representative members of each class of compounds. Acetic acid, for example, could be our prototype carboxylic acid and we can predict the behaviour of undecanoic acid, say, simply by *analogy* with acetic acid, at least as far as the carboxylic acid group is concerned. If you are not fluent in the typical reactions of functional groups, you

Figure 1.1.

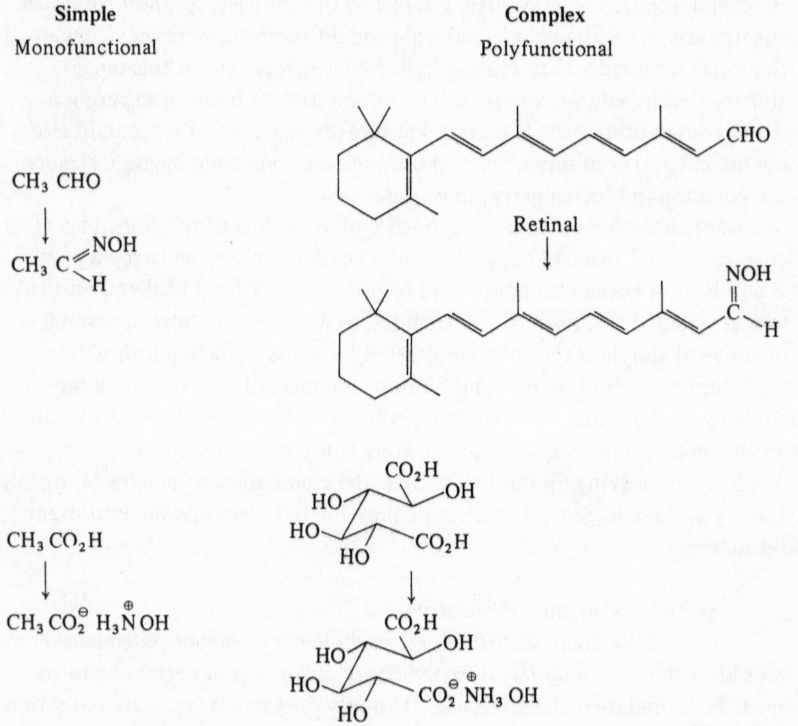

Simple	Complex
Monofunctional	Polyfunctional

should consolidate your knowledge by referring to a conventional organic textbook as you go along. Morrison & Boyd (1973) and Allinger *et al.* (1976) are recommended.

What can be said about cases like retinal, where in addition to the aldehyde there are also a number of carbon–carbon double bonds? How can we understand the chemistry of a polyfunctional compound such as this? Obviously it will show some of the typical reactions of each functional group independently, but it is also likely that some of the functional groups may interact with each other. The electronic theory of organic chemistry provides a rationalisation for the interaction of functional groups. It can describe in more or less pictorial terms why each functional group behaves as it does, and uses the same arguments to develop an understanding of how the functional groups affect each other (Suckling, Suckling & Suckling, 1978). The language used frequently refers to **effects**, inductive effects, resonance effects and steric effects, that change the properties of the functional group. These effects have been singled out chiefly as a result of experimental evidence, but they find some support in quantum-mechanical calculations. If you understand how to manipulate these effects, you can systematise much further information within them. For example, an electron-withdrawing effect exerts an acid-strengthening influence. Furthermore, you can make logical predictions about the behaviour of complex molecules or you can rationalise their observed behaviour. Such predictions are naive by the standards of modern computational chemistry but usually they are remarkably successful.

A third useful classification exists; reactions can be grouped together according to common features of **reaction mechanism**. These reaction types overlap with systematisation by functional groups and electronic theory because they employ concepts and methods from both. Thus, the above examples of aldehydes and carboxylic acids (Figure 1.1) illustrate a condensation reaction and the formation of a salt respectively. It is possible for condensation reactions to take place with esters, for example, and for salts to form with phosphoric acids as well as carboxylic acids. Further, we shall come across cases where we are not dealing with a simple recognisable functional group, although we can define the class of reaction occurring. Chapters 3 to 8 are built around relatively few single reaction types, but there is another equally important reason for this organisation. It is that we can bring the cohesive influence of reaction mechanism into our systematisation.

A reaction mechanism is best regarded as any description of the course of the reaction, however sophisticated or naive this description may be. It is a model of a reaction that guides us to make predictive leaps in the dark. Manipulating mechanism as an intuitive tool of chemists is very similar to manipulating electronic theory and we hope that you will develop a valuable facility in both as you proceed with this book.

1.3. Structural formulae and molecular models

When we think of a functional group, we bring to mind not only its reactions but also its structure. The crux of chemical argument is to relate the structure of a compound to its chemical properties and vice versa. To communicate structures between chemists it is customary to employ drawings or structural formulae. These pictures can be considered as stylised models of the molecules in question: they are constructed within the limits of fairly strict conventions. Structures symbolise not only molecular shape, but also chemical behaviour.

Structural formulae have limitations. In particular, it is necessary to employ conventional projections to represent the three-dimensional structure of molecules on a flat page or blackboard. One way round this is to use molecular models that allow us to hold and manipulate a more readily acceptable representation of a three-dimensional system (Mislow, 1966). All students of chemistry should be familiar with such models. Living beings, too, are three-dimensional and stereochemistry, the understanding of which is greatly aided by molecular models, is an essential part of the chemistry of living things. Molecular models also have their limitations, but it will divert us too much from our aim to consider them at length now. Where the representation of a compound shows an important feature, we shall point it out. Let us meanwhile bear in mind the essence of communication through structures as distilled by Robinson who as early as 1917 wrote concerning arguments over the structure of the alkaloid morphine (Robinson, 1917a): 'This formula [Figure 1.2] has been adopted especially since the formulae which it is suggested should replace it cannot without hesitation be accepted as superior representations of the properties of the substance.'

1.4. How organic chemists became interested in biochemistry

For many reasons, some commercial, some scientific and some philosophical and pedagogic, molecular science has been tending to ignore interdisciplinary boundaries in the last two decades. It is now common to

Figure 1.2.

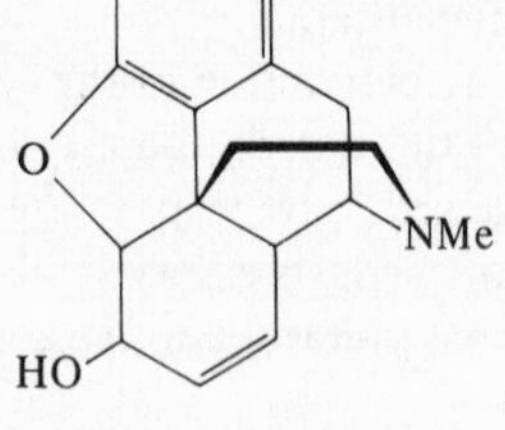

Morphine

hear scientists described as hybrids of conventional disciplines, such as the bioinorganic chemist. Some hybrids have become an established and prolific strain, molecular biologists for example. But before about 1940, it was unusual for a scientist of one discipline to pay more than polite passing attention to a colleague's work in a different but related discipline. The eminent German organic chemist Hans Fischer, whose life's work was expended upon the chemical synthesis and structure determination of the important naturally occurring compounds porphyrins, typified the situation when he wrote in 1937 'So far, we can only speculate upon the physiological significance of porphyrins in the plant and animal kingdoms' (Fischer & Orth, 1937). However, some twelve years earlier, Keilin described the identification of the porphyrin-containing proteins, the cytochromes; Fischer appears to have been unaware of them.

In the nineteenth century, disciplines were not so conventionally defined and men of genius, like Pasteur, were able to embrace in their careers all fields of natural science from physics to microbiology, and to contribute to each. The whole of modern chemical and biochemical stereochemistry is foreshadowed in Pasteur's work on optical isomerism (Robinson, 1974), not so much in its scientific precision and elegance but principally in the vision with which he assesses the significance of his results.

The twentieth century too had its prophets. Robinson, an organic chemist, interpreted the probable biosynthetic origins of alkaloids from the structures that he had deduced for them chemically (Robinson, 1917a). He demonstrated that the alkaloid tropinone could be prepared in the laboratory from the probable biological reactants under mild conditions (Robinson 1917b); (Figure 1.3). These early achievements were not surpassed until the skill of chemists had developed further over thirty years. Eventually (1945), it became possible to test the ideas of Robinson and others concerning biosynthesis by the use of radioactive tracers. Also, chemists learned to synthesise many of the complex naturally occurring molecules and began to study their behaviour in terms of chemical mechanism both *in vivo* and *in vitro*.

Technique and methodology, therefore, in chemical and biological sciences steadily reached a maturity that was compatible and apt for marriage. At the same time, the intuitive and theoretical concepts of organic chemistry matured,

Figure 1.3. A simple biogenetic-type synthesis.

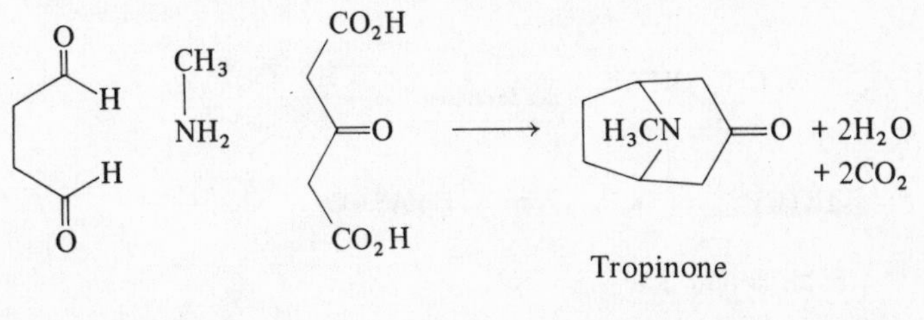

Tropinone

and chemists began to set themselves more severe tests posed by the complex molecules of nature. It was natural for organic chemists to be the first to be influenced by biochemists but, today, a chemist of any specialisation will readily apply his expertise to a biological problem. Complex problems in molecular science can now be approached with a formidable array of precedent, principle and perception that is remarkably self-consistent.

1.5.   Chemistry assimilates larger and more complex systems

Our knowledge of the biosynthetic pathways of organisms is chiefly the result of the combined efforts of chemists and biochemists during the two decades up to 1965. This was the classical era of biosynthesis (Bu'lock, 1965; Hendrickson, 1965) and we shall draw on it for examples.

A chemist confronted with this large body of information will perceive that in reality there is less to learn than at first sight seemed the case. Most of the transformations that biosynthetic studies have discovered are readily absorbed into the existing body of systematic chemical fact. For example, mevalonic acid, the key intermediate of steroid and terpene biosynthesis, is built up in nature from acetic acid in a series of steps. The individual steps are analogues of the chemically familiar reactions, the Claisen and aldol condensations (Figure 1.4).

Relatively few organic reactions simply transform A into B without the intervention of a catalyst; Claisen ester condensations, for example, require an alkoxide ion as catalyst. Uncatalysed reactions occur in both chemistry and biochemistry, but are especially rare in the latter. They are often spontaneous decarboxylations or dehydrations, usually occurring at room temperature in the laboratory. Most other reactions require more extreme conditions easily obtainable in the laboratory but not available to living systems. Reactions in living organisms must, of course, be carried out under the mild ambient conditions prevailing in the organism. This means that catalysis is almost invariably essential, and it is provided by enzymes.

Figure 1.4.

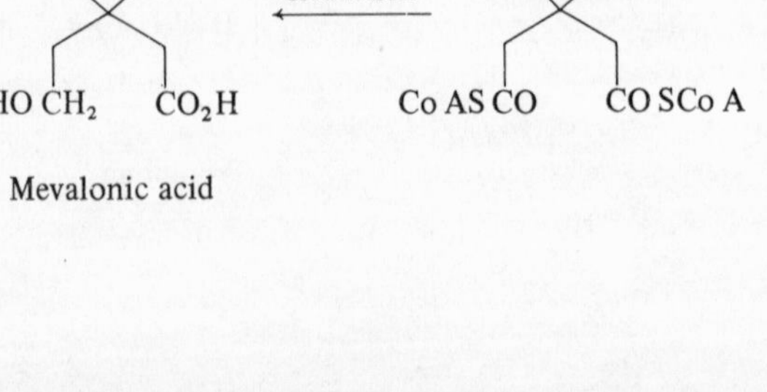

$$2\ CH_3\ CO\ SCoA \xrightarrow{\text{`Claisen'}} CH_3\ CO\ CH_2\ CO\ SCoA$$

'Aldol' | $CH_3\ CO\ SCoA$

$$\xleftarrow{\text{Reduction}}$$

Mevalonic acid

Enzymes are proteins whose molecular weights range from 10^4 to 10^6 and they catalyse metabolic reactions very efficiently. Chemists have been greatly intrigued to discover how enzymes achieve their catalytic efficiency under such mild reaction conditions. Accordingly, an immense research effort has been expended upon chemical, biochemical and crystallographic methods aimed at determining both the chemical structures of enzymes and the relationship of their structures to the mechanism of action. Both are exceedingly complex problems which can be solved satisfactorily only by using a combination of techniques. When answers are available, the organic chemist wants to see whether they fit in with his model of chemical behaviour. By and large, enzyme-catalysed reactions can be understood using modern organic chemical logic, especially if attention is concentrated upon the **substrate** of the enzyme.

Much has been written (Gray, 1971; Bender & Brubacher, 1973), about the mechanism of action of enzymes, but attention has usually focused on the enzyme through a physical approach to the organic chemistry involved. This approach is necessary, although a great deal must be known about the basic chemistry of the system before the often involved arguments can be appreciated. The topic then becomes one only for advanced undergraduate and postgraduate study.

On the other hand, if we consider the substrates, we can discuss the new chemistry and the related biochemical consequences at the same time. It makes little difference whether the substrate is attacked by a hydrolysing enzyme or by hot caustic soda solution, since its chemical properties govern what reactions may occur. By use of the qualitative but powerful logic of chemistry, we introduce new classes of compound as developments of simple aliphatic and aromatic functional-group chemistry. We can then study the implications of the chemistry in the biological field as a direct consequence of the inherent chemical reactivity of the substrate.

1.6. A chemical approach: amide hydrolysis, chymotrypsin

When a chemist plans a reaction, he examines the structure of the starting material and assesses its inherent chemical reactivity. He then selects the reagents and the reaction conditions that are most likely to give the desired result. Many attempts may be necessary in which the conditions are successively modified to achieve improvements. Nature has had more experience than chemists in handling molecules, and has developed in the course of evolution enzymes that are very efficacious. So precise is the relationship of an enzyme's chemical function to its substrate that a study can reveal aspects of the chemistry of the substrate that may well have lain hidden even to a most thorough-minded chemist.

Consider the hydrolysis of an amide bond as a simple example of what we can find out. What can chemistry tell us? Experimentally, if we simply compare the conditions required to hydrolyse an ester and the corresponding amide, we discover that the ester is more easily hydrolysed. An organic chemist would rationalise this by explaining that the contribution of the canonical form **B** (Figure 1.5*a*) to the resonance hybrid is much greater in the case of the amide because nitrogen is less electronegative than oxygen and therefore more readily donates its electrons. This fact is also apparent in the infrared stretching frequencies of the carbonyl groups. Typically an amide absorbs at about $1680\ \mathrm{cm^{-1}}$ and an ester close to $1740\ \mathrm{cm^{-1}}$, this reflects the weaker carbon–oxygen double bond in the amide, again a consequence of the large contribution of canonical form **B**.

Hydrolysis occurs by attack of a nucleophile (water or hydroxide ion) upon the carbon of the carbonyl group of the ester or amide. The resonance effect in the amide means that the attacking nucleophile senses a much weaker partial positive charge on the amide carbonyl group than the ester and, accordingly, more energy is required to form the intermediate adduct (Figure 1.6). Hydrolysis of amides and esters thus are examples of nucleophilic addition to carbonyl groups.

There is one other important consequence of the resonance interaction in amides. The canonical form **B** represents the donation of the lone pair of the nitrogen atom to interact with the p-orbital of sp^2-hybridised carbon in the

Figure 1.5. Bonding in amides and esters.

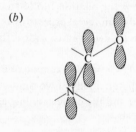

carbonyl group. The most stable situation will be reached when the orbital overlap is at its greatest, and this occurs when oxygen, carbon and nitrogen are all effectively sp^2-hybridised and lie in the same plane (Figure 1.5b). This simple piece of chemistry has very significant implications concerning the three-dimensional structure of proteins and polypeptides.

Both ester and amide hydrolysis are acid/base catalysed. We can easily understand qualitatively how this comes about. In aqueous base, negatively charged species such as hydroxide ion are present, and they are obviously better nucleophiles than dipoles like water. In acid, the substrate becomes protonated on the carbonyl group and this leads to an increase in the effective positive charge of the carbonyl group. Attack by a nucleophile is consequently much easier.

This is the state of affairs represented by a vast body of chemical precedent (O'Connor, 1970). Knowing the normal, we are then well placed to recognise

Figure 1.6.

Basic solution

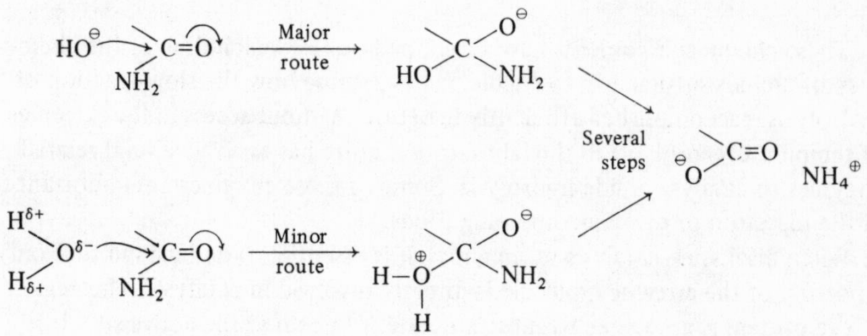

Acidic solution

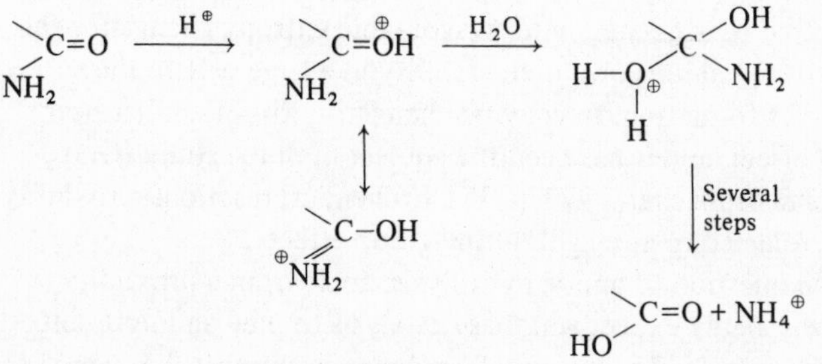

the unusual. Consider the more complex amide represented by the bicyclic structure in Figure 1.7 (Pracejus, Kehlen, Kehlen & Matschiner, 1965). The geometry of the aza-[2,2,2]-bicyclooctane system prevents the nitrogen, oxygen and carbon components of the amide lying comfortably in the same plane (make a model to convince yourself). Hence the resonance interaction that has such important chemical consequences in amides is destroyed. The i.r. stretching frequency of the carbonyl group of this amide is extraordinarily high at 1762 cm^{-1}, and this confirms that the amide is a special case. We can predict that this amide hydrolyses very much more easily than either an ester or a normal amide in a polypeptide chain.

Figure 1.7.

$$\nu_{C=O} = 1762 \text{ cm}^{-1}$$

These chemical arguments have developed our understanding of the chemistry of amides sufficiently to enable us to examine how the slow reaction of hydrolysis is accomplished efficiently in nature. Without access to the extremes of temperature available in the laboratory, nature has evolved several related enzymes to catalyse amide hydrolysis. Some of these enzymes are important in the digestion of protein-containing foods.

When discussing catalysis by enzymes, it is essential to understand that only a portion of the enzyme molecule is directly involved in catalysis. This region of the protein is accessible to substrates and is known as the **active site**. It is composed of the side-chains of some of the amino acids that make up the polypeptide chain of the enzyme, and the functional groups of the side-chains together with bound substrate and solvent molecules (water) are the reactants in the biological reaction. This molecular proximity gives the enzyme a big catalytic advantage. Effectively, the enzyme reaction then becomes intramolecular, a property that is associated with a favourable entropy change for the reaction. Chelates of metal ions owe their stability in a large part to the favourable entropy of formation. In enzyme chemistry, this effect has been called a proximity effect, and it has been the subject of much study (Gray, 1971, p. 28; Bender & Brubacher, 1973, p. 37). In chemical reactions, proximity effects are usually referred to as neighbouring-group effects.

Returning to the question of amide hydrolysis, apart from a proximity effect, we might reasonably expect acid/base catalysis to play an important role in the enzymic reaction. There is much evidence to support this view, particularly from kinetic studies (Gray, 1971, p. 52; Bender & Brubacher,

1973, p. 61). However, from the point of view of the substrate some deductions based upon X-ray crystallographic studies and work with inhibitors suggest that in the case of the enzyme chymotrypsin, the substrate amide is bound at the active site in such a way as to coerce the amide bond out of its favoured planar configuration (Powers, Baker, Brown & Chalm, 1974). We know from the properties of the bicyclic amide (Figure 1.7) that such a distortion will lead to very much easier hydrolysis.

The mechanisms drawn in Figures 1.5 and 1.6 use chemical symbols, including several types of arrow. All of these are well-defined and their usage should be scrupulously restricted to the correct circumstances:⟶ shows a reaction step,⟷ links canonical forms and⤳indicates the movement of an electron pair from tail to head of the arrow. Consult your textbook and Appendix 2 if you are hazy on this topic.

1.7. Developments of the chemical approach

The hydrolysis of amides illustrates the insight that can be gained by thinking about the chemistry of the substrate, and it introduces us to a number of other important subjects for discussion. Spectroscopy and crystallography were used in this example to provide detailed information about the molecules.

The idea of deliberately designing simple organic molecules to reproduce some of the features of enzyme catalysis or the reactions of coenzymes was developed during the 1950s with the aim of placing the newly discovered biological reactions upon a firm chemical footing. We shall look more closely at model compounds, both their good and bad points, in Chapter 3, but for now it is enough to say that there exists a laboratory analogy (model) for almost every biological reaction. Such models are not merely formal paper analogies, but are closely related to their natural prototypes in terms of detailed chemical mechanism.

One of the attractions of studying the science bordering on chemistry and biochemistry is that it cuts across disciplinary boundaries. This fact is equally true within the conventional field of chemistry, because inorganic chemists have adopted the same sort of approach in studying those natural reactions that are mediated by enzymes or coenzymes containing chelated metal ions Some examples are discussed in Chapters 3 and 9.

The prophetic work of Robinson mentioned earlier has been developed recently into an exciting area of chemical research. Now the argument is put the other way: if we understand some of the chemical processes of nature and we have discovered some novel things, can we not therefore design new chemical procedures based upon some of the tricks of nature? Some spectacular successes have recently been achieved with polycyclic steroids and terpenes following reactions found in nature (Chapter 5). However, a chemist hopes to learn from nature how to match the available catalytic and other chemical interactions to the reactants in order to achieve conveniently the required transformation in optimum yield. In fact a synergic state has currently been reached, where a perceptive chemist can learn much about the behaviour of small molecules from studies of large complex molecules like enzymes. Of course a better understanding of chemistry necessarily improves an understanding of biochemistry.

1.8.   Topics for discussion

Since one of the prime aims of this book is to develop your chemical intuition and understanding, we do not present an exhaustive survey of biological chemistry or of enzymic catalysis. These can be found elsewhere (Gray, 1971; Bender & Brubacher, 1973; Metzler, 1977; Scrimgeour, 1977). We have selected some of the common chemical reaction types and have associated them where appropriate with a major technique which has been important in that case. Each chapter is concerned with the study of one type of reaction and some related experimental techniques, but you will be encouraged to find interrelationships between the various sections and to see for yourself how different techniques and arguments may be applied more widely through suggested reading, problems and structured cases for further study.

What types of chemistry are to be expected and what techniques are important? Natural products contain representatives of most structural types, including the less usual cyclopropanes, cyclobutanes, polyacetylenes and organometallic compounds. Almost every field of known chemistry has a bearing upon biological chemistry. Similarly, we can use a great many techniques. Kinetic studies are valuable in both chemical and biochemical reactions because they reveal what molecules are involved in rate-determining steps. In some cases, the key to the problem lies in stereochemical studies through optical or enzymic methods, often with the aid of isotopic labels. Always we shall endeavour to adhere to the accepted body of chemical knowledge, either using simple analogy or, if no analogy is available, fitting in the missing links in the logical chain of reasoning with specially designed model compounds.

A structured problem: lysozyme

To get your ideas moving, work through the steps of this problem as they are presented. They will lead you to an understanding of some aspects of the mechanism of action of the enzyme lysozyme (systematic name *N*-acetyl-muramide glycanohydrolase EC 3.2.1.17)*. Lysozyme catalyses the hydrolysis of certain amino-polysaccharides which are constituents of the cell walls of bacteria. A section of the substrate is shown in Fig. 1.8.

(*1*) Notice that the functional linkage that is hydrolysed is an acetal.

(*2*) Write down the mechanism of acid-catalysed hydrolysis of the simple acetal (1.8.2), paying particular attention to the charged intermediates. If you cannot do it, look up your text book.

---

*The EC number is taken from a systematic numerical catalogue of enzymes according to the reaction they catalyse.

(3) Consider the same process in the case of a methyl glucoside (1.8.3): this cyclic acetal is similar to the polysaccharide substrate of lysozyme. Again concentrate upon the charged intermediates. (4) If lysozyme is to be a catalyst, it must somehow lower the energy barrier (activation energy) for the hydrolysis reaction. This it does by stabilising (lowering the energy) of one of the charged intermediates which are special kinds of **carbenium ions**. This can be shown on a simple reaction profile (1.8.4).

Figure 1.8. The mechanism of hydrolysis catalysed by lysozyme: a problem.

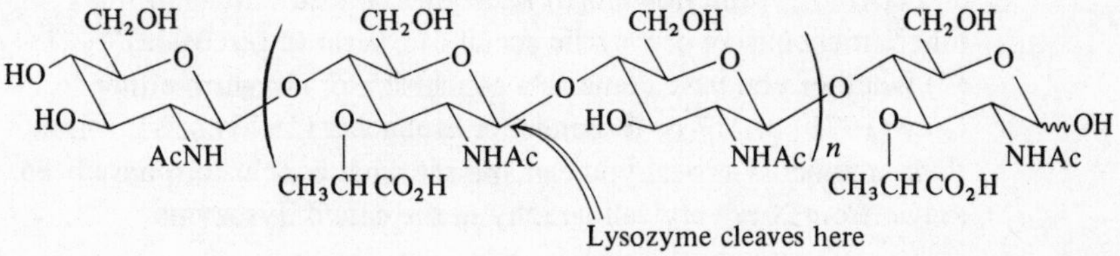

Lysozyme cleaves here

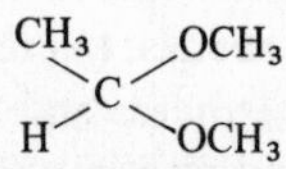

1.8.2 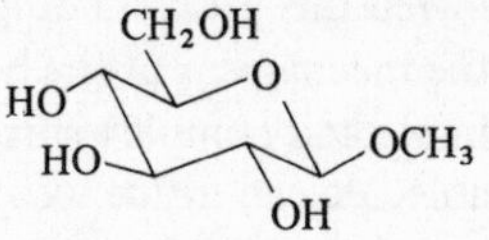   1.8.3

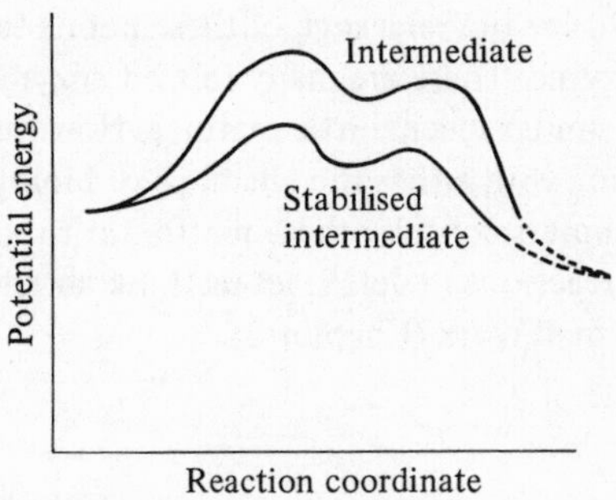

Reaction coordinate

1.8.4

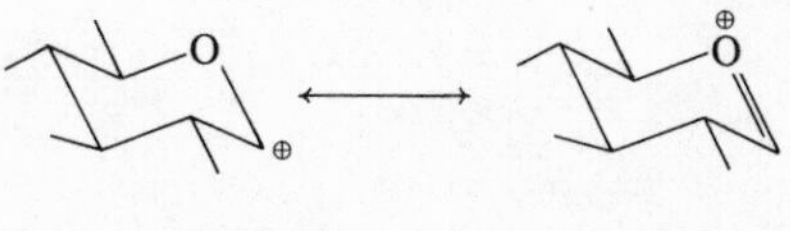

1.8.5

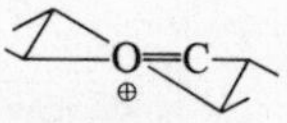

Skew boat

(5) What is the normal arrangement of the orbitals of carbon in a carbenium ion? Would you expect the charged intermediate from the glucoside to have this configuration of orbitals – if so why? If not, how could the molecule change its conformation to increase its stability? It is drawn here in the chair form. Note that none of the other functional groups need be considered at this stage, and they have therefore been omitted from the drawings (1.8.5).

(6) You should have argued that the intermediate ion in the hydrolysis of a cyclic acetal is more stable in the skew boat conformation than in the chair. Remembering what we said about amide hydrolysis and chymotrypsin, how might lysozyme take advantage of this inherent chemistry of a cyclic acetal of a sugar (a glycoside)?

(7) Whether you have come to a conclusion or not, turn either to Gray (1971) p. 217 or to Bender & Brubacher (1973) p. 81, follow their arguments as best you can and see what conclusions have been drawn from X-ray crystallography in the case of lysozyme.

If you have solved this problem diligently, you should now: (a) understand the mechanism of the hydrolysis of acetals; (b) recognise an acetal in both simple and complex surroundings; (c) understand how the geometry of a molecule can influence its stability and reactivity; (d) appreciate that the geometry of a molecule can be manipulated by an enzyme, thereby contributing to catalysis; (e) appreciate that we can learn much about chemistry and biochemistry by a careful examination of the chemistry of the substrate. In particular, you should see the relevance of these points to catalysis by chymotrypsin and lysozyme. There are many related enzymes (trypsin, the glucosidases) that have similar mechanistic features. However, you may have found difficulty in coping with either the chemical or biological aspects of this chapter. Subsequent chapters consider these matters at a more leisurely pace, but before considering reactions in detail, let us think about the biological systems with which we shall work (Chapter 2).

2   Biological chemistry and the character of biological systems

2.1.   The hierarchy of biological systems

Many students of biochemistry and biology are frequently sceptical about how far a specifically chemical and molecular approach can be applied to living systems. Perhaps this scepticism is based on subconscious remnants of the theories of vitalism, and it is certainly true that it becomes harder to use a purely molecular description of a living system as one examines more and more complex systems. However, molecular reasoning can help at all levels, and we hope to show the power of this reasoning in biological sciences. There is hardly a field in modern biology in which biochemical techniques and concepts are not used. An understanding of the molecular basis of these concepts and the ability to handle them is, therefore, essential for all workers in the biological sciences.

Chemists, on the other hand, are often bewildered by the variety and complexity of living systems. As in all mature sciences, biochemistry contains many unifying principles. In this chapter, we aim to illustrate how molecular questions can be asked of the many types of system studied by biochemists and to show that by carefully defining the system under study much of the apparent imprecision disappears and the questions which remain to be asked are left in sharper focus.

In order to make progress in understanding any apparently complex system every scientist accepts the need to make simplifications. In fact he will attempt to define quasi-independent systems which can be examined in more detail by the techniques and concepts available to him. The stepwise approach of isolating successively simpler quasi-independent systems is nowhere more apparent than in biochemistry (Suckling, Suckling & Suckling, 1978). The range of systems available to examine includes a complete organism, perhaps including its environment, and continues through specific organs and tissues to complete cells. The cells themselves can be fractionated to give various types of organelles and subcellular fractions, and from these fractions enzyme systems can be isolated and individual proteins with one catalytic function purified. All these systems can be described to some extent in molecular terms.

Progress in biochemistry has always been dependent on the techniques available. The rate-limiting step in biological research is frequently not conceptual but technical. Often the biological concepts reflect the background of the technique by which they were studied, and are expressed not in a biological language but in a language appropriate to the particular experimental method. We discuss the physical and chemical basis of biochemical techniques in Chapter 11.

Biochemistry aims to answer two general questions about living systems in order to understand their function in molecular terms. Firstly, biochemists want to know the chemical structures of the compounds which they are examining. Then they try to understand how the chemical structure is suited to its biological function and how this function is performed. This second stage involves combining results from related systems in order to describe a larger biochemical unit.

Let us examine the biochemical systems that can be isolated and defined for study. The hierarchy of systems increases in size and complexity from single molecules to whole organisms. The biochemist must be familiar with the characteristics of all of them and be able to move from one level of the hierarchy to another without confusion. The hierarchy of biological systems is summarised in Table 2.1.

Some of the subcellular components described in the table may be unfamiliar to readers with a chemical background. Figure 2.1 is a schematic drawing

Table 2.1. *Biological systems*

System	Size	Molecular weight	Important chemical and biological phenomena studied
(1) Model systems	Various, often small molecules		Mechanism of catalysis. Chemistry of incompletely understood biological structures.
(2) Single small molecule	0.5–1 nm	50–250	Chemistry of functional groups. Behaviour in different biological phases.
(3) Macromolecule, e.g. an enzyme	4–7 nm	10^3–10^6	Structure and function in cell. Mechanism of catalysis. Control of activity.
(4) Enzyme complex		10^6–10^9	As for 3.
(5) Membrane	About 9 nm across		Structure of protein and lipid components. Their relationships to each other and to the phases on either side of the membrane.
(6) Organelle	1–5 μm (nuclei, mitochondria)		Structure and integration of enzymic activities. Function with respect to the rest of the cell.
(7) Cell	1–20 μm		Integration of 2–7 into a unified picture of activity. Specialised functions and their relations with other cells and tissues.

Fig. 2.1. The 'average' eukaryotic cell. This composite drawing shows the principal organelles of both animal and plant cells, approximately to the correct scale. Abbreviations: BM, basement membrane; ER, endoplasmic reticulum (rough, with ribosomes attached; smooth ER is depicted nearer the nucleus and on the right side of the cell); Dl, deep indentation of plasma membrane; Gl, glycogen granules; Gap, space *c.* 10–20 nm thick between adjacent cells; M, mitochondrion; Mb, microbody; L, lysosome; D, desmosome; SG, secretion granule; Tj, tight junction; Mv, microvilli; C, cilium; G, Golgi apparatus; V, vacuole; CW, cell wall (of a plant); Ct, centrioles; P, plasmodesmata; N, nucleus; Nu, nucleolus; Cp, chloroplast; St, starch granule. Adapted from a drawing by Michael Metzler; reproduced from *Biochemistry: Chemical Reactions of Living Cells* by D. E. Metzler (Academic Press).

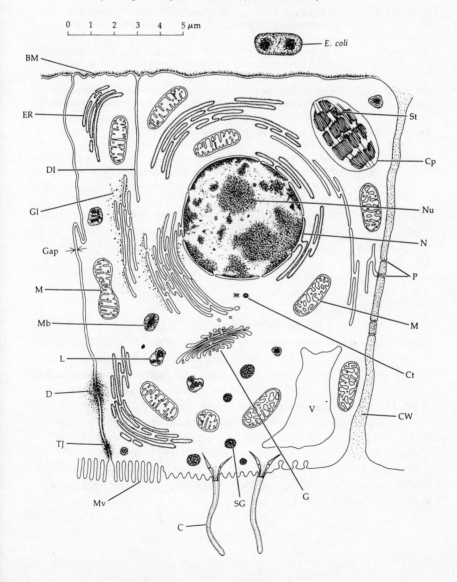

Table 2.2. *Some eukaryotic cell organelles discussed in this book*

Nucleus	Bounded by nuclear membrane. Contains DNA, the genetic material complexed with basic proteins known as histones. Histones and DNA organised in chromosomes 4–6 μm in diameter. Synthesis of messenger RNA takes place in nucleus. Nuclei are absent in prokaryotic cells (e.g. bacteria).
Mitochondrion	Separated from the cytoplasm by a double membrane. The two membranes have distinct functions, the inner membrane being associated with the processes of electron transport and oxidative phosphorylation (ATP production). Reactions of the citric acid cycle and β-oxidation of fatty acids take place in the matrix of the mitochondrion. About 1 μm in diameter.
Endoplasmic reticulum	A single-membraned system forming channels throughout the cell. Contains enzymes associated with the metabolism of drugs (liver) and sterol synthesis.
Ribosomes	Particles on which protein synthesis takes place. Frequently associated with endoplasmic reticulum (association known as rough endoplasmic reticulum).
Golgi apparatus	A membrane system associated with the secretion of proteins from cells and modifying proteins synthesised on the rough endoplasmic reticulum.
Cell membrane	Cell boundary. Selectively permeable for nutrients required by cell. Contains sites for recognition of cell e.g. by hormones.
Cytoplasm	The non-particulate so-called soluble part of the cell contains many important functions, e.g. energy production by breakdown of glucose (glycolysis).

of a typical **eukaryotic cell** (a cell with a nucleus such as is found in all higher organisms) and Table 2.2 summarises the functions of the components of the cell. For a more detailed discussion the reader should consult a basic text (e.g. Metzler, 1977).

2.2. Biological systems

2.2.1 Small molecules

Let us examine the perspective of biological systems by starting with the smallest – single molecules of low molecular weight. Such molecules have many functions: they can be primary energy sources, such as glucose; or they may as their main function be building-blocks for macromolecules, as amino acids are for proteins. Small molecules are not difficult to isolate and can be studied by the classical techniques of organic chemistry. If we are to understand how glucose can be used as an energy source, for example, we must understand how it behaves as a molecule. We need to know what kind of chemical reactions it can undergo, what its solubility in solvents of different polarities is, because all these properties are those on which the natural function of the molecule will depend. Most of the first part of this book deals with understanding these properties.

2.2.2.   Macromolecules

It is not a big step to enlarge the biochemical system to consider the behaviour of **macromolecules**. One of the characteristics of living systems is that they are substantially built up from polymers or associations of relatively simple subunits which are capable of a great diversity of properties. Amino acids, for example, are polymerised in a specific sequence to form a given protein. Such polymers, in contrast to many prepared chemically such as nylon, are not random but have a unique molecular weight. The defined structure of proteins is essential to their biological function whether it be catalysis (enzymes) or any other function, such as transport of a specific lower molecular weight species. Other macromolecular systems include polysaccharides (energy stores and cell wall components), nucleic acids (information stores) and associations of lipid molecules (leading to the membrane systems which we shall discuss shortly).

An advantage of studying a purified macromolecular system is that, once isolated, it forms a completely defined quasi-independent system. In Chapter 12 we consider how such systems can be exploited in order to understand the mechanism of catalysis by an enzyme. In a system of this size it is meaningful to ask what the three-dimensional structure of the enzyme is, and how this structure relates to the biological function. We can examine how the small molecules which are interconverted by the enzyme interact with it, firstly by binding to it and then by bond-making and -breaking reactions, such as are already understood from the laboratory chemistry of small molecules.

With all the biological systems we choose to define, we must be careful about two points. Firstly, we must be aware of what meaningful questions can be asked of the system. It is reasonable, for example, to expect to be able to make a detailed kinetic analysis of a pure enzyme catalysing a specific reaction, but it may be totally misleading to attempt a similar analysis with only a cell-free extract (a preparation from a cell which has not been purified beyond breaking open the cell and removing large particles). In such an unpurified system many other uncontrolled factors might operate. Secondly, we must know how the quasi-independent system differs from the prototype system *in vivo*. Evaluating the relevance of studies of simpler systems to the situation *in vivo* is not easy, and cannot always be done with certainty. Attempts to do this must be made, however, if studies of simplified systems are to have any value at all.

Consider the case of a purified enzyme. Normally it would be studied in a buffer solution at the pH which gives the maximum catalytic activity. The components of the buffer are usually chosen for practical convenience and often bear little relation to the species which act as buffers inside the cell. It is not easy to determine the pH within a cell (indeed it may vary in different parts of the cell) so there can be relatively little certainty about whether the pH of the isolated system and the pH within the cell are similar. The concentrations of substrates used in enzyme studies are frequently those which give measurable rates of reaction, and these concentrations may be orders of magnitude different from those which exist in the cell, or cell compartment, within which the enzyme is found, as may the relative concentrations of enzyme and substrate.

The activity of many enzymes is altered by small concentrations of other small molecules found within the cell (Chapter 13). These modifiers may not be present in the isolated system and will have to be added if their effect is to be studied. Again, like the substrates, the modifiers should be added at their physiological concentration, if this is known.

These are just a few of the clear differences between an isolated enzyme in the test tube and its natural situation in the cell. Each worker must decide whether these differences are significant for the purposes he has at hand. There would be little progress in a molecular understanding of cellular processes if simplifications were not made. It is work at this molecular level which particularly requires simplifications, even to the extent of examining laboratory analogies of living systems rather than preparations extracted from the living organisms themselves. An example of such studies, which is the use of model systems, will be found in Chapter 3.

There are undoubtedly great benefits obtained from the study of simple systems. With a pure enzyme it is possible to understand its three-dimensional structure in detail and to develop a picture of how it catalyses a chemical reaction. The concepts derived from the studies on simpler systems can be applied to more complex, less well-defined systems such as those which follow in this chapter. It is this transferability of concepts which makes it essential for those who work with the less-defined systems to be able to apply ideas derived from studies on defined systems to their own problems. In later chapters in this book we deal with situations like this which cannot at present be understood in great molecular detail. We hope to show how molecular thinking in these areas can lead to new hypotheses to test and aid in the definition of the problems to be tackled.

2.2.3. Enzyme complexes and enzyme systems

Enzymes do not, as we have already emphasised, exist in isolation within a cell. They are all part of the overall function of the cell in which energy sources are broken down to provide the energy for cellular activity. Individual reactions are joined together in pathways which are the biochemist's functional description of the overall chemical activity of the cell*. Some enzymes are physically so closely associated with their neighbours in a pathway that they form **multi-enzyme complexes**. These are associations of proteins which catalyse a sequence of reactions in a defined order. Examples include

*Schemes of the important pathways discussed in this book will be found in Appendix 1 and at appropriate places in the text.

the enzyme system that synthesises fatty acids, and the system that catalyses the oxidative decarboxylation of 2-oxoacids. Such systems can be described in very similar terms to those used for individual enzymes, although they may be several times larger. We discuss examples in more detail in Chapter 13.

Many enzyme systems are not organised in such close physical contact, and their component enzymes can be considered separately as distinct catalytic units. A cell is a self-regulating system and may be able to modify the activity of the whole pathway in a number of ways. The ways in which the activities of enzymes and pathways can be controlled within a cell can be described at a molecular level (Chapters 13 and 16).

In addition to the regulation of the activity of enzyme molecules, cells adopt many other means of control. The function of a tissue is determined by the specific genes which are expressed. Enzymes are synthesised and degraded in a controlled way, sometimes showing a characteristic fluctuation in the rates of synthesis and degradation according to the time of day. Foreign compounds (Chapter 15) and cells (Chapter 16) are recognised and removed.

2.2.4. *Organelles and membranes*

When we move to the more complex assemblies of Table 2.1 we enter an important new area. The cell **organelles**, the next most complex cellular systems, are large enough to be visible by microscopy. Many can only be seen by electron microscopy (for example the ribosomes and endoplasmic reticulum) but others, such as the nuclei of eukaryotic cells, can be seen with the light microscope. Provided that there are no artefacts introduced by the techniques of microscopy (Chapter 11), it is possible to get a good idea of what the organelles look like. This gives a gross structural basis on which a more detailed molecular structure can be built.

Each organelle has a specific function associated with it. Ribosomes conduct the synthesis of proteins, mitochondria the synthesis of ATP (amongst other things) and so on. Many of the organelles, in particular the larger ones, have one functional component in common. This is the membrane which divides them from the rest of the cell. Membranes are of such importance in modern biology that we devote a whole chapter to a discussion of the current molecular thinking about their structure and function (Chapter 14).

The larger the system, the more difficult it becomes to think of its behaviour in solely molecular terms. Very often we must use other ways of expressing ideas about the function of organelles. An example is a current approach to understanding the mechanism by which ATP is synthesised in mitochondria. This process is the main way by which ATP is produced in aerobic eukaryotic cells. It allows the reduced coenzymes, such as NADH, formed during the oxidation of energy sources to be re-oxidised so that they can be re-used, and at the same time, at specific stages in this process, molecules of ATP are synthesised from ADP and inorganic phosphate. The chemical significance of

these compounds is discussed in Chapter 3, it is sufficient at present to say
that the biological function of ATP is to be the molecular link between energy-
producing reactions and energy-requiring processes (Figure 2.2, *top*).

The way in which the oxidation of coenzymes is coupled to the production
of ATP is not understood in purely molecular terms, but some idea of what
may be happening can be gained by using an electrical analogy (Figure 2.2,
bottom). As the reduced coenzymes are re-oxidised, protons are ejected from
the mitochondrion. This creates a difference in the concentration of hydrogen
ions between the inside and the outside of the inner mitochondrial membrane
(which is not permeable to protons). Clearly this situation is a high-energy
state, since it is not at equilibrium. The free energy contained in this proton
gradient is thought to be used up by the protons re-entering the mitochondrion
at the specific site where ATP synthesis takes place, providing the driving force
for the condensation reaction. There is thus a proton current through the

Figure 2.2.

Electron transport
membrane system

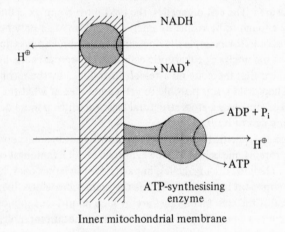

Inner mitochondrial membrane

Equivalent proton circuit
Protonmotive force≡electron transport

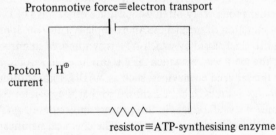

resistor≡ATP-synthesising enzyme

mitochondrial membrane. The proton-ejecting system can be regarded as a
chemical cell which generates a protonmotive force, and the enzymes which
capitalise on the proton current to synthesise ATP can be regarded as resistors
in the proton carrying circuit (Hinkle & McCarty, 1978).

In situations such as this it is not possible at present to use the same precise
molecular thinking that we can apply to single enzymes. Molecular ideas here
become less precise and confident, and it is this tendency towards a functional
rather than a molecular description which dismays many chemists.

It is obviously important for the biological chemist to be familiar with both
extremes of the biochemical approach. He must be molecularly literate but he
must also be able to appreciate defined systems within a cell and to envisage
how these systems might interact. Above all he must also be aware that a cell
is organised not only in reaction sequences but also in space. Reactions can
take place both in a purely chemical direction (making and breaking of covalent
bonds) and in a spatial direction. Substrates can be moved from one compartment of a cell to another, for example from the cytoplasm to the mitochondria,
or, in the case of the steroid hormones (Chapter 16) from the exterior of the
cell right through to the nucleus. The proton circuit we have just considered
is another example of a vectorial biochemical process. Such vectorial processes
have great significance in the integration and control of the function of various
parts of cells. Figure 2.3 illustrates the layout of some important processes
within cells which are considered in other parts of this book.

2.2.5. Cells and tissues

If we now move to the study of whole cells and tissues, the molecular
emphasis again changes. The main questions that can be asked now relate to
the overall function of the cell, such as how fuel is converted into energy and
how the energy is used to produce work or the specific products of the cell as
they are required. Techniques that were used to study single reactions cannot
usually be applied to these systems in unmodified form. Frequently the cells,
whether they be present as a preparation of single cells or aggregated in a tissue,
must be treated as a continuous chemical reactor. The reasoning used can be
similar to that used in industrial chemistry for very much larger but much less
complex reactors.

This does not mean that molecular considerations are absent. A study at
the level of the intact cell greatly benefits from a knowledge of the more
microscopic levels of simplification. The chemical structures of the various
types of biological molecules still define the reactions which are possible, and
allow us to predict how cell components will behave in various cell compartments. In fact at the level of the whole cell we need to integrate information
obtained from all the earlier levels of simplification. Thus the understanding
of individual enzymes is elaborated into the concept of metabolic pathways.
Pathways are controlled by internal stimuli which are themselves under the

influence of external stimuli (the effect of the environment of a cell on its activity). The sum of these processes defines the molecular activity of the cell.

Obviously the level of precision of the molecular description of all these events will differ. In some cases precision will be very high, but in others the information about the structure and chemical nature of the system and its environment will be more speculative. Here the worker must use a special kind of insight which will be more powerful if it is based on a molecular understanding of the properties of biomolecules and how they behave in more completely understood biological systems. We present in the next few chapters some molecular systems which can be understood in terms of modern chemical

Figure 2.3. Some pathways of fatty acid metabolism and the parts of the cell in which they might be found. Note that not all of the steps shown would take place in the same cell at the same time.

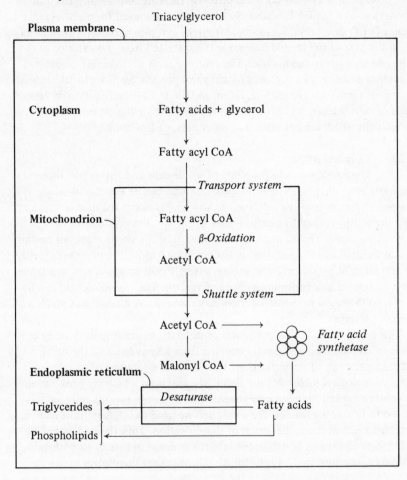

theory. After a pause to consider how chemical and physical methods can be applied to biological systems, we examine the various levels of simplification that we have just outlined in turn. We hope that the reader will appreciate the power and limitations of molecular thinking in many different fields and be prepared to tackle a problem using a molecular approach combined with appropriate concepts derived from less microscopic sources.

3 Polarised double bonds: derivatives of carboxylic, phosphoric and sulphonic acids

A polarised double bond is found whenever two elements of different electronegativity are joined together by a double bond. In biologically significant compounds this includes not only double bonds to carbon but also doubly bonded sulphur and phosphorus. To emphasise the generality of the chemical arguments, we shall take the unusual course of deferring discussion of $C=O$ in aldehydes and ketones until the next chapter. Instead, we shall begin by considering compounds that contain polarised double bonds and that are acidic because of the presence of hydroxyl groups associated with the double bond. These include carbonic acid derivatives (RO.CO.X) carboxylic acid derivatives (R.CO.X) phosphoric acid derivatives (RO.PO$_2$H.X) and derivatives of sulphonic acids (R.SO$_2$.X) (where X = NR$_2$, OR, SR, OP, OH etc.). To begin, let us consider the electron distribution in multiple bonds to carbon in a very elementary way that allows a ready comparison with the sulphur and phosphorus analogues.

3.1. The carbonyl group: revision summary

The carbonyl group, $C=O$, is probably the most important functional group in organic chemistry, especially from a synthetic or biosynthetic point of view. It is no accident that our introductory examples in chapter one featured the carbonyl group in two of its guises, an amide and an acetal.

The familiar way of describing a carbonyl group is typical of a polarised double bond. The more electronegative oxygen atom bears a greater share of the electron density between itself and the less electronegative carbon atom, and this can be depicted either as a polarisation using $\delta+$ and $\delta-$ symbols, or as a resonance hybrid of two canonical forms (Figure 3.1a). However you choose to look at it, the chemical reactivity of the carbonyl group is such that nucleophiles will attack the positively polarised carbon atom but electrophiles will bond with the more electron-rich oxygen atom. Thus a proton (an electrophile) attacks oxygen and an amine (a nucleophile) attacks carbon (Figure 3.1b). In this example, the amine *adds* across the double bond. A very large number of reactions can be classified as additions to polarised double bonds (see also the following chapter) and, in principle, the same holds for reactions of any compounds in which an element less electronegative than oxygen is double bonded to oxygen.

Figure 3.1.

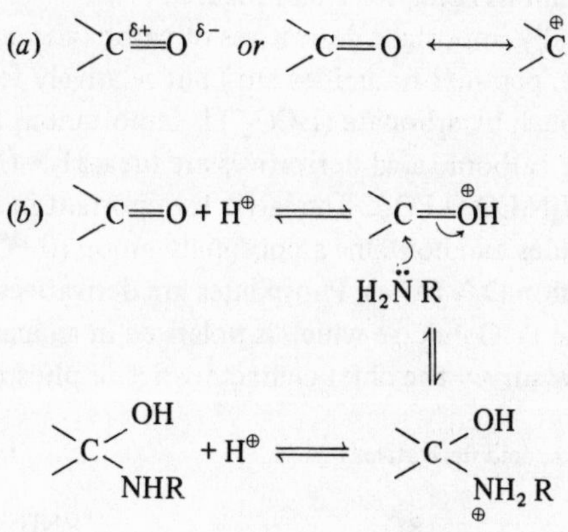

(a) $\quad \overset{}{>}C \overset{\delta+}{=\!=} O^{\,\delta-} \quad or \quad >C\!=\!O \longleftrightarrow \overset{\oplus}{>}C\!-\!\overset{\ominus}{O}$

(b) $\quad >C\!=\!O + H^{\oplus} \rightleftharpoons \;>C\!=\!\overset{\oplus}{O}H$

$$H_2\ddot{N}R$$

$$\overset{OH}{\underset{NHR}{>C<}} + H^{\oplus} \rightleftharpoons \overset{OH}{\underset{\underset{\oplus}{NH_2R}}{>C<}}$$

3.1.1. Carbonic and carboxylic acids and their derivatives

In additions to carbonyl groups, the carbon atom acts as an electron acceptor from an external nucleophile. It can also accept an electron pair from an adjacent atom bonded to it. (Figure 3.2). In each case, the negative charge

Figure 3.2. Charge delocalisation in some anions.

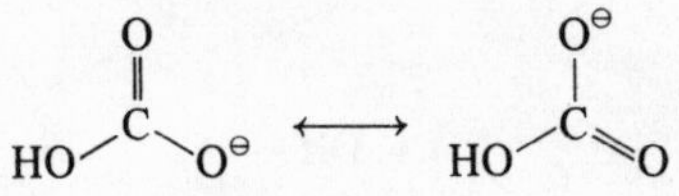

$$HO\overset{\displaystyle \overset{O}{\|}}{\underset{}{C}}O^{\ominus} \longleftrightarrow HO\overset{\displaystyle \overset{O^{\ominus}}{|}}{\underset{}{C}}\!\!=\!\!O$$

Carbonate anion

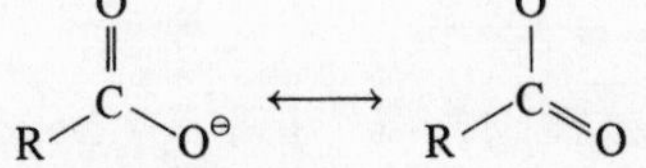

$$R\overset{\displaystyle \overset{O}{\|}}{\underset{}{C}}O^{\ominus} \longleftrightarrow R\overset{\displaystyle \overset{O^{\ominus}}{|}}{\underset{}{C}}\!\!=\!\!O$$

Carboxylate anion

$$R\overset{\displaystyle \overset{O}{\|}}{\underset{}{C}}CH_2^{\ominus} \longleftrightarrow R\overset{\displaystyle \overset{O^{\ominus}}{|}}{\underset{}{C}}\!\!=\!\!CH_2$$

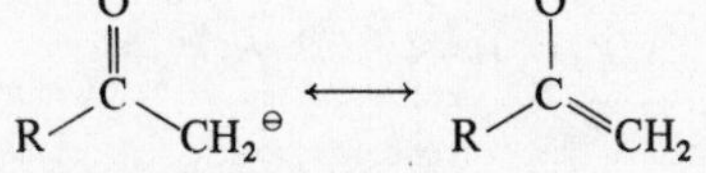

Enolate anion

is spread over all three atoms. When a charge is delocalised in an ion like this, the ion is stabilised. Such familiar arguments are used to rationalise the acidity of carbonic and carboxylic acids in water and many reactions of aldehydes, ketones and esters (Chapter 4). Similarly, a resonance contribution to the electronic structures of esters and amides can be invoked if the non-bonded electrons of nitrogen or oxygen are donated to the carbonyl group, as the negative charge was in the anions (Chapter 1 and Figure 3.2).

There are many biologically important derivatives of carboxylic acids (e.g. acetate, succinate, pyruvate, peptides or amides etc.) but relatively few derivatives of carbonic acid, although bicarbonate (HCO_3^-) is important as a buffer ion in blood. Chief amongst carbonic acid derivatives are urea, $H_2N.CO.NH_2$, and carbamyl phosphate, $H_2N.CO.O.PO_3^-$. The latter is important in the biosynthesis of purine nucleotides and contains a phosphate group ($O.PO_3^-$) in addition to the amide function ($>N.CO-$). Phosphates are derivatives of phosphoric acids and contain the $P=O$ linkage which is polarised in an analogous manner to $C=O$. Let us now survey the chief characteristics of phosphates.

Figure 3.3. Phosphoric acid derivatives.

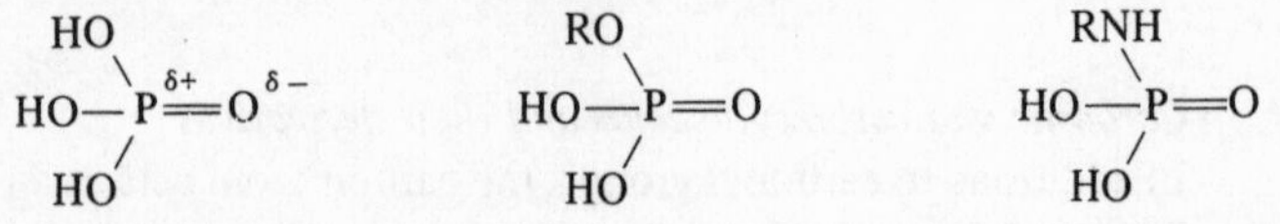

Phosphoric acid Phosphate ester Phosphoramide

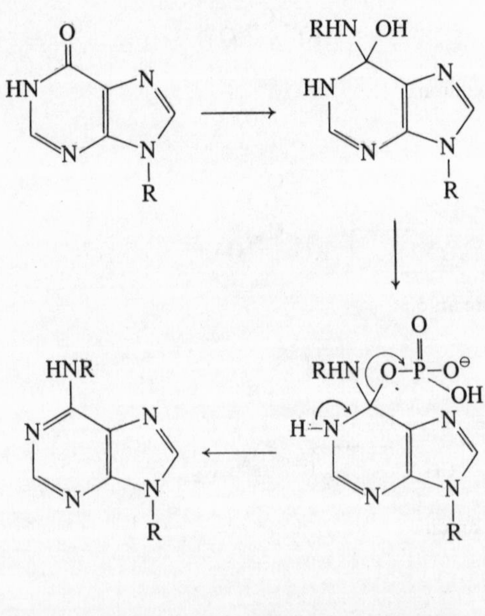

3.1.2. Phosphoric acids, phosphate esters and their derivatives

Phosphoric acids and their derivatives are of great significance bio-chemically, chiefly as a consequence of the chemical versatility of phosphate esters. In natural products, they can function either as structural units or they can behave as activating and leaving groups in metabolic reactions (Figure 3.3). Phosphoramides, on the other hand, are not widespread in nature and their chemistry is chiefly of interest to the synthetic chemist. Many of the reactions of phosphates are essentially addition reactions analogous to those of car-boxylic and carbonic acids, but the substitution reactions in which phosphates act as leaving groups have no parallel in carboxylate chemistry, except in extreme cases; we shall consider why shortly. Further differences exist in the mechanism of addition reactions of phosphates. The chief distinctions are

Figure 3.3. (continued) Some naturally occurring phosphate esters.

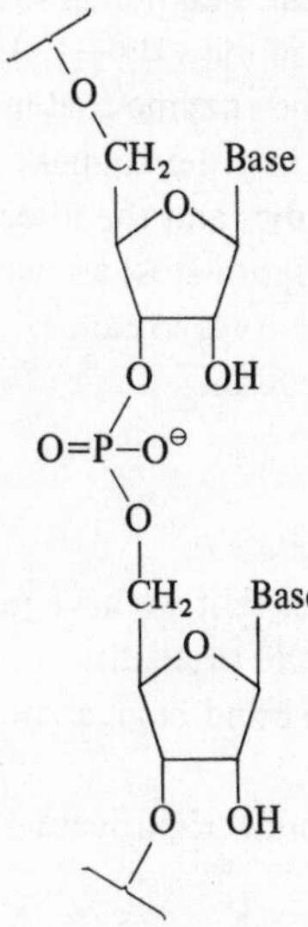

Ribonucleic acid

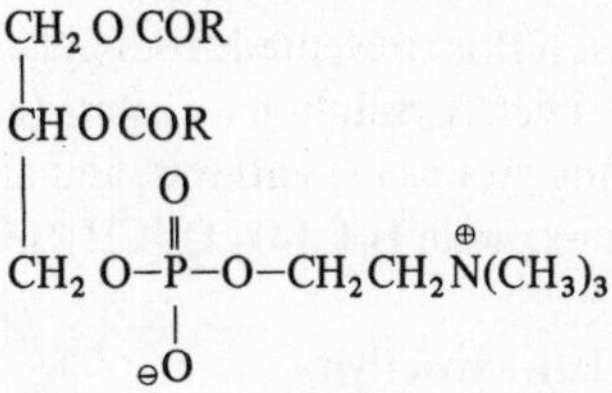

Lecithin

$R = C_{14} - C_{18}$ saturated chain

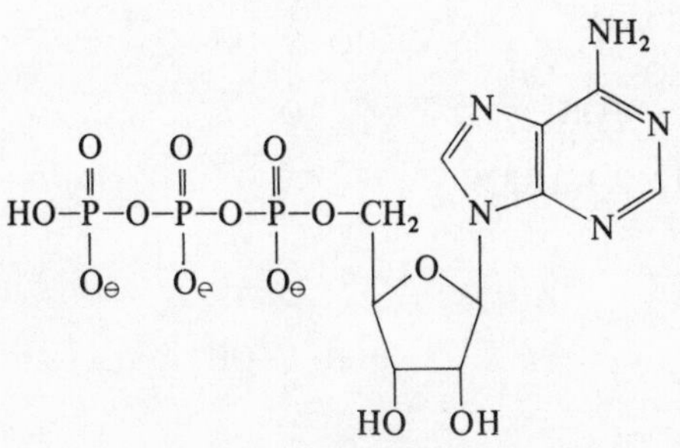

Adenosine triphosphate (ATP)

consequences of the fact that phosphorus is bonded to four oxygen atoms compared with two for carbon in a carboxylate derivative.

3.1.3. Sulphonic and sulphuric acids and their derivatives

The trend towards substitution rather than addition reactions established moving from carboxylate to phosphate esters is continued in the derivatives of sulphonic acids. Sulphonic acids are directly analogous to carboxylic acids, sulphuric acid is related to carbonic acid (Figure 3.4). In nature, sulphates occur as esters in situations where a negatively charged group is permanently required, for example in bile salts, in some tyrosine-containing peptides and in polysaccharides found in connective tissue. Sulphonic acids contain a C–S bond; naturally occurring examples include taurine which is a component of taurocholic acid (Figure 3.4) and is significant in brain. The important antibacterial drugs, sulphonamides, (see p. 315) are analogous to the natural metabolite *p*-aminobenzoic acid, which is involved in the biosynthesis of dihydrofolic acid; folic acid is an essential vitamin for man. Sulphonamides are active antibacterial agents because the $-SO_2N$ group mimics the $-CO_2^-$ of the natural *p*-aminobenzoic acid, competing with it for the enzyme that makes dihydrofolic acid. The bacterium also requires folic acid in order to make nucleic acids; if it is prevented from making folic acid it dies and the infection is cured. Chemically, sulphonate esters (e.g. *p*-toluene sulphonates) are widely used as leaving groups in synthesis, and also as alkylating drugs in cancer therapy, for example $H_3C.SO_2.O.(CH_2)_4.O.SO_2.CH_3$: Busulfan.

3.2. Relative reactivity
3.2.1. The relative reactivity of C=O, P=O and S=O in acids

The mimicking of carboxylate by sulphonamides that we have just discussed leads us to consider the relative reactivities of the three different polarised double bonds. To do this, we place each double bond in an analogous chemical environment and investigate what happens.

Consider firstly the series of acids in Figure 3.5. The most significant

Figure 3.5. Strengths of some biologically significant acids.

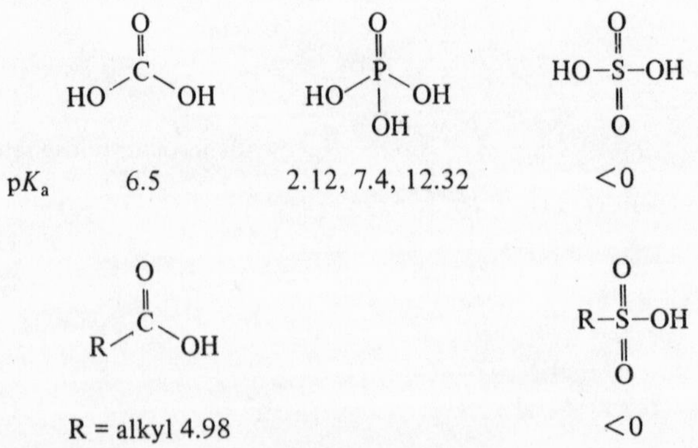

pK_a 6.5	2.12, 7.4, 12.32	<0

R = alkyl 4.98

<0

Figure 3.4. Sulphuric acid derivatives.

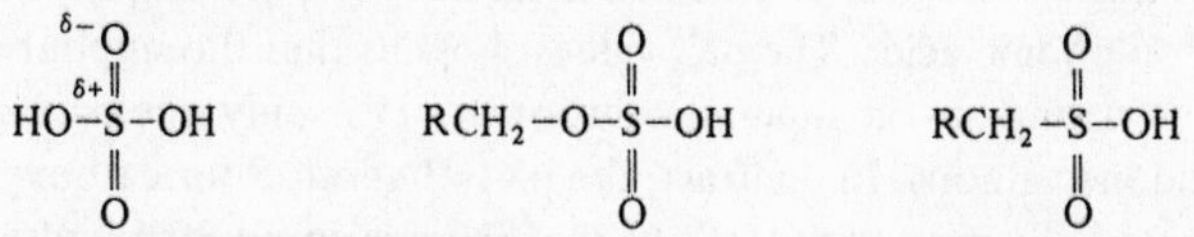

Sulphuric acid   A sulphate ester   A sulphonic acid

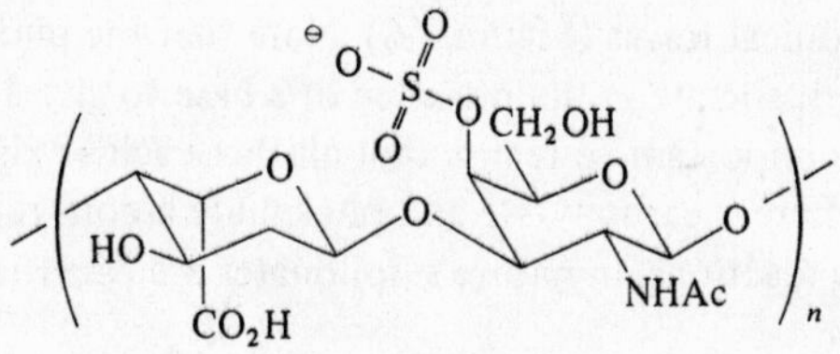

Dermatan sulphate

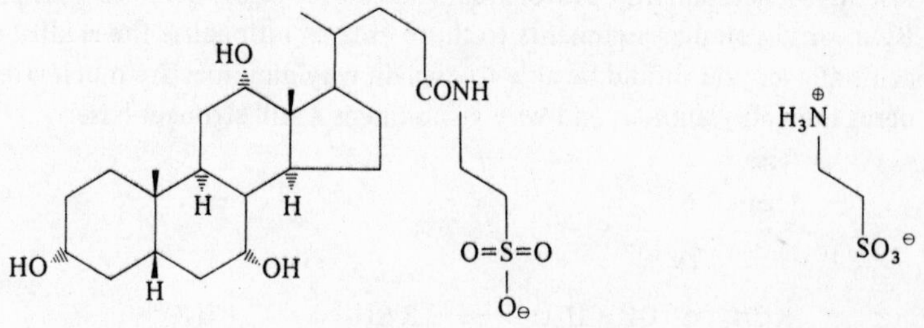

Taurocholic acid

Taurine

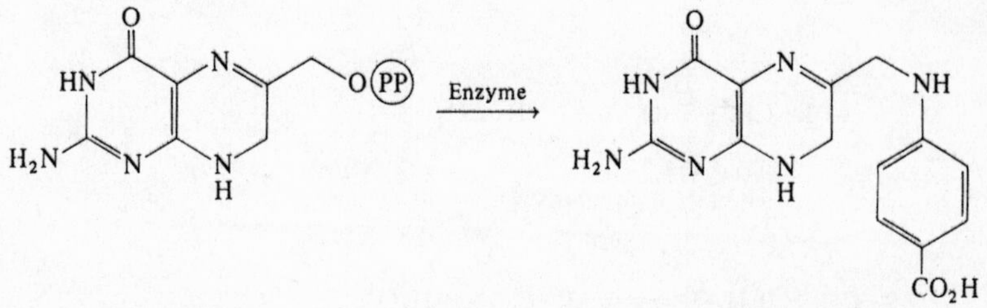

Dihydropteroic acid

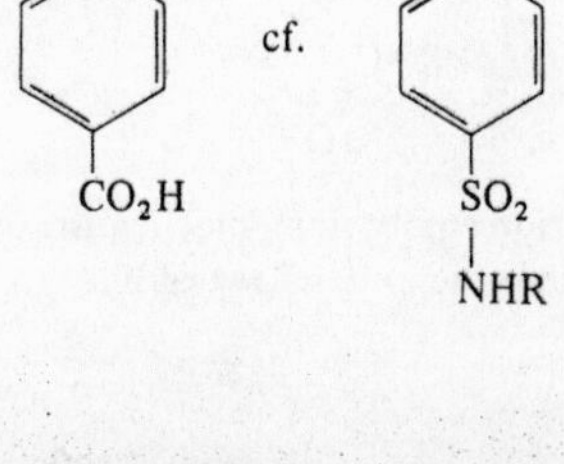

cf.

comparisons from the biological point of view are between the carboxylic, phosphoric and sulphonic acids. The pK_a values show us that the sulphur acids are by the far the strongest – in aqueous solution at pH 7, only one part in 10^7 is not dissociated into anions. In contrast, the pK_a of about 5 for carboxylic acids (unsubstituted alkanoic acids) tells us that there is about 99% ionisation in a solution at pH 7. It is not difficult to find a rationale for this behaviour in the descriptive conceptual language of organic chemistry. The sulphonate anion is more highly delocalised than the carboxylate or phosphate anion, as is illustrated by the canonical forms (Figure 3.6). Note that the phosphate monoanion can further dissociate in the presence of a base to give a dianion and a trianion. It is very important to realise that all these acids exist as anions at physiological pH*. Whereas carboxylate and phosphate anions readily undergo reprotonation during reactions, in nature sulphonates are essentially permanent negative charges.

At the other end of the spectrum of acidity in biologically important compounds is the guanido group, found in the amino acid, arginine. It is essentially permanently protonated, and the conjugate acid has a pK_a of 12.5. By means of similar arguments to those used to rationalise the acidity of the acids above, you should be able to explain why amidines are much stronger bases then alkylamines, and why guanidine is a still stronger base.

Figure 3.6.

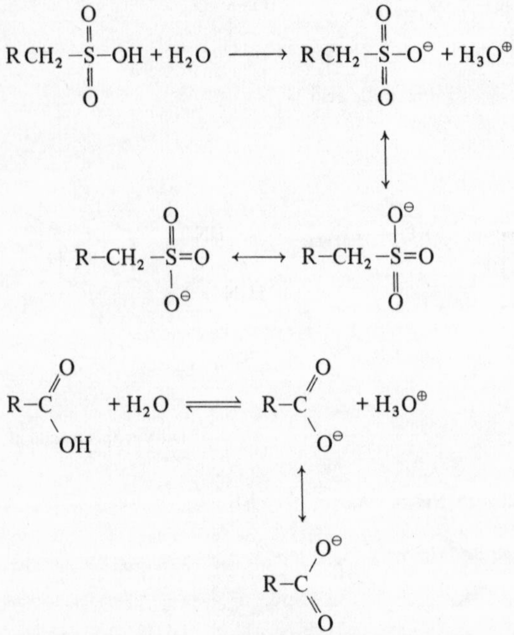

*Because of the ionisation equilibrium, biochemists often use the name of the acid and the name of the anion interchangeably.

Think about charge delocalisation in the cations coresponding to the three bases (Figure 3.7).

Figure 3.7.

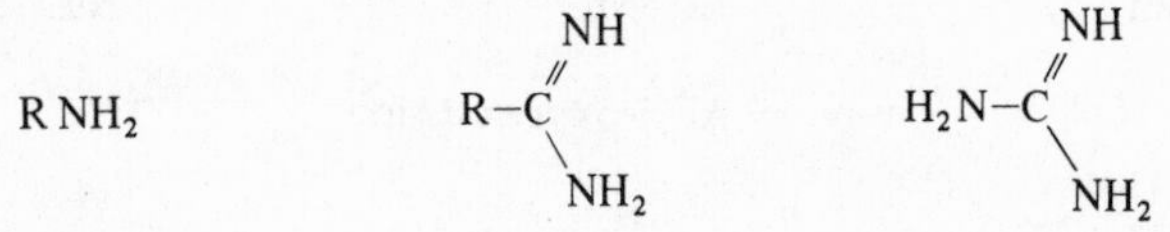

$$R\,NH_2 \qquad\qquad \underset{\displaystyle NH_2}{\overset{\displaystyle \overset{NH}{\|}}{R-C}} \qquad\qquad \underset{\displaystyle NH_2}{\overset{\displaystyle \overset{NH}{\|}}{H_2N-C}}$$

An alkylamine   An amidine   Guanidine

3.2.2. *Reactivity of C=O, P=O, and S=O derivatives in esters – addition versus substitution*

Now let us turn to the ester derivatives of our acids and investigate what happens to each in the presence of a nucleophile (Nu⁻). Again it is helpful to concentrate firstly upon the two extreme cases, carboxylates and sulphonates. From what we have already seen of the behaviour of polarised double bonds, the strong possibility exists that the nucleophile will attack the positive end of the X=O dipole. The adduct thus obtained can either revert to starting materials or decompose to yield the products. If the nucleophile is hydroxide anion or water, hydrolysis of the ester occurs. The carboxylate esters almost always fit in with this expectation – the bond between the carbonyl group and the ester oxgyen atom is broken. This is termed **acyl–oxygen fission** (Figure 3.8). However, in sulphonate esters a different mechanism is found. Instead of acyl–oxygen fission, the bond between the ester carbon and the ester oxygen breaks. This cleavage is referred to as **alkyl–oxygen fission**. It is possible to distinguish between these two mechanisms by labelling either the ester oxygen or the hydroxide nucleophile with ^{18}O. As Figure 3.8 shows, the location of the label in the products reveals the mechanism by which hydrolysis occurred.

What explanations can be advanced for the different behaviour of the sulphonate? Two factors may be invoked. Firstly the positively charged sulphur atom is closely guarded by three electronegative oxygen atoms and the nucleophile, which of course also is rich in electrons, must pass the three in order to add across S=O. Because like charges repel, there is a high energy barrier to this direction of attack and so the nucleophile seeks another route. This is easy to find because the three oxygens bonded to sulphur together exert a powerful electron-withdrawing inductive effect, which induces a positive charge on the carbon derived from the alcohol. Accordingly, the nucleophile attacks here causing the C–O bond to break. The liberated sulphonate anion is extensively charge delocalised and accordingly is stable in polar solvents. Thus in the sulphonate ester, nucleophilic substitution at carbon occurs in preference to nucleophilic addition to the sulphonyl group. We say that the sulphonate anion is a good leaving group; anions of strong acids are usually good leaving groups.

Figure 3.8.

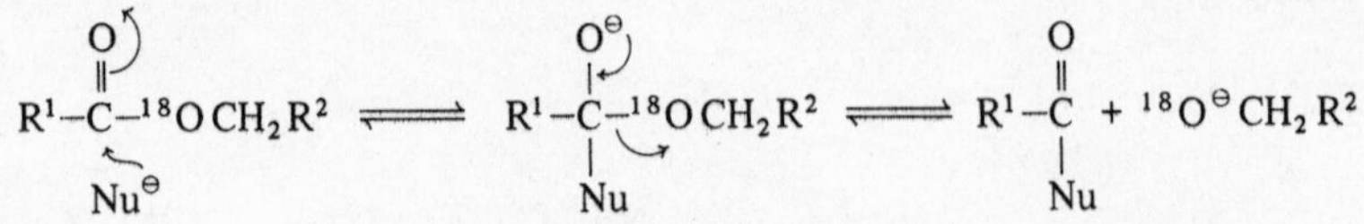

$$R^1-\overset{\overset{O}{\|}}{\underset{\underset{Nu^\ominus}{|}}{C}}-{}^{18}OCH_2R^2 \rightleftharpoons R^1-\overset{\overset{O^\ominus}{|}}{\underset{\underset{Nu}{|}}{C}}-{}^{18}OCH_2R^2 \rightleftharpoons R^1-\overset{\overset{O}{\|}}{\underset{\underset{Nu}{|}}{C}} + {}^{18}O^\ominus CH_2R^2$$

Acyl–oxygen fission

$$R^1-\overset{\overset{O}{\|}}{\underset{\underset{O}{\|}}{S}}-{}^{18}O-CH_2R^2 \quad Nu^\ominus$$

$$\nearrow \times \quad R^1-\overset{\overset{O}{\|}}{S}-{}^{18}OCH_2R^2 \quad {}^\ominus O \diagdown Nu$$

$$\searrow \quad R^1-\overset{\overset{O}{\|}}{\underset{\underset{O}{\|}}{S}}-{}^{18}O^\ominus + NuCH_2R^2$$

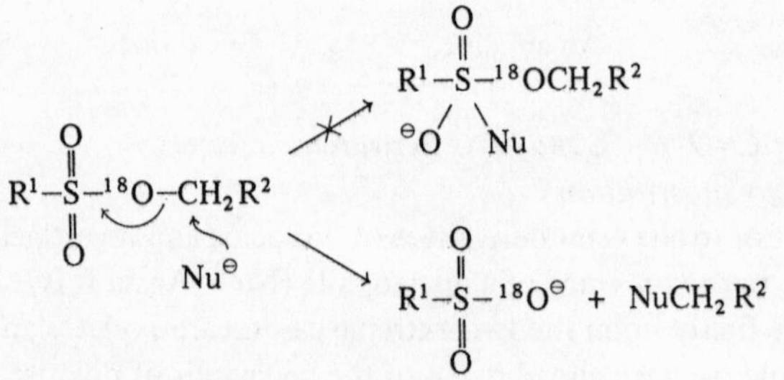

Alkyl–oxygen fission

With the contrast of carboxylate and sulphonate in mind, how do phosphate esters behave? As might be expected from the pK_as of the acids, phosphates display features of both analogues. What happens depends both upon the precise structure of the phosphate and upon the reaction conditions. It is essential for many metabolic reactions that phosphate esters are able to play this dual role. By an addition process, a phosphate ester can **phosphorylate** a nucleophile. Usually in nature, phosphorylations are accomplished by phosphoric anhydride derivatives, as we shall see shortly. Alternatively, if a substitution reaction occurs the nucleophile can be alkylated (Figure 3.9). The biosynthesis of dihydrofolic acid that we looked at earlier is an example of this type of reaction.

Figure 3.9.

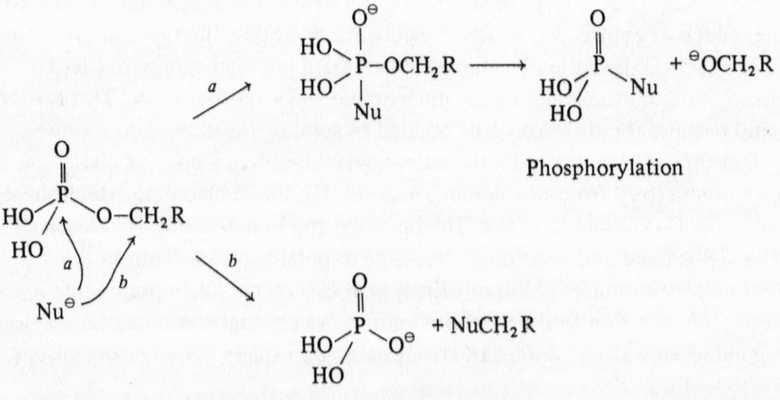

$$\overset{a}{\nearrow} \quad \underset{HO}{\overset{HO}{\diagdown}}\overset{\overset{O^\ominus}{|}}{\underset{\underset{Nu}{|}}{P}}-OCH_2R \longrightarrow \underset{HO}{\overset{HO}{\diagdown}}\overset{\overset{O}{\|}}{\underset{}{P}}\diagdown Nu + {}^\ominus OCH_2R$$

Phosphorylation

$$\underset{HO}{\overset{HO}{\diagdown}}\overset{\overset{O}{\|}}{P}\diagup O-CH_2R$$

$$Nu^\ominus \quad \overset{a}{\diagup}\overset{b}{\diagup}\overset{b}{\searrow}$$

$$\searrow^{b} \quad \underset{HO}{\overset{HO}{\diagdown}}\overset{\overset{O}{\|}}{P}\diagdown O^\ominus + NuCH_2R$$

Alkylation

There are multitudinous examples of both types of reactions in nature and we shall need to consider how the reaction course can be controlled (Chapter 12). Try the following structured problem to develop your ideas.

Structured problem 3.1
(1) The carboxylate anion is not such a good leaving group as phosphate or sulphonate. Suggest a reason for this.
(2) A carboxylate cannot undergo a nucleophilic addition reaction across C=O. Why?
(3) What modification to the carboxylate anion must be made in order to permit addition? Answer this firstly in descriptive terms and then suggest a suitable chemical method for carrying out your proposed modification (it is very simple).
(4) Describe points of similarity between phosphate and carboxylate anions.
(5) By analogy with your answer to point (2), suggest how the direction of attack on a phosphate ester by a nucleophile might be controlled.
See Section 3.3 for the answer.

The following sections describe chemistry of carboxylate and phosphate derivatives in relation to this problem. So far, we have concentrated upon rather elementary conceptual descriptions of simple chemical reactions; to help you extend your ideas into the biological realm we offer this second structured problem the solution of which relates closely to the first.

Structured problem 3.2
The enzyme carboxypeptidase catalyses the hydrolysis of peptide bonds as shown below.

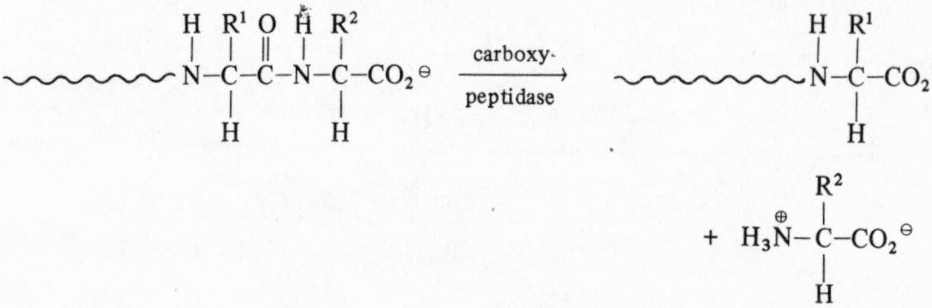

A Zn^{2+} ion is essential for its activity.
(1) What type of a reagent is Zn^{2+}? (Hint: anhydrous $ZnCl_2$ is a useful mild catalyst for Friedel-Crafts reactions.)
(2) Hence how might the presence of a divalent metal ion at the active site of carboxypeptidase contribute to catalysis by the enzyme?
(Gray, 1971, p. 180; Bender & Brubacher, 1973, p. 202.)

We shall discuss in a subsequent chapter how such information is obtained and how a composite picture of the mechanism of action of some enzymes can be established with the aid of results from many different types of experiments.

3.3. Catalysis of addition reactions to polarised double bonds

The first of the above problems emphasises that a nucleophile, which is anionic in character, must penetrate a negative charge cloud in order to attack the carbon of the carbonyl group. This is obviously unfavourable, as is the analogous addition to the phosphoryl group (Figure 3.10, *top*). The simplest mechanism involves neutralisation of the negative charge with a positive charge – at its simplest, a proton. Consequently, most additions to polarised double bonds are catalysed by protons or by metal cations, as the example of carboxypeptidase shows. Instead of protonating the amide car-

Figure 3.10.

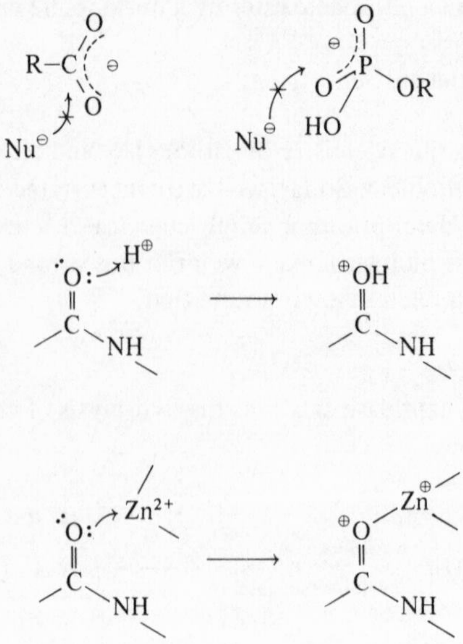

bonyl group, the carbonyl group coordinates with Zn^{2+} (Figure 3.10). The metal ion is functioning as a Lewis acid, that is an electron acceptor, and the polarisation of C=O is enhanced. Both proton-donating acids (Brønsted acids) and Lewis acids are effective catalysts for the preparation and hydrolysis of carboxylate esters, as the following equations illustrate.

$$C_{15}H_{32}CO_2H + MeOH \xrightarrow[H_2SO_4]{1 \text{ drop}} C_{15}H_{32}CO.OMe + H_2O$$

$$C_{15}H_{32}CO_2H + BF_3(MeOH) \longrightarrow C_{15}H_{32}CO.OMe + H_2O.BF_3$$

The situation with phosphate esters is similar, although, because the structures of carboxylate and phosphate esters differ, there are significant mechanistic differences (Figure 3.11).

Figure 3.11. Mechanisms for hydrolysis of phosphate esters.

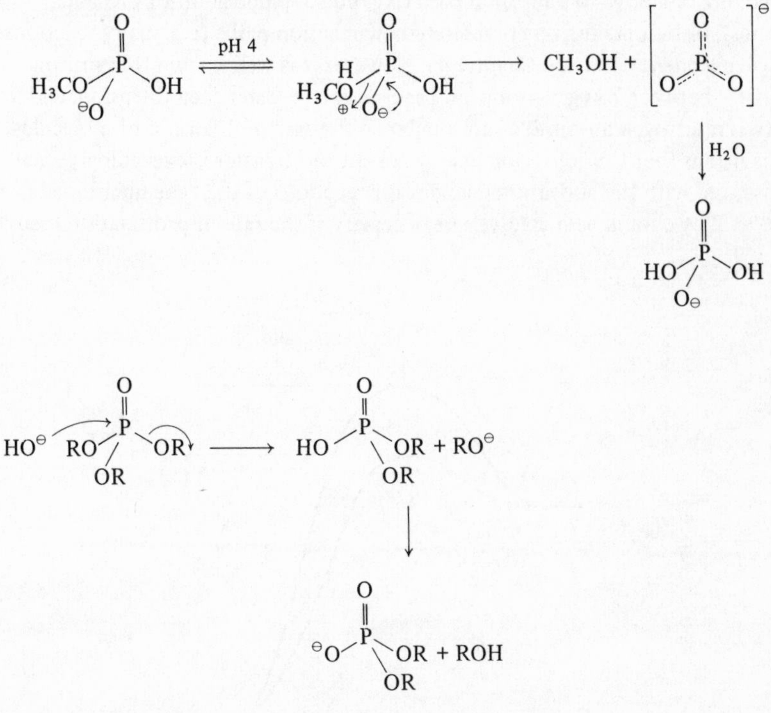

3.3.1. *The significance of acid/base catalysis in chemical reactions and in catalysis by enzymes*

Before looking at details of some of the examples in the above survey, we shall pause to take a closer look at acid/base catalysis. Acid/base catalysis is, as the structured problems and examples show, intimately involved in the mechanisms of the reactions of polarised double bonds. What follows in this section is brief and if you wish to study this topic more deeply, consult either Bender & Brubacher (1973, p. 37) or at a more advanced level, Jencks (1969). The physical chemical principles apply equally to reactions in later chapters.

To bring acid/base catalysis into clearer focus, we should firstly recall that the function of a catalyst is to lower the activation energy ($E^{\ddagger}$) for a reaction, thereby permitting the reaction to proceed more quickly. This can be illustrated by a reaction profile (Figure 3.12); the corresponding mathematical formulation is

$$\text{Rate constant } k = A \exp(-E^{\ddagger}/RT)$$

where A is a pre-exponential factor, R is the gas constant per mole and T is the absolute temperature. The rate constant, of course, is determined by the *slowest* step of the reaction; obviously this must involve the catalyst. The catalyst may achieve a lowering of activation energy by promoting an existing reaction path (e.g. H^{+}-catalysed addition reactions to carbonyl groups), by removing obstacles to a reaction path (e.g. nucleophilic additions to phosphates), or by opening up completely new reaction paths (e.g. many reactions involving coenzymes, see Chapter 4). However it is well known that proton transfer between oxygen atoms proceeds at a rate faster than diffusion, and between nitrogen atoms at a rate similar to the rate of diffusion of molecules. This means that the activation energy for proton transfer is exceedingly small compared with the activation energies for reactions of the type mentioned above. Why should acid catalysis be necessary if the rate of protonation is so

Figure 3.12.

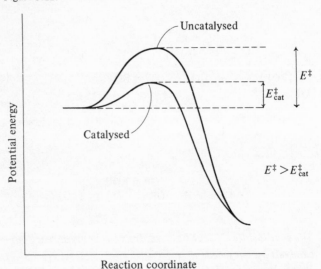

high? The answer is that frequently the proton transfer will lead to a species that is unstable or on the unfavoured side of an equilibrium. The conjugate acid of an ester is an example. It is easy to force the reaction to the right by adding an excess of acid which lowers the pH. An enzyme, however, operates at a constant pH in the organism of which it is part and the protons that it requires for catalysis must be provided very specifically and stoichiometrically at the active site. The enzyme must in effect provide a local pH change, and the change must be reversible if the enzyme is to catalyse the transformation of a further molecule of substrate. Consequently, acid/base catalysis is extremely important in enzyme-catalysed reactions.

3.3.2. The detection of acid/base catalysis

It is not always easy to prove that acid/base catalysis is operating in a reaction. This is because there may be many acids or bases involved and the distinction between the several possible mechanisms that can usually be written requires very careful kinetic studies (Jencks, 1969). The most simple method available is to show that the rate of the reaction depends upon the pH of the solution. Thus in Figure 3.13, the pH–rate curve for reaction (1) represents acid catalysis only occurring in a reaction (e.g. acetal hydrolysis).

Figure 3.13. pH–rate profiles for typical reactions.

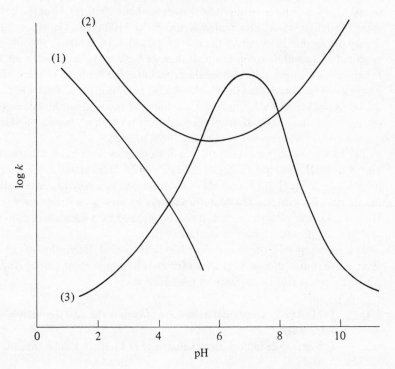

Similarly, reaction (2) is catalysed by both acid and base (eg. alkyl ester hydrolysis).

Such pH–rate profiles can be obtained for enzyme-catalysed reactions also, but because the microenvironment of the enzyme's active site can greatly modify the pK_a of an ionisable group, the mechanistic interpretation of such data is hazardous. Reaction (3) shows the appearance of a typical case.

An excellent way of detecting acid/base catalysis is to slow down the hydrogen-transfer step without altering the concentration of reactants so that an acid-catalysed reaction would show a decrease in reaction rate. This effect can be achieved by exchanging all the transferable protons of a reactant for the heavier isotope of hydrogen, deuterium. In general, the cleavage of a carbon–deuterium bond takes place 2–7 times more slowly than a corresponding carbon–protium bond. The same goes for bonds between hydrogen and elements close to carbon in the periodic table, like nitrogen and oxygen. If the slowest step of the reaction requires the cleavage of a bond to hydrogen then the reaction as a whole will be slowed down by exchanging protium for deuterium. We call this phenomenon a **kinetic isotope effect** and it is an invaluable tool for elucidating reaction mechanisms (see also Chapter 4).

We can appreciate why substitution of deuterium for protium should cause a decrease in rate by looking at the infrared spectra of both labelled and un-labelled molecules. The unlabelled species containing protium shows a C–H stretching vibration for an aliphatic system at about 2950 cm^{-1} but the corre-sponding deuterium-labelled molecule absorbs at 2100 cm^{-1}. Transition-state theory of reaction rates relates the rate of a reaction to the frequency of vibration of a bond or group of bonds that are broken in the slowest step of the reaction. Consequently, we would expect that the breaking of the C–D bond would be slower than the C–H bond because its infrared stretching fre-quency is smaller. For a rigorous understanding of the origin of kinetic isotope effects, some knowledge of quantum mechanics is required. Jencks provides a general discussion of the topic (Jencks, 1969, p. 243).

Returning to acid/base catalysis, it is obvious from what has just been said that if an N–H bond breaks during catalysis, an N–D bond will break many times more slowly. For enzyme-catalysed reactions, it is easy to convert all the transferable protons into deuterons simply by dissolving the enzyme in D_2O. If acid/base catalysis occurs, it will be revealed by a decrease in rate of reaction in D_2O with respect to the rate in H_2O.

The preceding paragraphs present a highly simplified description of complex phenomena, but it should be clear nevertheless how important kinetic studies are in the determination of reaction mechanisms.

3.4. Typical and exceptional behaviour of carboxylic acid derivatives – a summary

The examples in both this chapter and in Chapter 1 show that the normal behaviour of a carboxylic acid derivative undergoing hydrolysis involves nucleophilic addition to the carbonyl group aided by acid/base catalysis. Usually acyl–oxygen fission occurs, and the reactions are **bimolecular**. Both nucleophile and carboxylate are involved in the rate-determining step. There is, however, an exceptional case for which no parallel has yet been found in nature. If an ester is dissolved in very strong acid such as 90% sulphuric acid, it becomes extensively protonated in a rapid equilibrium (Figure 3.14). There follows a slower step in which the protonated ester undergoes **unimolecular** acyl–oxygen fission to give an acylium ion which is then rapidly trapped by water, affording the protonated acid.

This example demonstrates that a change in reaction conditions may cause a profound change in mechanism. It is very important to bear in mind possible differences in reaction conditions whenever enzyme-catalysed reactions and their laboratory analogues are compared in detailed mechanistic terms.

Figure 3.14.

$$R^1 CO{-}OR^2 + H^\oplus \xrightleftharpoons{\text{Fast}} R^1 CO{-}O\!\!\overset{\displaystyle R^2}{\underset{\oplus}{\diagdown}}{}_{H}$$

$$R^1 CO{-}O\!\!\overset{\displaystyle R^2}{\underset{\oplus}{\diagdown}}{}_{H} \xrightarrow{\text{Slow}} R^1 CO^\oplus + R^2 OH$$

$$R^1 CO^\oplus + H_2O \xrightarrow{\text{Fast}} R^1 CO\, \overset{\oplus}{O}H_2$$

$$R^1 CO\, \overset{\oplus}{O}H_2 \xrightleftharpoons{\text{Fast}} R^1 CO_2H + H^\oplus$$

3.5. The ambivalent behaviour of phosphate esters

The greater number of oxygen atoms bonded to phosphorus in a phosphate ester than to carbon in a carboxylate ester adds the further dimension of substitution reactions to the chemistry of phosphate esters. Because both addition and substitution reactions are so widespread in nature, a discussion of each of them is important.

3.5.1 Substitution reactions of phosphate esters

Consider the phosphate ester dimethylallyl pyrophosphate, which is an important intermediate in the biosynthesis of isoprenoid compounds including steroids (Figure 3.15). The name 'pyrophosphate' refers to the diphosphoric acid in which one oxygen atom joins two phosphorus atoms.

Note in passing that the PO.O.PO grouping is part of the phosphoric anhydride group, a topic for discussion later in this chapter.

Suppose that there is no catalysis available and that dimethylallyl pyrophosphate must react just as it is. A nucleophile would find great difficulty in attacking either phosphorus atom through the protective sheath of negative charge and therefore it takes the alternative pathway; it attacks carbon. Substitution takes place in which the pyrophosphate (diphosphate) anion is the

Figure 3.15. Examples of alkylation by allylic and benzylic pyrophosphates.

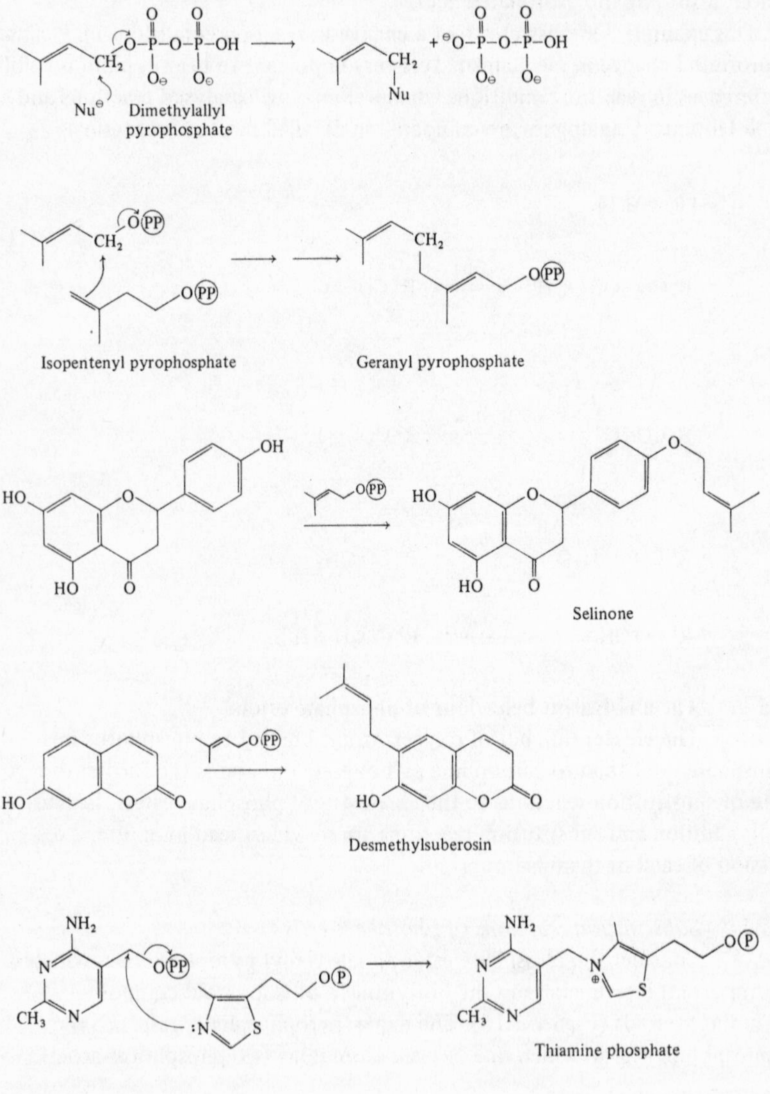

Dimethylallyl
pyrophosphate

Isopentenyl pyrophosphate   Geranyl pyrophosphate

Selinone

Desmethylsuberosin

Thiamine phosphate

leaving group. This anion is analogous to the halide anions that are the leaving groups of the common alkylating agents, alkyl halides. From the point of view of the nucleophile, dimethylallyl pyrophosphate is behaving as an alkylating agent.

There are many biosynthetic reactions in which dimethylallyl pyrophosphate behaves in this manner (Figure 3.15). One of the most important is the reaction with its isomer, isopentenyl pyrophosphate, a reaction by which the carbon skeleton of terpenes and steroids is constructed (see Chapter 5). It can also act as an alkylating agent for phenols, either on oxygen or by electrophilic substitution on the aromatic ring (see Chapter 7). We have already noted in the introduction to this chapter (Figure 3.4) that p-aminobenzoic acid is alkylated by a diphosphate ester. In this case, the leaving group is α to a carbon–nitrogen double bond. A similar situation is found in the biosynthesis of the vitamin thiamine, where the pyrophosphate is α to an aromatic (pyrimidine) ring. All of these alkylations by phosphate esters take place with systems that are activated to substitution; 'allylic' and 'benzylic' halides are well known to be highly reactive as alkylating agents, and the phosphate esters follow their example. In the absence of catalysis, the preference for allyl and benzyl compounds to undergo substitution reactions tips the balance for the phosphate esters away from addition at the phosphorus–oxygen double bond. Phosphates are also the leaving groups in a wide range of reactions, principally dehydrations, examples of which will be found later in the book.

3.5.2. *Addition reactions at phosphorus in phosphate esters*

The analogy between carboxylate and phosphate esters leads us to suggest the mechanism for the hydrolysis of a phosphate monoester shown in Figure 3.16, (1). However, the negatively charged oxygen atoms surrounding phosphorus make this unfavourable, as we have seen. The drawback applies equally to the alternative single-step mechanism (2), which is a substitution reaction at phosphorus. The distinction between these two mechanisms and substitution at carbon is, in principle, simple to make if the oxygen that joins phosphorus and carbon is labelled with ^{18}O.

In the first two cases the ^{18}O turns up in the product alcohol because P–O bond cleavage takes place. On the other hand, when substitution at carbon occurs, the label is found in the phosphate anion itself due to C–O bond cleavage. In the case of enzyme-catalysed hydrolyses of phosphate esters P–O cleavage usually occurs adjacent to the group for which the enzyme has a specific binding site (Figure 3.16). Consequently, the enzymes must have some way of getting around the problem of the negatively charged oxygen atoms shielding phosphorus; some sort of catalysis must be operating.

Figure 3.16.

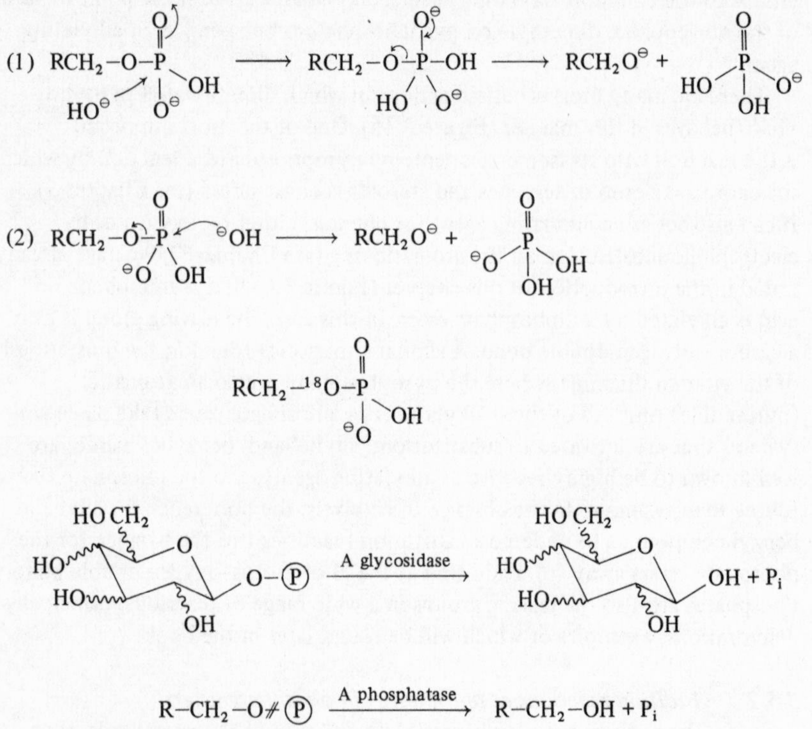

$$(1)\quad RCH_2-O-\underset{\underset{O^\ominus}{|}}{\overset{\overset{O}{\|}}{P}}-OH \longrightarrow RCH_2-O-\underset{\underset{O^\ominus}{|}}{\overset{\overset{O^\ominus}{\|}}{P}}-OH \longrightarrow RCH_2O^\ominus + \underset{\underset{OH}{|}}{\overset{\overset{O}{\|}}{P}}(HO)(O^\ominus)$$

$HO^\ominus$

$$(2)\quad RCH_2-O-\underset{\underset{OH}{|}}{\overset{\overset{O}{\|}}{P}}\,{}^\ominus OH \longrightarrow RCH_2O^\ominus + {}^\ominus O-\underset{\underset{OH}{|}}{\overset{\overset{O}{\|}}{P}}-OH$$

$${}^\ominus O$$

$$RCH_2-{}^{18}O-\underset{\underset{O^\ominus}{|}}{\overset{\overset{O}{\|}}{P}}-OH$$

$$\xrightarrow{\text{A glycosidase}}$$ (sugar structures) $+ P_i$

$$R-CH_2-O\!\!\!/\;\textcircled{P} \xrightarrow{\text{A phosphatase}} R-CH_2-OH + P_i$$

Model systems

The enzyme-catalysed process is complex because of the size of the enzyme molecule, and it is therefore often useful to study the fundamental chemistry pertaining to the enzyme-catalysed reaction with the aid of simplified models of the enzyme situation. To design a model, we isolate what we believe to be the important chemical features of the enzyme system and study them thoroughly in isolation from the whole system. With the aid of suitable model systems it is possible to investigate the factors that favour the various hydrolysis mechanisms for phosphate esters that we have outlined. The same is true for the hydrolysis of carboxylate esters and amides. Let us set up some simple chemical model systems for reactions of naturally occurring phosphate esters and see what can be learned from the models' behaviour. A detailed discussion of the problems associated with the construction and operation of models is beyond the scope of this chapter: you will find further examples throughout this book, and a more extensive specialised discussion can be found in Suckling, Suckling & Suckling, 1978.

When designing a model for an enzyme-catalysed reaction, the chemist firstly looks to see what enzymologists can tell him about the chemical requirements of the enzyme. In the case of alkaline phosphatase, he will be told that divalent zinc cations are essential for enzyme activity. A functional role for such a cation in a model system is therefore essential. Another factor, always important in enzyme-catalysed reactions, is the ability of the enzyme to bring the reactants together into proximity at the active site ready to react. This property produces a favourable entropy of activation for the reaction, and it can be compared with chelation and with neighbouring-group effects as we saw in Chapter 1. Let us take this enzyme property first and set up a model for the intramolecular hydrolysis of a phosphate ester.

Intramolecular hydrolysis

Look at the structures of the phosphate esters of 1,2-diols (Figure 3.17); both are **diesters** of phosphoric acid. The simpler structure is an ester of glycerol and the more complex is a part structure of a ribonucleic acid molecule. These esters of diols also contain a hydroxyl group that is suitably positioned to attack the phosphorus atom, either by addition or by substitution. The proximity of the hydroxyl group to the phosphorus, held together as they are in the same molecule, is sufficient to overcome the repulsion due to the negatively charged oxygen atoms that surround the phosphorus atom. This type of catalysis, in which a nucleophile (hydroxyl group) forms a covalently bonded intermediate, is often referred to as **nucleophilic catalysis**. Nucleophilic catalysis can occur either inter- or intramolecularly and it is a subclass of the group of mechansims known as **covalent catalysis**. The other main subclass, **electrophilic catalysis**, is well illustrated by the reactions of a number of coenzymes that are discussed in the next chapter. Nucleophilic catalysis is an extremely powerful effect, and can significantly alter the chemistry of a compound. Consider for yourself how nucleophilic catalysis might account for the lability of RNA and the stability of DNA to alkaline hydrolysis. Part structures of DNA and RNA are shown in Figure 3.18, p. 47.

This leaves the question of whether intramolecular hydrolysis occurs via substitution at phosphorus or via addition followed by elimination. The answer to this is complex and it must suffice to say that in most cases, including the reaction catalysed by the enzyme ribonuclease, it has been found that the substitution mechanism (the second that we mentioned in this section) is favoured whenever nucleophilic catalysis can occur.

Metal ions

Phosphate ions can coordinate with many metal ions. For a diphosphate or a triphosphate (like ATP) there are two or three charged groups available through which the phosphate can chelate with the metal ion. This is very important in the chemistry of ATP, as we shall see shortly. Once again, it is possible to demonstrate the effect of a metal ion in a model system (Cooperman & Hsu, 1976; Figure 3.19). In the reaction of the imidazolium phosphoramide with the oxime of pyridine 2-carboxaldehyde, the positively charged imidazolium is a good leaving group. In the absence of a metal ion, no reaction occurred, but when zinc cations were added, the phosphoramide hydrolysed apace. Figure 3.19 indicates the probable mechanism. The nucleophile is

Figure 3.17. Intramolecular hydrolysis of phosphate esters.

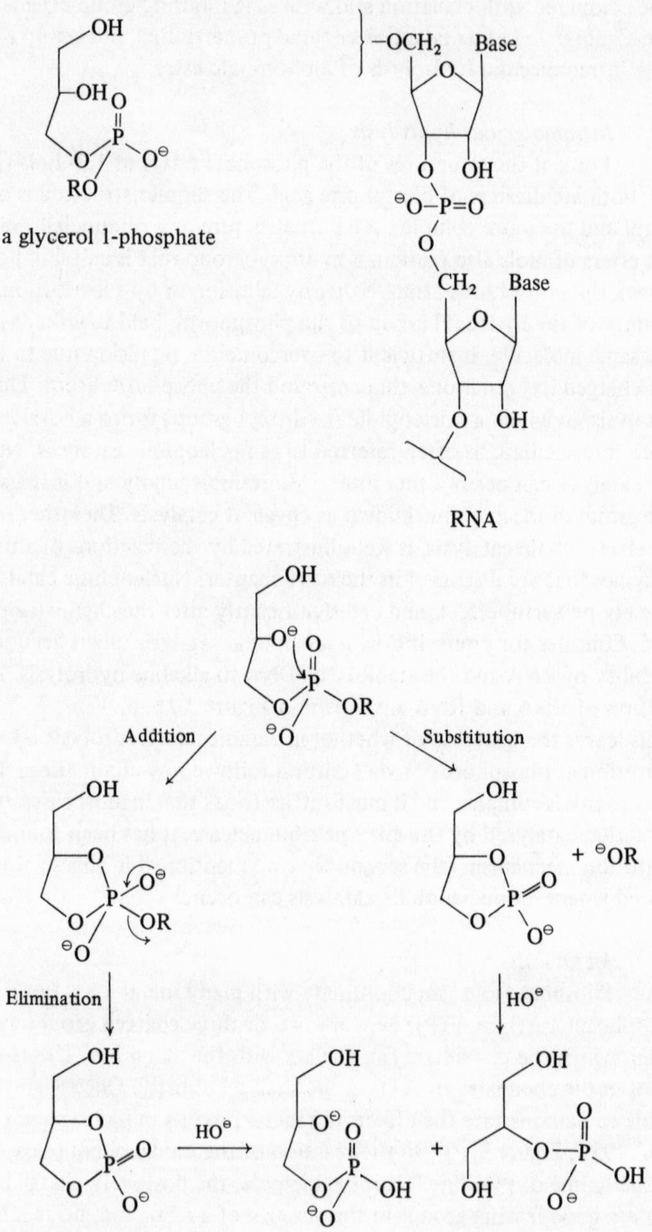

Figure 3.18.

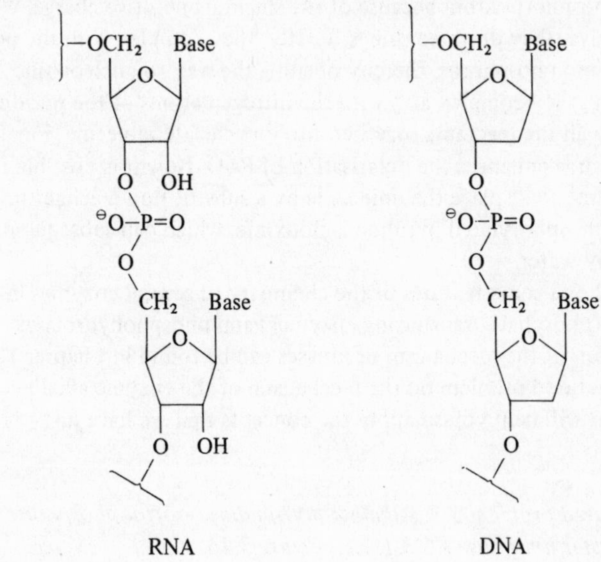

RNA   DNA

Figure 3.19.

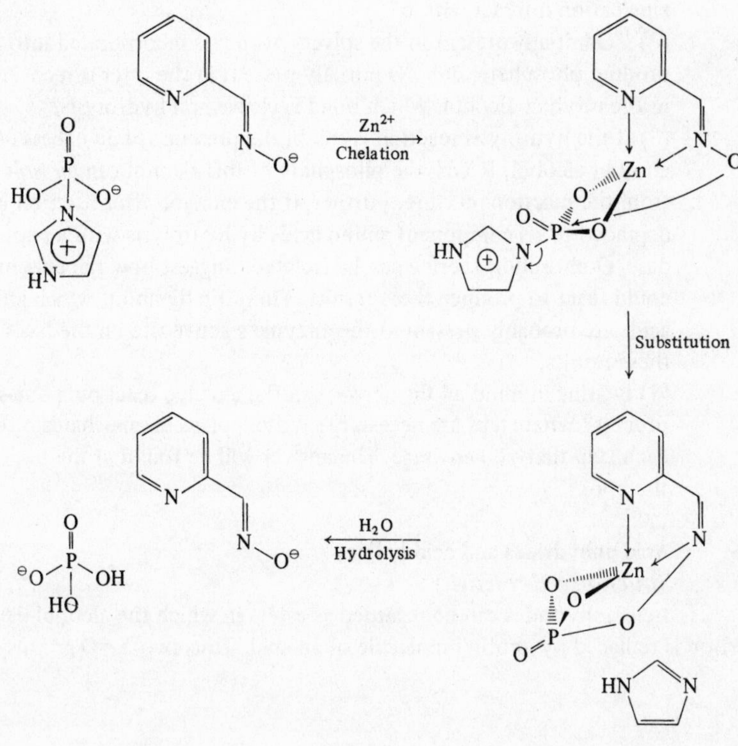

the O⁻ on the pyridine oxime and, in the absence of zinc cations, it cannot approach the phosphorus atom because of the shielding negative charge. When the zinc ions arrive, they do three things. Firstly, they complex with the negative charges around phosphorus, thereby opening the way to nucleophilic attack. Secondly, they complex also with the nitrogen atoms of the pyridine oxime and bring all the reactants together into one chelate molecule. Finally, their positive charge enhances the polarisation of P=O. Now it is possible for the pyridine oxime to displace the imidazole by a substitution mechanism and to yield *O*-phosphorylated pyridine 2-aldoxime, which can subsequently be hydrolysed by water.

This model shows some features of the chemistry of several enzymes including kinases (phosphate-transferring enzymes) and phosphohydrolases. Further discussion of the mechanism of kinases can be found in Chapter 12.

Here is a structured problem on the mechansim of the enzyme alkaline phosphatase that will help you to apply the concepts that we have just described.

> *Structured problem 3.3. Alkaline phosphatase – orthophosphoric acid monoester hydrolase EC 3.1.3.1 (Figure 3.16, p. 44)*
>
> (1) The enzyme contains Zn^{2+}, and functions at a pH optimum of 8. In what form is a phosphate ester present at pH 8? How could the zinc cation interact with it?
>
> (2) ^{18}O initially present in the solvent, water, is incorporated into the product phosphate, but ^{18}O initially present in the ester is recovered in the product alcohol. Which bond is cleaved on hydrolysis?
>
> (3) If the hydrolysis reaction is run in the presence of an excess of another alcohol, R^2OH, the phosphate of this alcohol can be isolated from the reaction mixture. Further, if the enzyme after reaction is degraded to its component amino acids by hydrolysis with a peptidase, *O*-phosphoryl serine can be isolated. Suggest how the enzyme could react to produce these results. Think firstly about which amino acids are probably present at the enzyme's active site on the basis of these results.
>
> (4) Bearing in mind all the above, can the enzyme react by a one-step process? What steps are necessary? Write a plausible mechanism for each step that you envisage. The answer will be found at the back of the book.

3.6. Acid anhydrides and acid halides
3.6.1 Structure and reactivity

Acid anhydrides can be regarded as esters in which the alcohol-derived portion is replaced by another molecule of an acid. Thus two X=O groups are attached to the same oxygen atom. Pyrophosphate can be considered as a phosphoric anhydride, although the typical chemical behaviour of phosphoric anhydrides is chiefly shown by triphosphates such as ATP, as we shall soon see. You should be familiar with simple anhydrides of carboxylic acids such as acetic anhydride. A very important class of anhydrides both in laboratory synthesis and in biosynthesis are mixed anhydrides formed from two different acids (Figure 3.20). Closely related in chemical reactivity to acid anhydrides are acid halides in which a halogen atom takes the place of the alkoxy group of an ester.

Figure 3.20. Acid anhydrides and acyl halides.

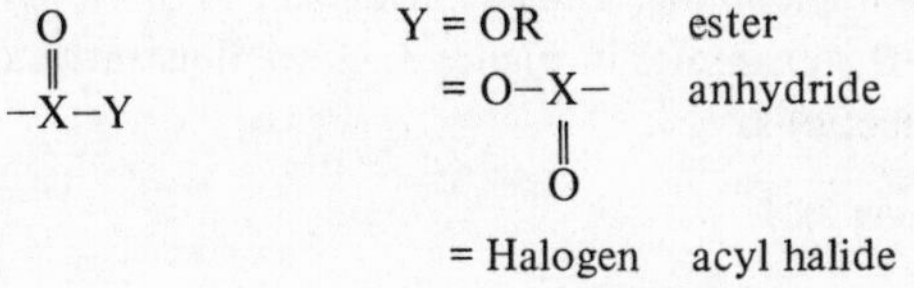

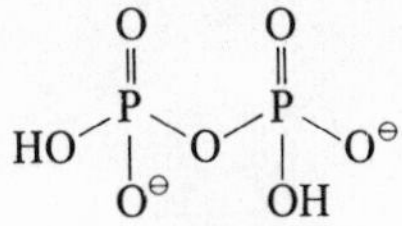

Pyrophosphate or diphosphate

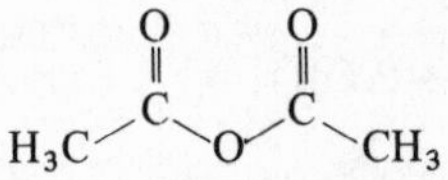

Acetic anhydride

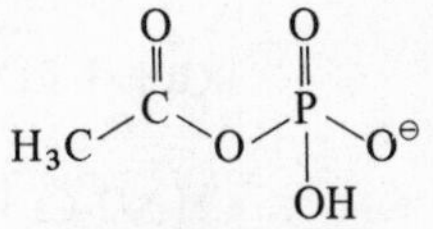

Acetyl phosphate

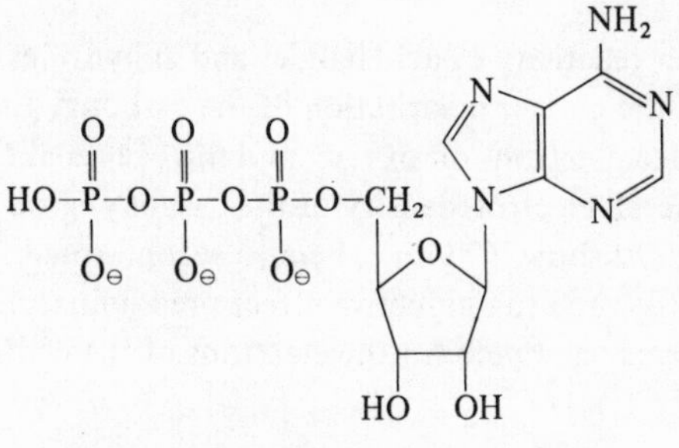

Adenosine triphosphate (ATP)

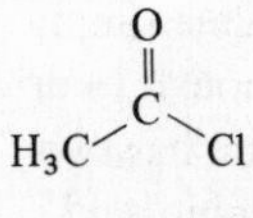

Acetyl chloride

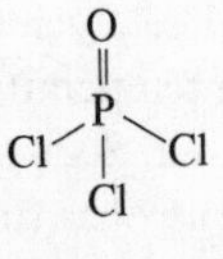

Phosphoryl chloride

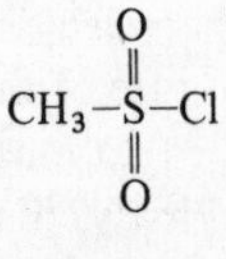

Methane sulphonyl chloride

The name anhydride signifies that water has been removed in order to form the compound. Conversely, if water is added to the anhydride, the parent acids will be formed. It is easy to understand conceptually why anhydrides should react more readily than esters of aliphatic alcohols with water: the product of hydrolysis of an anhydride is an ion that can be stabilised by resonance, whereas an alkoxide ion has a charge localised upon one oxygen atom. In other words, the acid anion is a better leaving group than alkoxide ion. Similarly, halide anions are better leaving groups than alkoxides and, accordingly, acid halides are very reactive towards nucleophiles, including water. We therefore expect that the typical reaction of both acid anhydrides and acid halides will be transfer of the acyl group, whether carbon-, phosphorus- or sulphur-derived, to a nucleophile. The acylation step is often part of a dehydration process and the examples in Figure 3.21 are illustrative of laboratory and biological reactions.

Figure 3.21.

$$CH_3CO.O.COCH_3 + RNH_2 \xrightarrow{\text{aq. NaOAc}} RNH.COCH_3 + CH_3CO_2H$$

$$POCl_3 + ROH \xrightarrow{\text{base}} P(OR)_3$$

$$CH_3SO_2Cl + RNH_2 \xrightarrow{\text{base}} CH_3SO_2NHR$$

$$CH_3CO_2H + ATP \xrightarrow[\text{kinase}]{\text{acetate}} CH_3.CO.O.PO_3 H + ADP$$

A further reason for the greater reactivity of acid halides and anhydrides compared to esters and amides is the greater polarisation of the carbonyl group. In esters or amides, the non-bonded electrons on oxygen and nitrogen make a significant contribution to the overall electron density of the carbonyl group, as the canonical forms in Figure 3.22 show. Chlorine, being a second-row element, can contribute only poorly, and the inductive effect predominates. In an anhydride, two carbonyl groups compete for the electrons of the central oxygen atom.

3.6.2. Synthesis of peptide bonds

Peptide bonds are the amide links formed between two amino acids, and they are the chief covalent linkage in molecules of proteins. It is possible to form a peptide bond between amino acids like glycine and leucine simply by heating together a mixture of the two to a temperature hot enough to drive out the molecule of water that must be removed. Such a procedure has two very serious drawbacks. Firstly, there is no control as to which amino acid reacts with which and how many times they react, and therefore a random co-polymer of the two would result. Secondly, many amino acids that are

Figure 3.22.

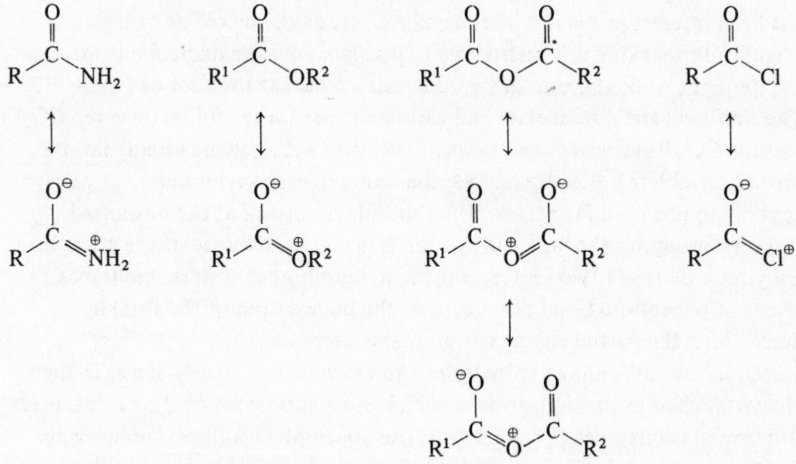

important in proteins are unstable to the vigorous conditions. For example serine dehydrates in the side chain (Figure 3.23, *top*).

For the above reasons, it is essential to have a controlled method of forming peptide bonds between amino acids. Both laboratory and biological processes

Figure 3.23. Activation of amino acids for peptide synthesis.

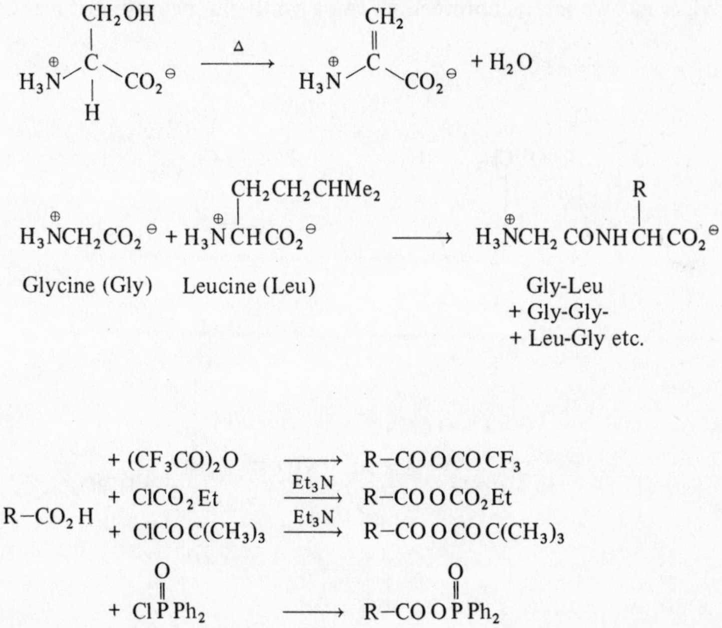

Glycine (Gly)   Leucine (Leu)

Gly-Leu
+ Gly-Gly-
+ Leu-Gly etc.

employ activating groups, which are very commonly mixed anhydrides. Traditionally, mixed carboxylic anhydrides have been used, but recently there has been interest in the use of phosphoric/carboxylic mixed anhydrides. Chemically speaking it is instructive to discover why the nucleophile attacks the desired carbonyl group in the four cases shown at the foot of Figure 3.23. The first two, trifluoroacetate and carbonate, are successful because the CO_2Et and the $COCF_3$ groups exert powerful electron withdrawing effects (cf. the previous problem), thereby making the required carbonyl group highly susceptible to nucleophilic attack. The third blocks attack at the unwanted carbonyl group by the bulk of the t-butyl group. In the case of the phosphoric anhydride there are two factors; the phenyl groups cause steric hindrance to attack at phosphorus, and also decrease the polarisation of the P=O by delocalising the partial charge within themselves.

Nature too uses mixed anhydrides. An example is the early stages in the biosynthesis of purine nucleotides, which are components of the nucleic acids and several coenzymes (Figure 3.24). The nucleophile is ribosylamine as its 5-phosphate, and it attacks an anhydride formed from the amino acid glycine and phosphoric acid. This mixed anhydride in turn is formed from ATP, as we shall see in a moment. It is highly significant that this enzyme-catalysed reaction proceeds quite specifically; none of the other nucleophiles, the hydroxyl groups of the sugar, attack the anhydride. In the absence of enzyme, we would have certainly obtained a complex mixture. In nature, the binding of the substrates to the enzyme's active site controls the specificity – the enzyme provides its own internal protection. But a synthetic chemist must make use of

Figure 3.24.

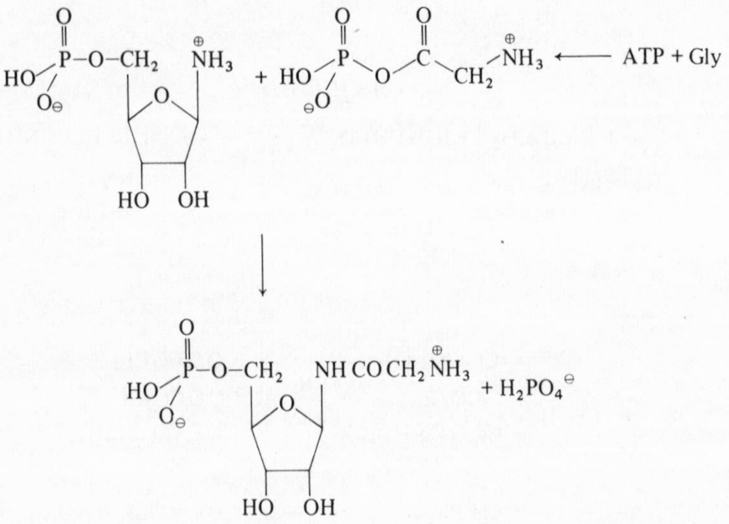

protecting or blocking groups in order to control his synthesis. In the case of nucleophiles like alcohols and amines it is customary to protect them as esters or as amides. Carboxylic acids are usually protected as esters which are, of course, less reactive than the anhydrides used in the peptide-bond-forming step. You should be able to recognise in the examples of protection and coupling to form peptides, the typical chemical reactivity of the reagents and reactants employed (Figure 3.25). All have been mentioned earlier in this chapter.

This section has not considered all the common methods for forming peptide bonds or of protecting reactive groups. You will find a discussion of these topics in Tedder, Nechvatal, Murray & Carnduff (1972). To complete the survey, however, attempt the following two questions; you will find answers at the back of the book.

Problem 3.4.

Why are the esters of the more acidic phenols (e.g. *p*-nitro-, pentachloro- and 2,4,5-trichlorophenols) useful reagents for forming peptide bonds? These compounds are known as 'active esters'.

Problem 3.5.

The scheme in Figure 3.26 illustrates the formation of a peptide bond by means of a very useful dehydrating reagent, a carbodiimide. Describe in words the reactions that are occurring at each stage and compare them with an example from earlier in this chapter or from your own experience.

Figure 3.26.

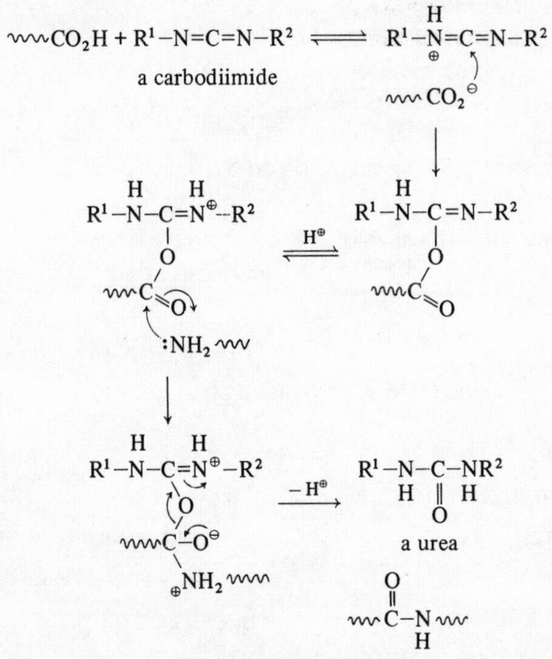

Figure 3.25.

Amine protection

$$RNH_2 + PhCH_2OCOCl \xrightarrow{NaOH} RNH\boxed{COOCH_2Ph}$$

Benzyloxycarbonyl   a urethane
chloride   (an amide ester of carbonic acid)

$$\downarrow H_2/Pd$$

$$RNH_2 \xleftarrow{-CO_2} RNHCO_2H + CH_3Ph$$

$$RNH_2 + Bu^tOCON_3 \longrightarrow RNH\boxed{COOBu^t}$$

$$\downarrow H^\oplus \text{ (HBr } or \text{ CF}_3CO_2H)$$

$$RNH_3^\oplus \xleftarrow{-CO_2} RNHCO_2H + (CH_3)_2C=CH_2$$

Acid protection

$$H_3\overset{\oplus}{N}CHRCO_2^\ominus + MeOH \xrightarrow{H^\oplus} H_3\overset{\oplus}{N}CHR\boxed{CO_2Me}$$

$$\downarrow HO^\ominus$$

$$H_2NCHRCO_2^\ominus$$

$$H_3\overset{\oplus}{N}CHRCO_2 + PhCH_2Cl \longrightarrow H_3\overset{\oplus}{N}CHR\boxed{CO_2CH_2Ph}$$

$$\downarrow H_2/Pd$$

$$H_3\overset{\oplus}{N}CHRCO_2^\ominus + H_3CPh$$

Coupling procedure

$$Bu^tOCONHCH_2CO_2H$$

$$\downarrow$$

$$\boxed{Bu^tOCO}NHCH_2CO\boxed{\begin{matrix}O\\\|\\OPPh_2\end{matrix}} + H_2NCHCO\boxed{OMe} \quad \overset{(CH_2)_4\overset{|}{N}H\boxed{COOCH_2Ph}}{}$$

Protecting group   Activating   Protecting groups
              group

$$\downarrow$$

$$(CH_2)_4\overset{|}{N}HCOOCH_2Ph$$

$$Bu^tOCONHCH_2CONHCHCOOMe$$

anhyd. HBr $\downarrow$ aq. OH$^\ominus$ $\downarrow$ H$_2$/Pd

$$(CH_2)_4\overset{|}{N}HCOOCH_2Ph \qquad\qquad H_2N(CH_2)_4$$

$$H_3\overset{\oplus}{N}CH_2CONHCHCO_2Me \qquad Bu^tOCONHCH_2CONH\overset{|}{C}H$$

Free α-amino   Free ϵ-amino $\overset{|}{C}O_2Me$

$$(CH_2)_4\overset{|}{N}HCOOCH_2Ph$$

$$Bu^tOCONHCH_2CONHCHCO_2^\ominus$$

Free α-carboxyl

3.6.3.   *Adenosine triphosphate*

The importance of adenosine triphosphate, ATP, (Fig. 3.28) in nature can scarcely be overemphasised since most metabolic or biosynthetic sequences require it at some stage. ATP provides the driving force for many reactions by supplying a good leaving group for carbon–carbon bond-forming reactions and elimination reactions. You will often see pseudo-thermodynamic discussions of the role of ATP in terms of the free energy of hydrolysis to adenosine diphosphate and adenosine monophosphate, but since very few biological reactions either involve these transformations directly or, more importantly, take place in the closed equilibrium systems to which thermo-dynamic arguments rigorously apply, it is better to think of the chemistry of ATP in terms of its function in the reactions.

We have already come across several examples of ATP acting as a donor of pyrophosphate (diphosphate) in this chapter. The pteridine diphosphate (Figure 3.4), dimethylallyl and isopentenyl diphosphates (Figure 3.15), and the pyrimidine diphosphate (Figure 3.15) all derive from the general reaction shown in Figure 3.4 at some stage in their biosynthesis (not necessarily the stage in which they themselves are formed). All of these phosphate esters and diphosphate esters are reactive compounds towards substitution by nucleophiles.

Figure 3.27. Phosphate derivatives as activating groups.

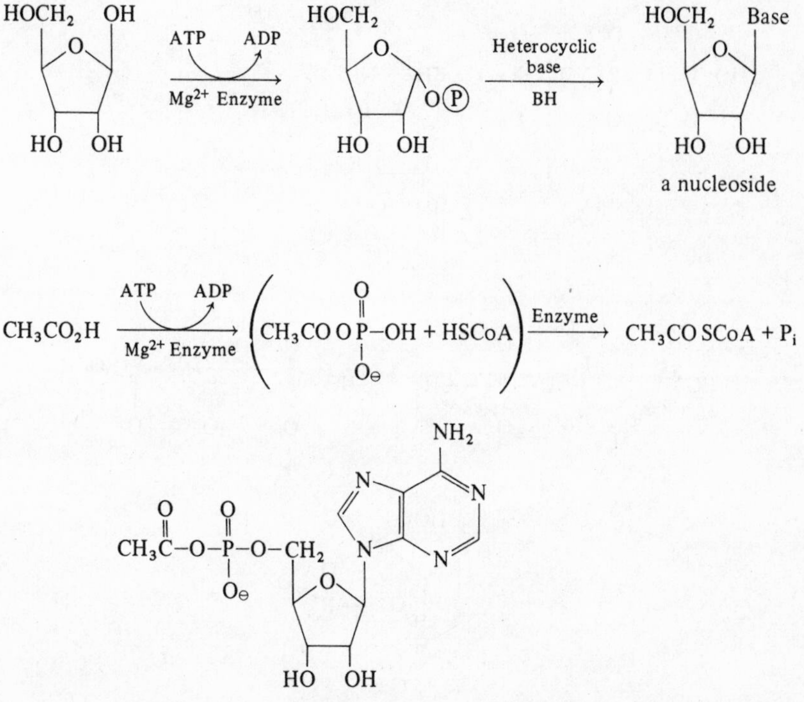

Acetyl adenylate

There exists one further method by which ATP activates molecules to substitution reactions – it can form a mixed anhydride with a carboxylic acid. A well-studied case is the biosynthesis of acetyl coenzyme A which is the subject of a structured problem at the end of this section. The key intermediate formed in bacteria is acetyl phosphate in contrast to eukaryotic cells in which acetyl adenylate is formed (Figure 3.27).

We need to be able to account for this multiplicity of reactions by considering the chemical structure of ATP. It is a double anhydride and so there are several possible positions for hydrolysis or for transferring phosphate to a nucleophile. (Figure 3.28 *top*). If mode (*a*) is followed, ATP transfers a phosphate and produces ADP. Mode (*b*) is the diphosphate-donating mode, mode (*c*) allows acetyl adenylate formation, and mode (*d*) occurs in only one known case, the synthesis of the important biological methylating agent, *S*-adenosyl methionine (see Chapter 6). Chemically, it would be very hard to predict which of these reaction modes would be preferred in any particular case but the enzyme-catalysed reactions are highly specific. It is thought that the enzyme controls which reaction of ATP takes place by chelating with the phosphates. All the enzymes that catalyse reactions of ATP in which phos-

Figure 3.28.

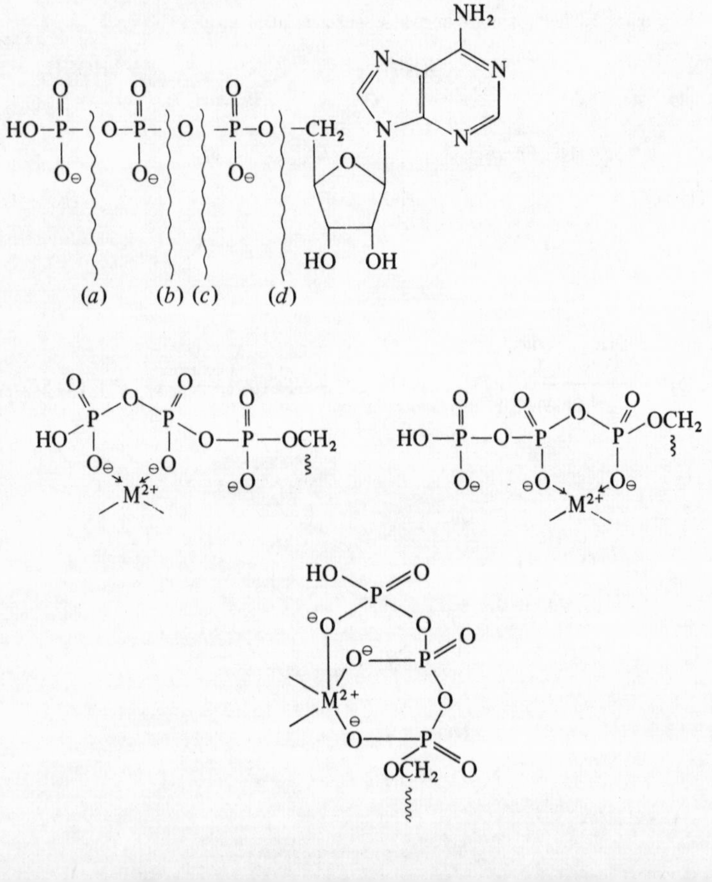

phate transfer occurs require for their activity the presence of a divalent metal cation, usually magnesium or manganese which have similar ionic radii. Not only will the metal ions provide control for the direction of the reaction, they will also catalyse phosphate transfer by shielding the negatively charged oxygen atoms from the incoming nucleophile and enhancing P=O polarisation in the usual way. Once again, nature gets everything she can from the inherent chemistry of the reactants. We can formulate some of these chelate compounds schematically (Figure 3.28, *bottom*).

There are a number of other nucleotides of other purines and pyrimidines that have important functions in biosynthesis. They behave in a similar way to the transfer of adenosine monophosphate (mode (*b*)), except that usually the nucleotide diphosphate is the group transferred. The transfer of glucose units aided by uridine triphosphate is an example. Cytidine nucleotides are involved in the biosynthesis of phosphatidyl cholines, guanine nucleotides are significant in protein biosynthesis, but inosine derivatives are rarely employed.

There are also a number of non-nucleotide phosphorylating agents in nature that phosphorylate ADP during the breakdown of hexoses to pyruvate, a pathway known as glycolysis (Appendix 1). 1,3-Diphosphoglyceric acid (Figure 3.29) contains a mixed phosphoric–carbonic anhydride, which we have seen earlier acting as an acylating agent. If the enzyme 1,3-diphosphoglycerate

Figure 3.29. ATP formation from phosphate esters.

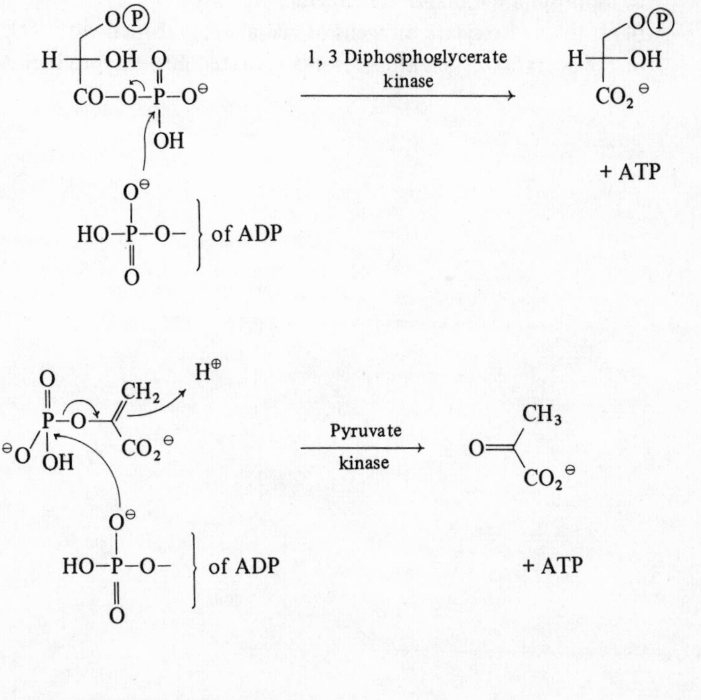

kinase directs ADP to react at phosphorus instead of carbon, 1,3-diphosphoglyceric acid can then behave as a phosphorylating agent. Phosphoenol pyruvate is a good phosphorylating agent because, like most enol esters, cleavage of the ester linkage affords the enol, which rapidly tautomerises to the more stable keto form. The keto form could be termed a good leaving group.

Structured problem 3.6. Acetyl coenzyme A synthase

From each piece of information presented below, make a deduction about the mechanism of action of this enzyme. Finally, place all the information in a coherent mechanistic scheme.

The overall equation for the reaction is:

$$CH_3CO_2H + ATP + CoASH \rightarrow CH_3CO.SCoA + AMP + H_2O$$

Coenzyme A is a complex molecule containing a thiol at one terminus. For the purposes of this problem and for most simple discussions of its chemistry, it can be regarded as CoA–SH.

Adenine–ribose–$\circled{P}$–$\circled{P}$–$CH_2C(Me_2)CH(OH)$ $CONH(CH_2)_2$ $CONHCH_2CH_2SH$

(1) The enzyme requires magnesium cations for activity.

(2) Acetyl adenylate (acetyl–AMP) is converted into acetyl CoA when treated with the enzyme and coenzyme A.

(3) Acetyl–AMP is converted into ATP and acetic acid in the presence of diphosphate ion and the enzyme.

(4) If the carboxylate oxygens of acetate are labelled with ^{18}O, one of the acetate oxygen atoms is incorporated into the product AMP.

4 Condensation reactions

4.1. Introduction: scope and definitions

When two molecules combine together with the concomitant elimination of a small molecule such as water, we say that a condensation reaction has taken place. Nature uses condensation reactions to construct a large number of molecules, both simple and complex, for example fatty acids, phenolic compounds and heterocyclic compounds (Figure 4.1). The chemistry of these biosynthetic reactions is easy to interpret if we have an understanding of simple laboratory condensation processes.

Figure 4.1.

'Claisen'-type condensation

$$CH_3 CO\,SACP + CH_2\,CO\,SACP \longrightarrow CH_3\,CO\,CH_2\,CO\,SACP$$
$$\overset{|}{CO_2H}$$

$$\downarrow$$

Fatty acids

'Aldol'-type condensation

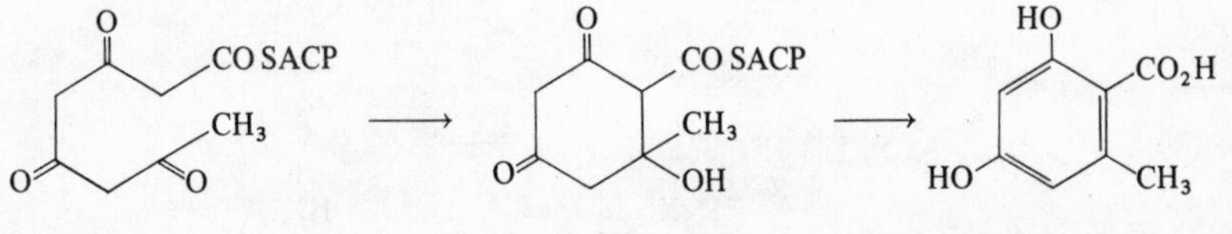

Phenolic compounds

Usually, there are many steps to one condensation reaction. The principal of these is addition of a nucleophile to a polarised double bond which precedes elimination of a small molecule. Familiar simple derivatives of aldehydes, ketones and esters such as oximes and phenylhydrazones are prepared by

Figure 4.1 cont.

Imine formation

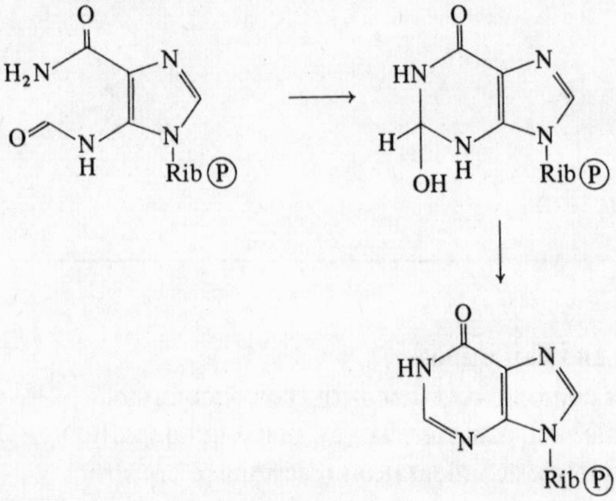

Inosylic acid

SACP = acyl carrier protein

Rib(P) = ribose 5′-phosphate

typical condensation reactions (Figure 4.2). The general mechanism shows that a condensation reaction forms C–X bonds, where X = C, N or O. It is in this property that the synthetic importance of condensation reactions lies.

Figure 4.2.

$$RCHO + H_2NNHPh \longrightarrow RCH{=}N{-}NH{-}Ph + H_2O$$

$$RCO_2Et + H_2NOH \longrightarrow R{-}C(OH){=}NOH + EtOH$$

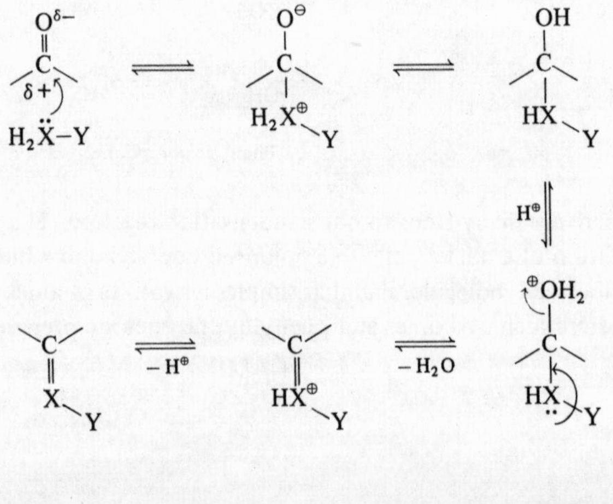

Note also in passing that the reactions of the nitrogen nucleophiles with the carbonyl compounds are usually acid-catalysed, involving protonation of the carbonyl group, thereby enhancing its polarisation; protonation also aids the departure of the small molecule. You will notice formal similarities between this mechanism for condensation reactions and one of the mechanisms that we discussed for ester hydrolysis.

4.1.1. Carbanions and carbon acids

The simple examples above all use the lone pair of nitrogen as a nucleophile to form a C–N bond. If we wish to form a C–C bond by a condensation reaction, obviously we require a carbon species that is a nucleophile. Carbon anions or **carbanions** meet the requirements. A carbanion can be formed simply by removal of a carbon-bound proton from a molecule, and the ease with which this can be accomplished depends greatly upon the structure of the compound. Recall from Chapter 3 that delocalisation of negative charge in the anions of carboxylic, phosphoric and sulphonic acids increases acidity (the anions are resonance stabilised). The same effect can be observed with carbon acids – the more easily the charge can be delocalised, the stronger the acid will be. You can see how this qualitative, but powerful rule of thumb works in the examples in Table 4.1. Write down the structures of some of these examples and see how delocalisation of the negative charge can be effected. Fluorene is drawn out as an example in Figure 4.3, p. 62.

Table 4.1. pK_a' *values of some carbon acids*

Compound class	Example	pK_a'
Hydrocarbons	Methane	50
	Benzene	35
	Fluorene	25
	Cyclopentadiene	16
Carbonyl compounds	Acetonitrile	25
	Ethyl acetate	24
	Acetone	20
Dicarbonyl compounds	Diethyl malonate	13
	Malononitrile	11
	Acetyl acetone	10
	Nitromethane	10

Clearly, very strong bases are required to prepare anions of hydrocarbons, with the notable exception of cyclopentadiene, and the usual procedure is to react an alkyl or aryl halide with lithium or with magnesium in a suitable dry solvent. The latter generates the well-known Grignard reagent. The anions formed can then be used in one of two ways – either as strong bases them-

Figure 4.3.

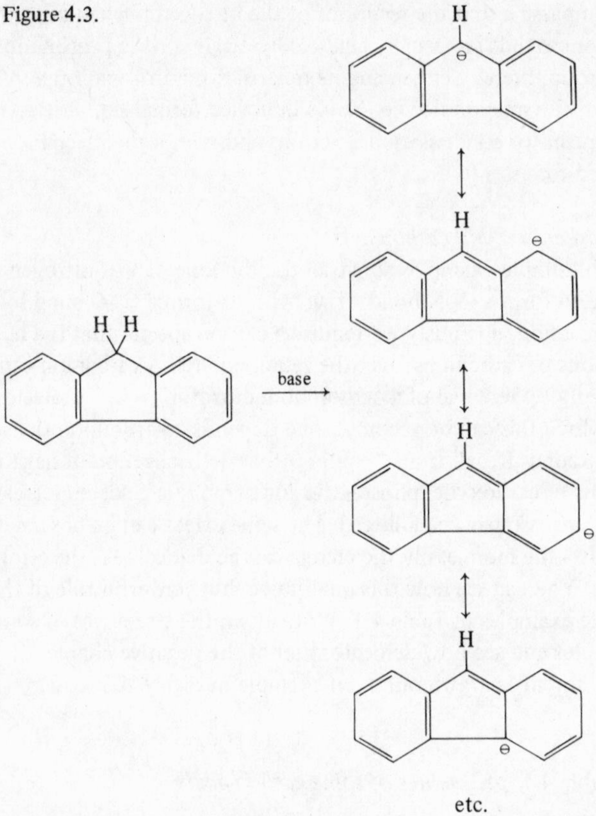

etc.

selves to give the anions of stronger acids than themselves, or as nucleophiles to form carbon–carbon bonds (Figure 4.4).

The fact that such strong bases are required to form useful carbanions places nature at a severe disadvantage, because enzyme-catalysed reactions take place in aqueous solution in which the strongest available base is hydroxide ion. Obviously some form of catalysis of carbanion formation will be required, and we shall see towards the end of this chapter how several coenzymes can provide the necessary delocalisation and stabilisation of the anion. The rate of removal of protons bound to carbon and the rate of protonation of carbanions are not always as fast as proton transfers between oxygen acids, which are virtually as rapid as the diffusion of molecules in solution. The step in a reaction in which a proton is removed from carbon is therefore a good candidate for the rate-determining step.

4.1.2. Condensation reactions of aldehydes and ketones

Whenever an aldehyde or a ketone that bears a hydrogen atom on the

Figure 4.4. Formation and reactions of some carbanions.

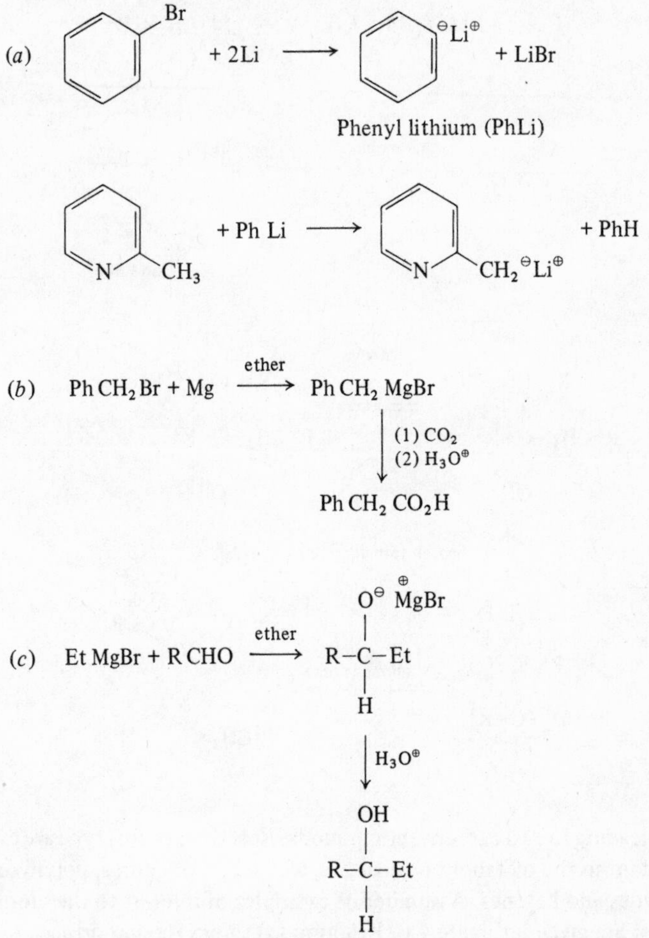

(a)

$$\text{PhBr} + 2\text{Li} \longrightarrow \text{Ph}^{\ominus}\text{Li}^{\oplus} + \text{LiBr}$$

Phenyl lithium (PhLi)

$$\text{(2-methylpyridine)} + \text{Ph Li} \longrightarrow \text{(pyridine-2-yl)} \text{CH}_2^{\ominus}\text{Li}^{\oplus} + \text{PhH}$$

(b) $\text{Ph CH}_2\text{Br} + \text{Mg} \xrightarrow{\text{ether}} \text{Ph CH}_2\text{MgBr}$

$$\downarrow \begin{array}{l}(1)\ \text{CO}_2 \\ (2)\ \text{H}_3\text{O}^{\oplus}\end{array}$$

$$\text{Ph CH}_2\text{CO}_2\text{H}$$

(c) $\text{Et MgBr} + \text{R CHO} \xrightarrow{\text{ether}}$

$$\begin{array}{c} \text{O}^{\ominus}\ \overset{\oplus}{\text{MgBr}} \\ | \\ \text{R}-\text{C}-\text{Et} \\ | \\ \text{H} \end{array}$$

$$\downarrow \text{H}_3\text{O}^{\oplus}$$

$$\begin{array}{c} \text{OH} \\ | \\ \text{R}-\text{C}-\text{Et} \\ | \\ \text{H} \end{array}$$

carbon α to the carbonyl group is treated with a sufficiently strong base, an acid/base equilibrium between the carbonyl compound and its enolate anion is set up. In an enolate, the negative charge is delocalised between carbon and oxygen (Figure 4.5). Enolate anions are effective nucleophiles and they add readily to carbonyl groups to form addition products known as aldol adducts. In some cases, the aldol adduct is isolable but very often it dehydrates to form an α,β-unsaturated carbonyl compound and thereby completes the condensation reaction.

Dehydration is especially favoured when the double bond that is formed is further conjugated to a benzene ring or to another double bond. All the reactions prior to dehydration are equilibria and this implies that a β-hydroxy aldehyde or ketone (an aldol adduct), can undergo the reverse reaction (retro-

Figure 4.5.

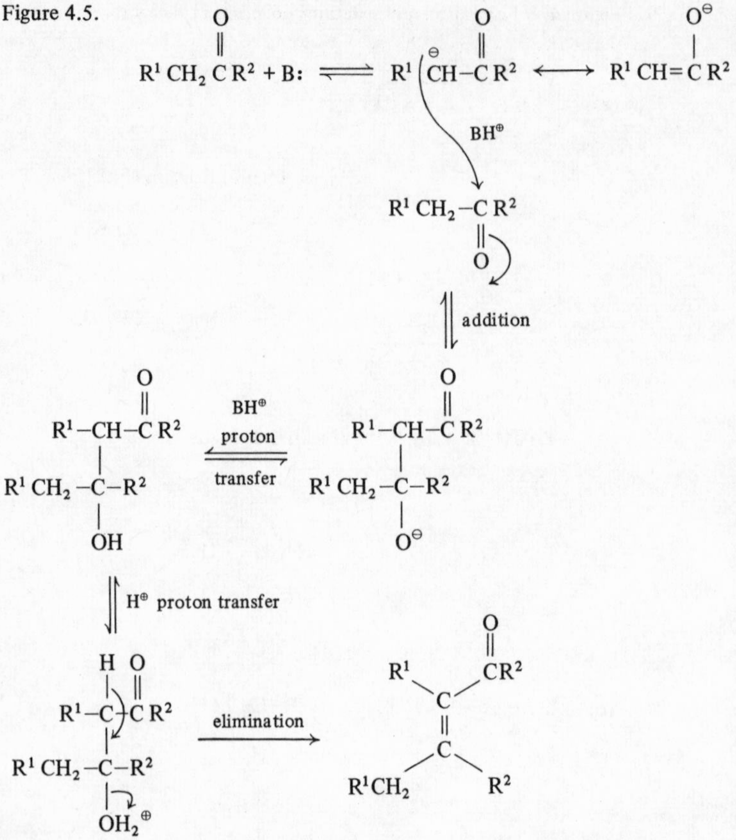

$$R^1 CH_2 C R^2 + B: \rightleftharpoons R^1 \overset{\ominus}{C}H - \overset{O}{\overset{\|}{C}} R^2 \longleftrightarrow R^1 CH = C R^2$$

$$\downarrow BH^\oplus$$

$$R^1 CH_2 - \overset{\|}{\underset{O}{C}} R^2$$

$$\big\Updownarrow \text{ addition}$$

$$\begin{array}{c} R^1 - CH - \overset{O}{\overset{\|}{C}} R^2 \\ | \\ R^1 CH_2 - C - R^2 \\ | \\ OH \end{array} \quad \overset{BH^\oplus}{\underset{\text{transfer}}{\overset{\text{proton}}{\rightleftharpoons}}} \quad \begin{array}{c} R^1 - CH - \overset{O}{\overset{\|}{C}} R^2 \\ | \\ R^1 CH_2 - C - R^2 \\ | \\ O^\ominus \end{array}$$

$$\big\Updownarrow H^\oplus \text{ proton transfer}$$

$$\begin{array}{c} \overset{H\ O}{\underset{| \ \|}{R^1 - C + C R^2}} \\ | \\ R^1 CH_2 - C - R^2 \\ | \\ OH_2{}^\oplus \end{array} \quad \xrightarrow{\text{elimination}} \quad \begin{array}{c} R^1 \diagdown \quad \diagup CR^2 \\ C \\ \| \\ C \\ R^1 CH_2 \diagup \quad \diagdown R^2 \end{array}$$

aldol) leading to two carbonyl compounds. Reactions of this type are especially important in the metabolism of sugars, which are, of course, polyhydroxy aldehydes and ketones. A number of examples of interest to the bioorganic chemist are given in Figure 4.6. Reaction (*a*) shows the importance of condensation reactions in synthesis. The cyclisation to fuse a second six-membered ring on to the existing one (an annelation) was first developed by Robinson and was until recently the mainstay of steroid and terpene synthesis. Recent developments have shown the use of cationic carbon intermediates to be advantageous (see Chapter 5). Formally analogous to the previous example is the cyclisation (Figure 4.6*b*) by which the polyketide or acetate-derived natural products orsellinic acid and 6-methylsalicylic acid are formed. Note the enolisation and dehydration steps to attain the stability of the aromatic product. The driving force for the retroaldol reaction (4.6*c*) is the relief of strain of the four-membered ring. A very common retroaldol reaction of carbohydrates is the base-catalysed cleavage of one sugar into two smaller molecules. In glycolysis, a reaction formally of this type is catalysed by the enzyme aldolase (Appendix 1).

Figure 4.6. Condensation reactions of aldehydes and ketones. (*a*), (*b*) aldol-type condensations; (*c*), (*d*) reverse reaction, or retroaldol; (*a*), (*b*) and (*e*) aldol condensations followed by dehydration.

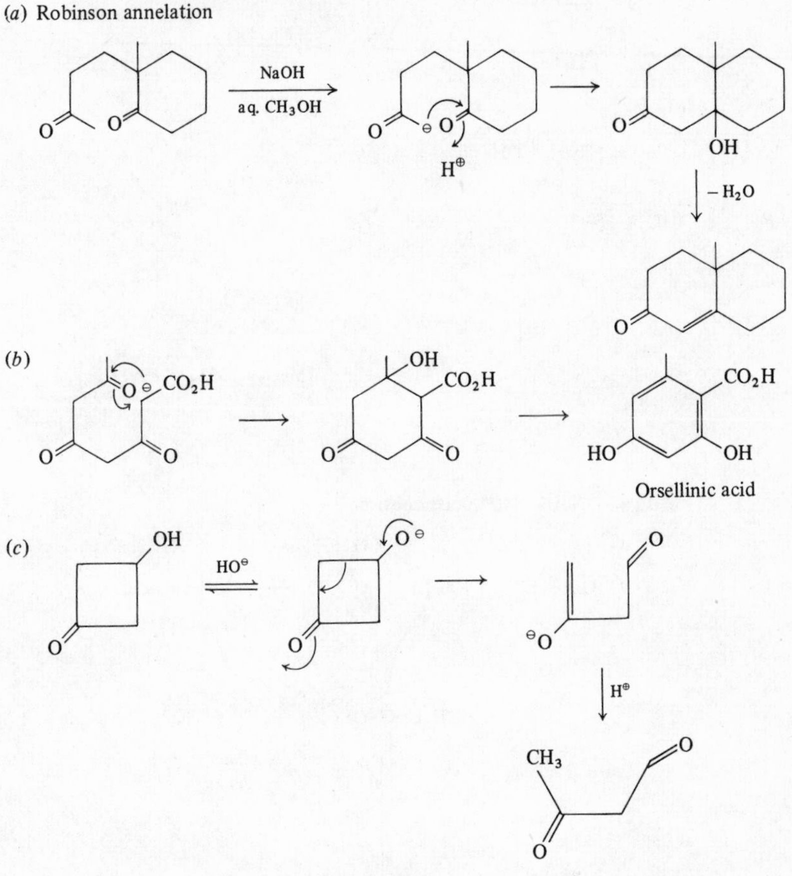
(a) Robinson annelation
NaOH
aq. CH₃OH
O
H⊕
OH
– H₂O
(b)
CO₂H
OH
HO
Orsellinic acid
(c)
HO⊖
H⊕
CH₃

Example 4.6*e* illustrates the condensation of a carbanion derived from a molecule other than an aldehyde or ketone, in this case from nitromethane. It is a reaction that has had wide application in the synthesis of compounds for the study of the biosynthesis of alkaloids based on the benzyl isoquinoline ring system (see Chapter 7 and later in this chapter).

4.1.3. *Condensation reactions of esters*

When esters undergo condensation reactions the carbanion-forming and addition steps are closely analogous to the reactions of aldehydes and ketones. The elimination step results in the formation of a β-keto-ester which is acidic (Table 4.1); it therefore reacts with the base to form an anion. The stability of this anion is due to charge delocalisation, and its irreversible formation provides the driving force to pull the initial equilibrium from its normal position far to the left over towards the reaction products (Figure 4.7).

In outline this reaction, which is known as the Claisen ester condensation,

Figure 4.6. (*d*) and (*e*).

(*d*)

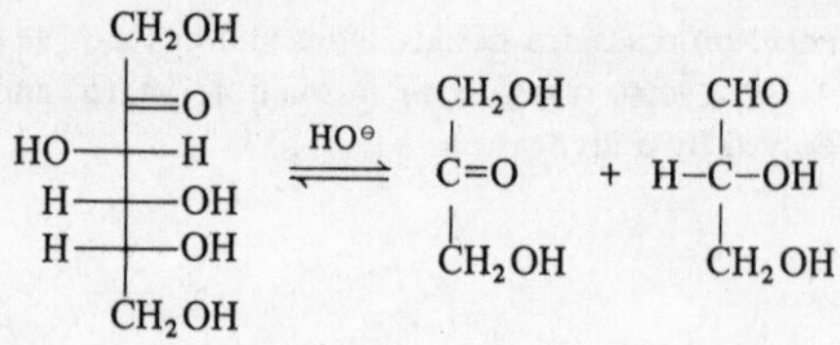

(*e*)

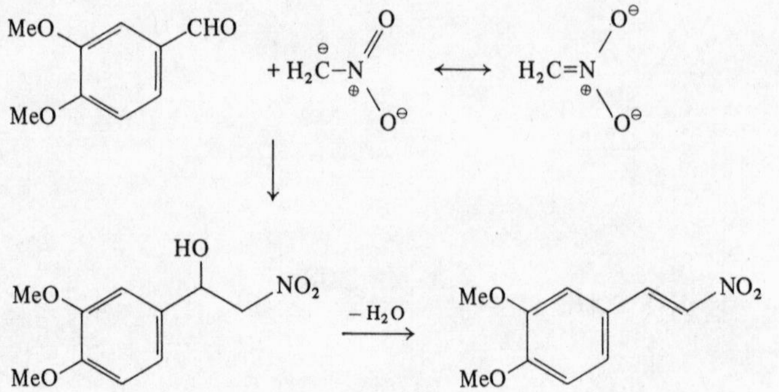

Figure 4.7. Claisen ester condensation.

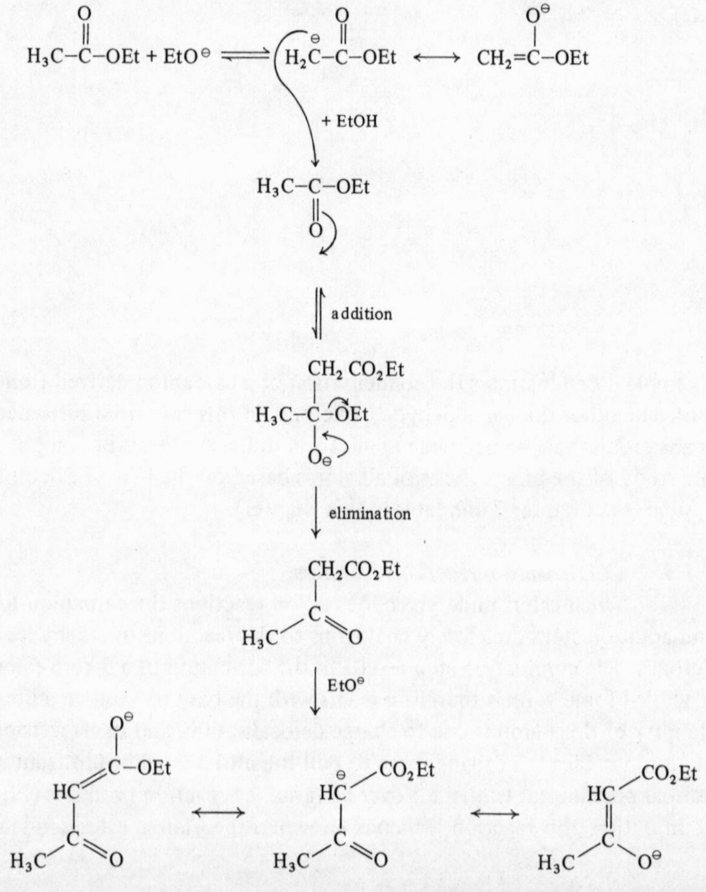

is just what is required to build up the β-dicarbonyl system that cyclised to orsellinic acid in the example in Figure 4.6b. Nature follows the condensation plan using esters of coenzyme A as reactants. Coenzyme A has a terminal thiol (SH) group and its esters (thiol esters) are especially reactive, as we shall see shortly. Thiol esters are also important in the biosynthesis of fatty acids and several condensation reactions in the tricarboxylic acid cycle. To illustrate the formal analogy between β-keto-ester synthesis by the Claisen condensation and the enzyme-catalysed process, follow through Figure 4.8. It shows a possible biosynthetic route to the natural product phloroacetophenone via four separate condensation reactions.

Figure 4.8.

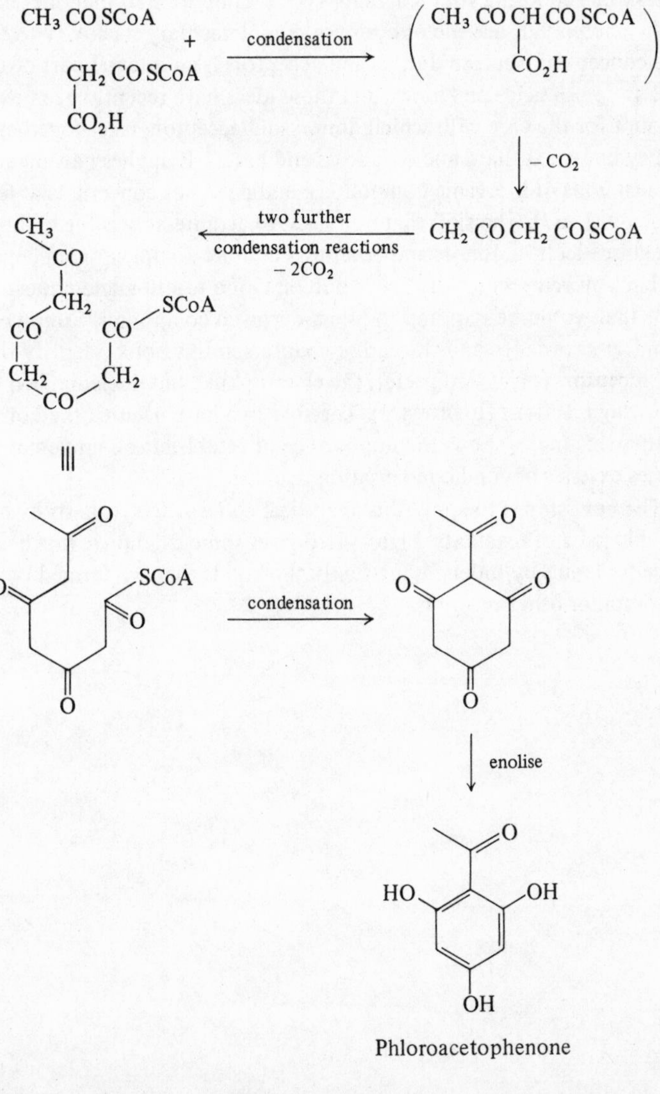

Phloroacetophenone

4.1.4. Condensation reactions generalised

The carbon atom of a carbonyl group is electrophilic because it bears a partial positive charge due to the polarisation of the carbon–oxygen double bond. In other words, the carbon atom can act as an electron **acceptor**. Similarly, nucleophiles like carbanions and nitrogen-containing compounds can be described as electron **donors**. In all of the above condensation reactions, bonds are formed between an electron donor and an electron acceptor. The first step in setting up a synthetic scheme using a condensation reaction is to ensure that a complementary arrangement of electron-pair donors (nucleophiles) and electron-pair acceptors (electrophiles) is available to form the bond required. Mechanistic details can be considered later. It matters little whether you prefer to think of donors and acceptors of electrons or of nucleophiles and electrophiles when choosing your reactants – what is important is that one component of the chosen pair has the opposite or complementary reactivity to the other. The concept of electron donors and acceptors is an integral part of the Lewis description of acids and bases, and these ideas have recently been extended to account for the ease with which donors and acceptors react together in a theory known as 'hard and soft acids and bases'. It applies particularly well to the reactions of inorganic compounds, and need not concern us at length here (Pearson, 1967). The skill that you need to acquire now is the ability to recognise electron donors and acceptors in quite complex organic molecules so that you can predict the likely outcome of a reaction and choose the reactants that would be required to prepare a given compound. From your general chemical knowledge and the earlier chapters in this book, identify the donors and acceptors (or, if you prefer, the electrophiles and nucleophiles) in the following reactions (Figure 4.9). The first two have been worked out for you. As these examples show, the importance of establishing complementary reactivities extends beyond condensation reactions.

The next step is to apply this analytical concept to synthesis by choosing suitable pairs of reactants. Figure 4.10 gives some examples. It is best to tackle these problems by indentifying firstly the bonds that are formed by the condensation or other reaction.

Figure 4.9.

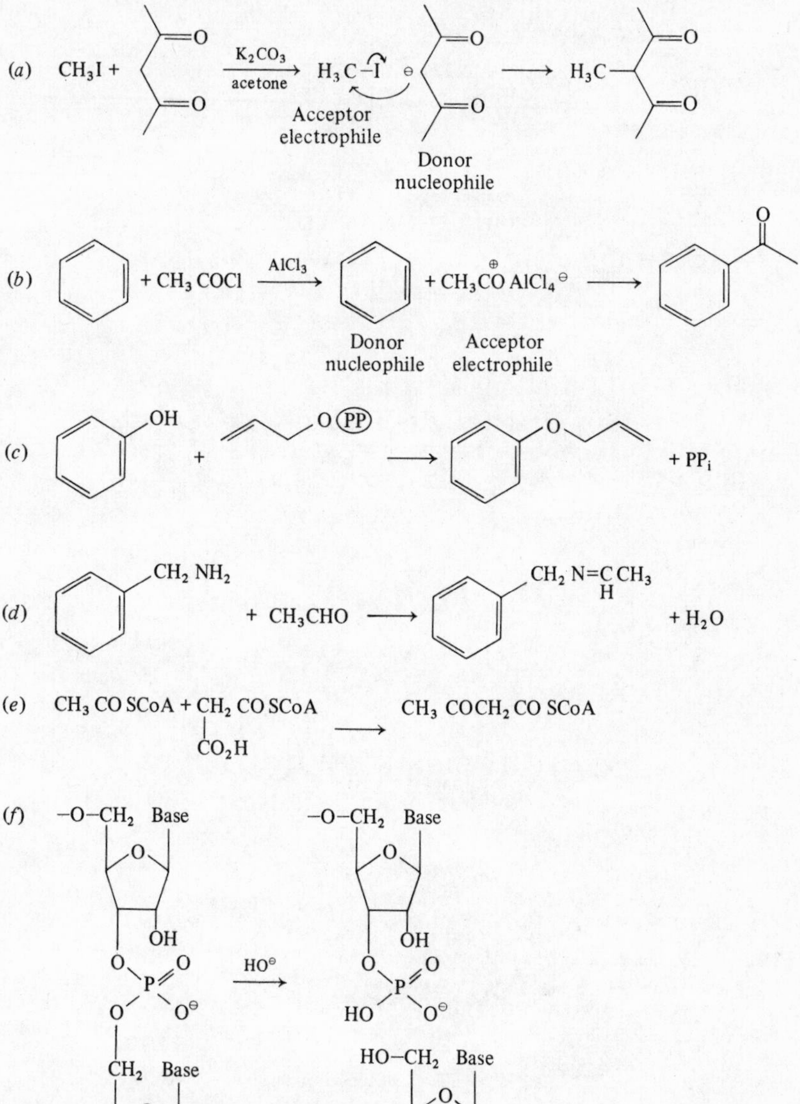

(a) CH_3I +

$\xrightarrow[\text{acetone}]{K_2CO_3}$ $H_3C\overset{\frown}{-}I$ $^\ominus$

Acceptor
electrophile

Donor
nucleophile

$\longrightarrow$ H_3C—

(b)

+ CH_3COCl $\xrightarrow{AlCl_3}$

+ $CH_3\overset{\oplus}{C}O\ AlCl_4{}^\ominus$ $\longrightarrow$

Donor
nucleophile

Acceptor
electrophile

(c)

OH

+

$O\ (PP)$

$\longrightarrow$

O

+ PP_i

(d)

CH_2NH_2

+ CH_3CHO $\longrightarrow$

$CH_2\ N=\underset{H}{C}CH_3$

+ H_2O

(e) $CH_3\ CO\ SCoA$ + $CH_2\ CO\ SCoA$
$\qquad\qquad\qquad\quad |$
$\qquad\qquad\qquad CO_2H$
$\longrightarrow$ $CH_3\ COCH_2\ CO\ SCoA$

(f)

—O—CH_2 Base

OH

$O\overset{O}{\underset{O^\ominus}{-P=}}O$

CH_2 Base

$\xrightarrow{HO^\ominus}$

—O—CH_2 Base

OH

$HO\overset{O}{\underset{O^\ominus}{-P=}}O$

$HO-CH_2$ Base

O OH

(g) $CH_3CO_2CH_2CH_2\overset{\oplus}{N}(CH_3)_3$ $\xrightarrow[\text{esterase}]{\text{acetylcholine}}$ $CH_3CO_2{}^\ominus$
$+ HO\ CH_2CH_2\overset{\oplus}{N}(CH_3)_3$

Figure 4.10.

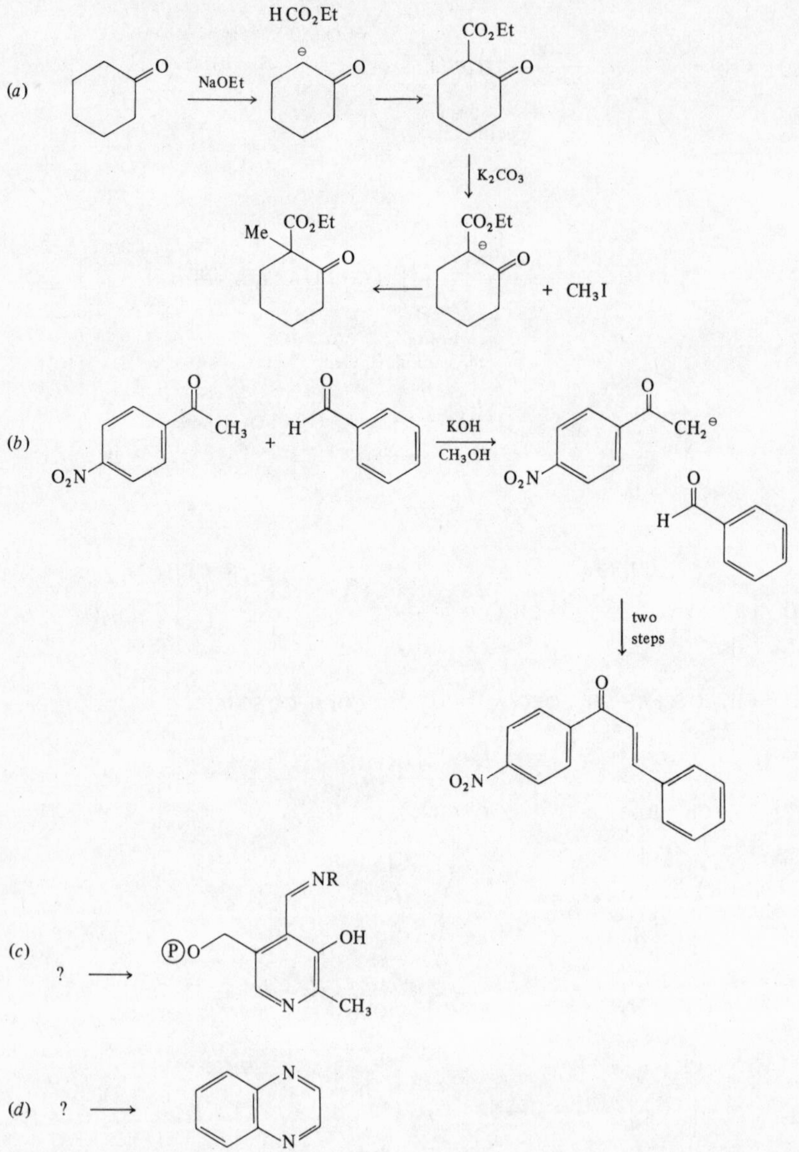

(a)

(b)

(c)

(d)

4.2. The synthesis of heterocyclic compounds by condensation reactions

Heterocyclic compounds contain rings made up of carbon and one or more other elements. In nature, they are found in the bases of nucleosides, in alkaloids, in the amino acids histidine and tryptophan, and in several coenzymes. All of the ring systems are biosynthesised by a sequence of condensation reactions.

Figure 4.11. Examples of the synthesis of heterocyclic compounds via condensation reactions.

(a) Pyridines

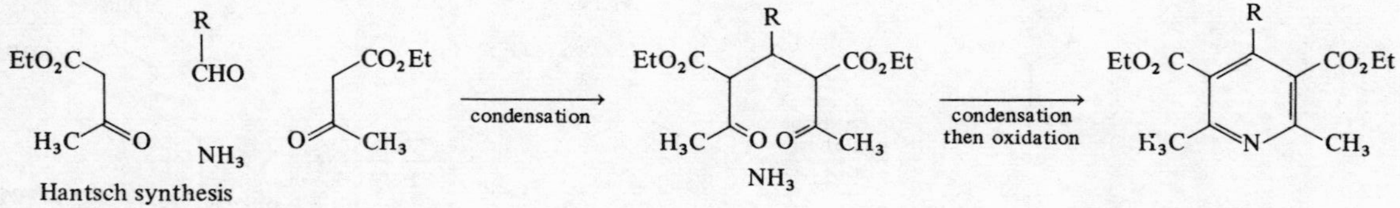

Hantsch synthesis

(b) Quinolines

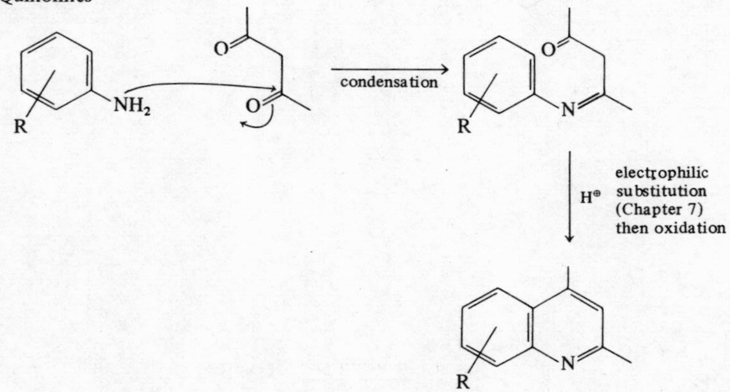

Figure 4.11. (continued).

(c) Isoquinolines

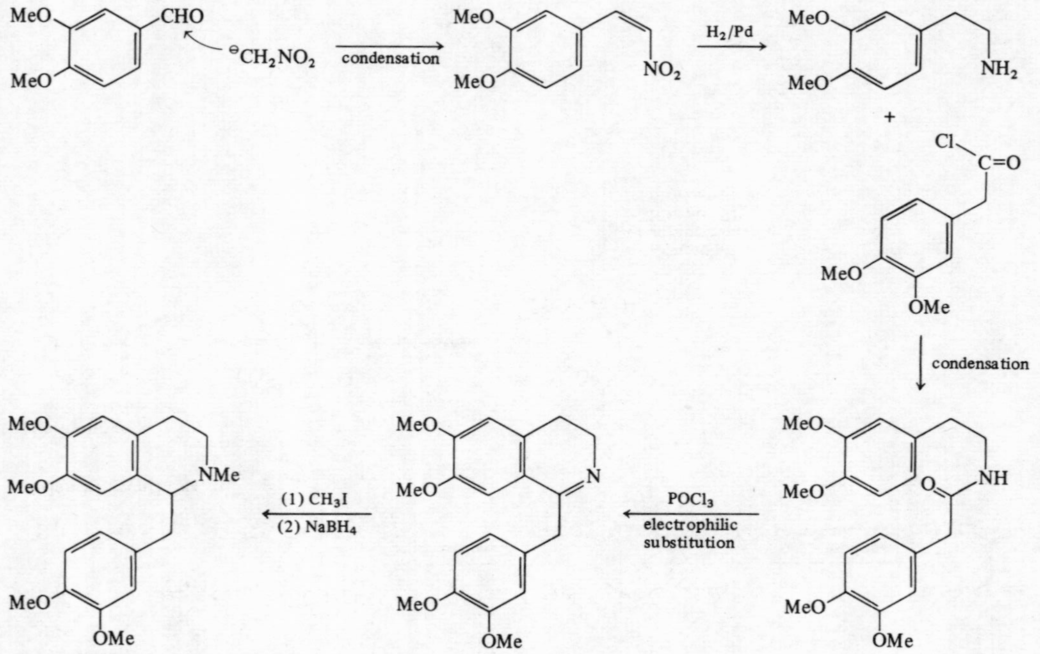

A benzyl isoquinoline alkaloid

(*d*) Pyrroles, furans and thiophenes

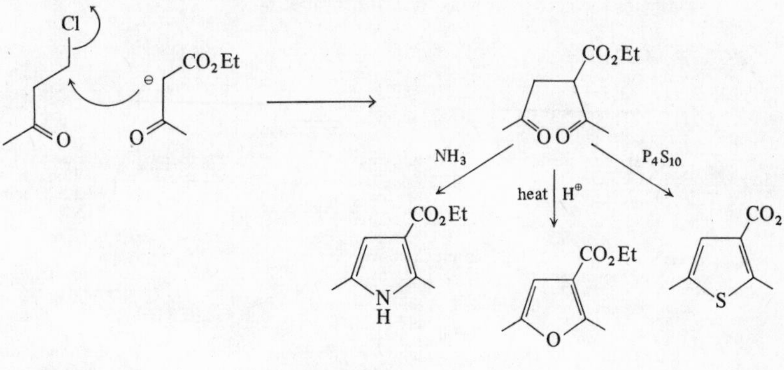

(*e*) Pyrroles

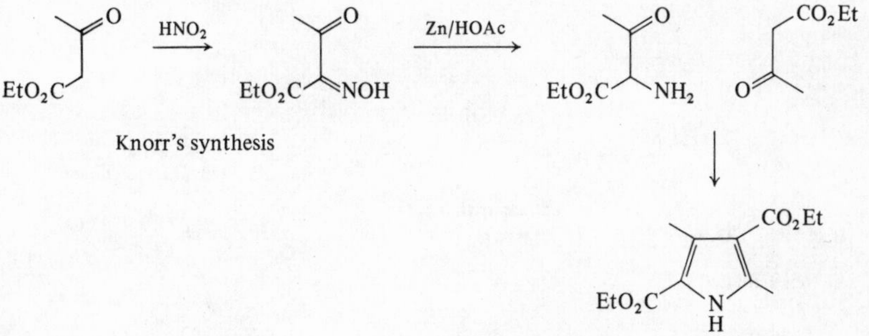

Knorr's synthesis

Figure 4.11 illustrates some important synthetic procedures for laboratory reactions. Notice that it is sometimes necessary to use a masked electron donor or electron acceptor. In principle, almost any donor/acceptor combination that you can think of will, under the correct conditions, give rise to a heterocyclic product. However, because the reactants are always polyfunctional, side-reactions will take place, affording mixtures of products that may be hard to separate. Details of the standard routes to various heterocycles are helpfully discussed by Joule & Smith (1978).

4.2.1. Porphobilinogen

Porphobilinogen (PBG) is a pyrrole and a key intermediate in the biosynthesis of porphyrins (Chapter 7). It is formed by a condensation reaction from two molecules of the keto-amino-acid, δ-aminolaevulinic acid (δ-ALA) as shown in Figure 4.12. δ-ALA is also the product of some complex condensation reactions involving the coenzyme pyridoxal phosphate; we shall look at this reaction shortly. The enzyme that catalyses PBG synthesis uses the side-chain amino group of the amino acid lysine, which is present at the active

Figure 4.12. Condensation reactions of porphobilinogen (PBG). Asterisk indicates the carbon bearing the tritium label.

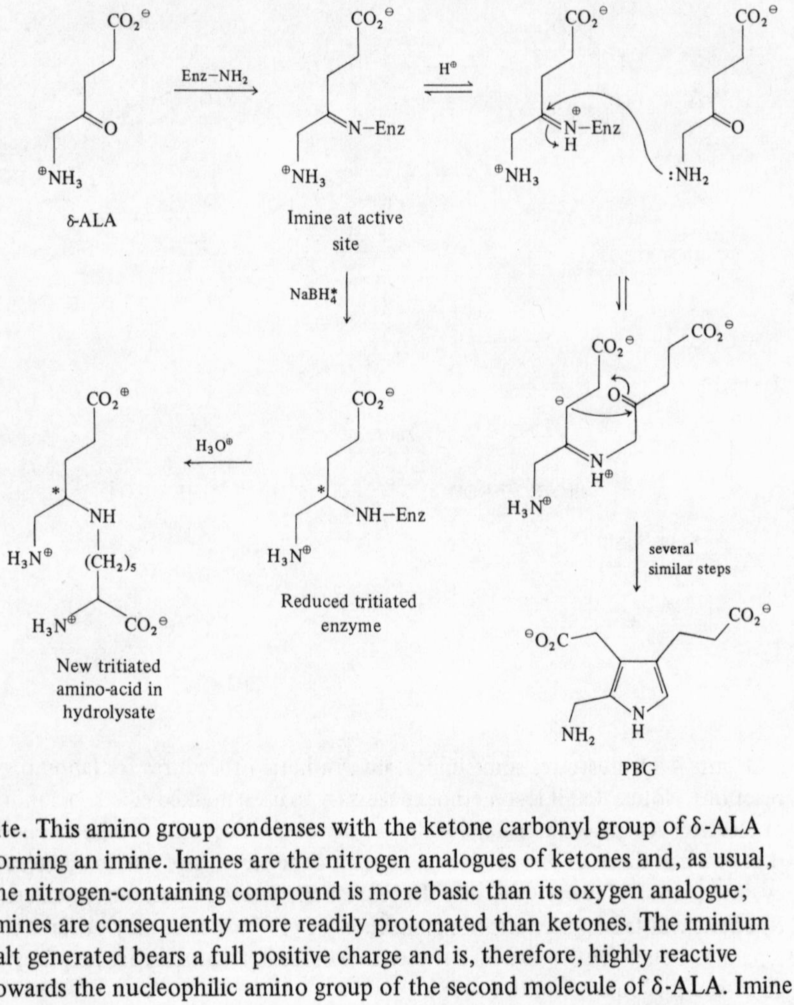

site. This amino group condenses with the ketone carbonyl group of δ-ALA forming an imine. Imines are the nitrogen analogues of ketones and, as usual, the nitrogen-containing compound is more basic than its oxygen analogue; imines are consequently more readily protonated than ketones. The iminium salt generated bears a full positive charge and is, therefore, highly reactive towards the nucleophilic amino group of the second molecule of δ-ALA. Imine formation is a very common biological method for catalysing condensation reactions, a fact particularly well shown by the chemistry of pyridoxal (see section 4.4). In general, protonated imines (iminium cations) are more reactive than ketones in condensation reactions.

In this case, the presence of an imine intermediate was demonstrated by adding tritiated sodium borohydride to the reaction mixture. The imines were reduced to amines, tritium was incorporated into the product and δ-ALA became irreversibly attached to the enzyme. It was found that when the enzyme was hydrolysed to its component amino acids, a new amino acid containing the rest of the δ-ALA molecule and tritium from the sodium borohydride was isolated (Jencks, 1969, p. 125).

Problem 4.1.

Glutamate dehydrogenase catalyses the reaction:

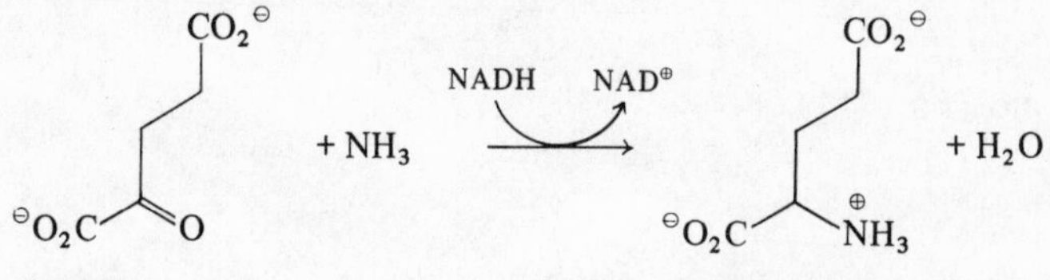

2-Oxoglutarate   Glutamic acid (Glu)

NADH is a redox coenzyme that effects reactions similar to those of $NaBH_4$ in the laboratory (see Chapter 8). Suggest a mechanism for the transformation. This reaction is a major route for the fixation of nitrogen in mammals.

4.2.2. Purine biosynthesis.

The biosynthesis *de novo* of purine nucleotides is a lengthy sequence (Figure 4.13), but at several of the ring-forming steps, we can identify donor–acceptor interactions in condensation reactions. Because purines and pyrimidines are components of nucleic acids and are essential for the division of cells and the growth of organisms, the inhibition of purine and pyrimidine biosynthesis has become a target for anti-cancer drugs (see Chapter 15). Through a sequence of amide-forming reactions involving ATP and derivatives of the coenzyme tetrahydrofolic acid, the first compound in Figure 4.13 is constructed. It is an ideal compound for cyclising to an imidazole, all that is required is a dehydrating agent and a catalyst. In the laboratory, we might reach for a hot-plate and a bottle of acid, but nature uses an enzyme with the aid of ATP. The sequence of events is probably as shown in Figure 4.13. The pyrimidine ring of the purine system is built on to the preformed imidazole (most laboratory syntheses work the other way round) and once again nature constructs the perfect precursor for cyclisation via condensation reactions.

Figure 4.13.

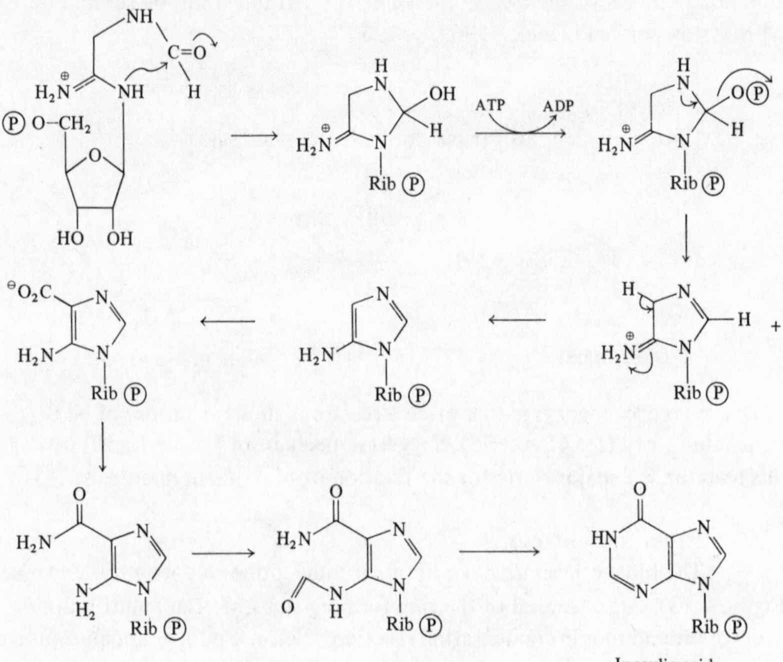

$\ominus$O$_2$C

H$_2$N

ATP   ADP

+ P$_i$

Inosylic acid

Problem 4.2: pyrimidine biosynthesis

The biosynthesis of the pyrimidine nucleotides follows a rather different course, the pyrimidine ring system being constructed before the ribose is added. The sequence of events is reproduced in Figure 4.14. Interpret the likely courses of the individual reactions in as much mechanistic detail as you can. You will need to use some of the material of Chapter 3 also. The fourth reaction has not yet been discussed – it is a dehydrogenation and comes into Chapter 8.

4.3.   Thiol esters

As we saw in the early part of this chapter, strong bases are often required to form the carbanions that take part in condensation reactions. One of the ways in which nature obviates the need for strong bases is to use thiol esters as intermediates. The replacement of oxygen by sulphur in an ester has two effects on the electronic structure of the ester. Firstly, the lower electronegativity of sulphur compared with oxygen means that the carbonyl group will be more ketone-like because the inductive effect of the oxygen is absent (Figure 4.15). Secondly, because sulphur is a second-row element, it is less effective at donating its non-bonded electron pairs in a resonance interaction with the carbonyl group. This also makes the polarisation of the carbonyl group greater than in an oxygen ester. A consequence of both these effects is that the protons on the carbon α to the carbonyl group should be more acidic

Figure 4.14.

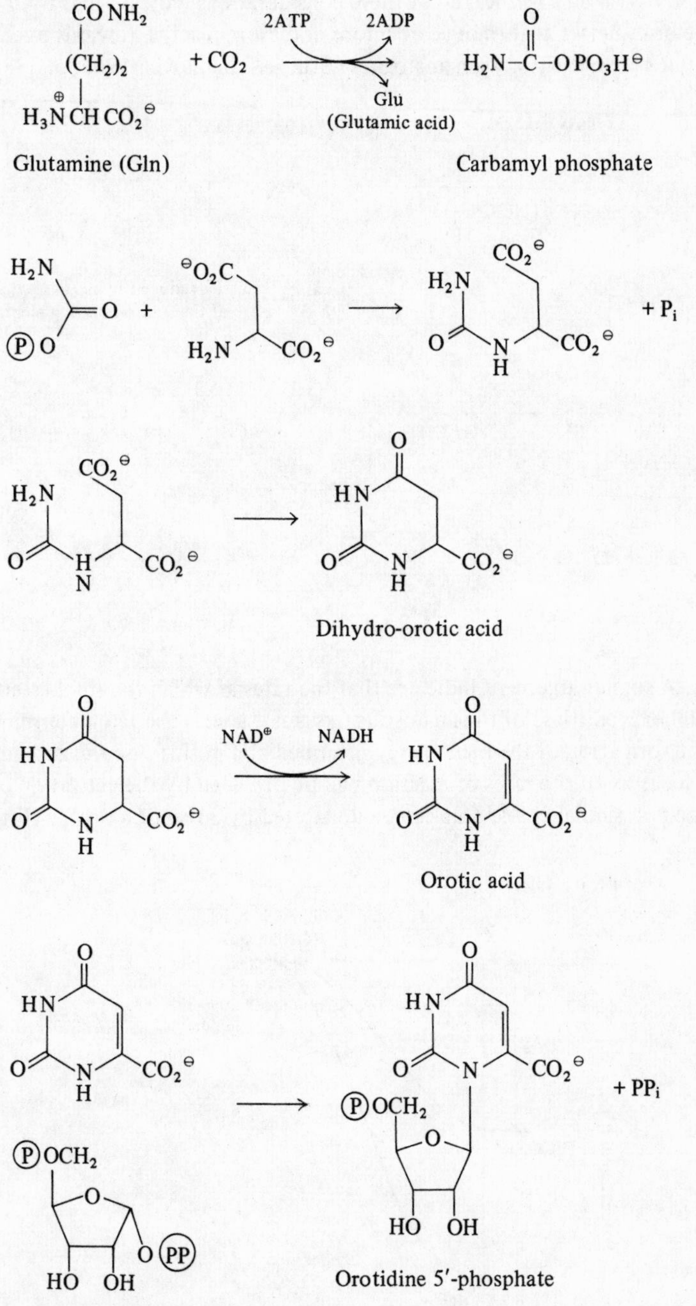
CO NH2
(CH2)2
H3N CH CO2
+ CO2
2ATP
2ADP
Glu
(Glutamic acid)
H2N–C–O PO3H
Glutamine (Gln)
Carbamyl phosphate
Dihydro-orotic acid
NAD
NADH
Orotic acid
+ Pi
+ PPi
Orotidine 5′-phosphate

than in the oxygen case. The canonical forms of Figure 4.15 illustrate these points. Taken together, all of these considerations indicate that thiol esters will be both better at forming carbanions and more reactive towards nucleophilic attack than their oxygen analogues (Bruice & Benkovic, 1966, p. 388).

Figure 4.15.

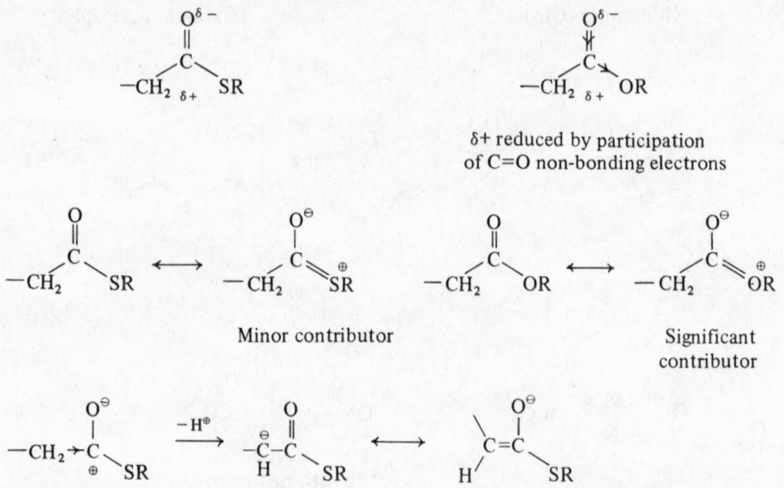

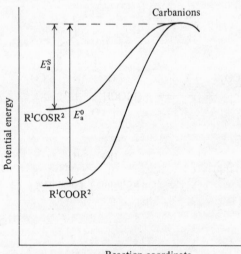

$\delta+$ reduced by participation of C=O non-bonding electrons

Minor contributor

Significant contributor

A similar argument indicates that the rates at which the thiol esters react will exceed those of the analogous oxygen esters. If the rate-determining step is the formation of the high-energy intermediate, in this case the carbanion, then a measure of the rates of reaction will be provided by the activation energy for the two similar cases. This can be illustrated by a reaction profile (Figure 4.16)

Figure 4.16.

Carbanions

E_a^S

R^1COSR^2 | E_a^0

R^1COOR^2

Potential energy

Reaction coordinate

in which the carbanions are placed at about equal energies. The resonance argument used above tells us that the potential energy of the thiol ester will be greater than the oxygen. Therefore the activation energy will be less for thiol ester carbanion formation ($E_a^S < E_a^0$), and the rate of reaction of the thiol ester will be correspondingly greater.

The reaction of thioethyl-4-nitrobenzoate with n-butylamine occurs at 25 °C (Figure 4.17, *top*), whereas the corresponding oxygen ester is inert under the same conditions. Thiol–malonate esters react four times faster than the oxygen analogues in condensations with benzaldehyde (Figure 4.17, *bottom*). Physical data (Table 4.2) also illustrate the electronic characteristics of the thiol ester group as being more like a ketone than an oxygen ester. These data are entirely consistent with there being little resonance interaction involving sulphur in thiol esters.

Figure 4.17.

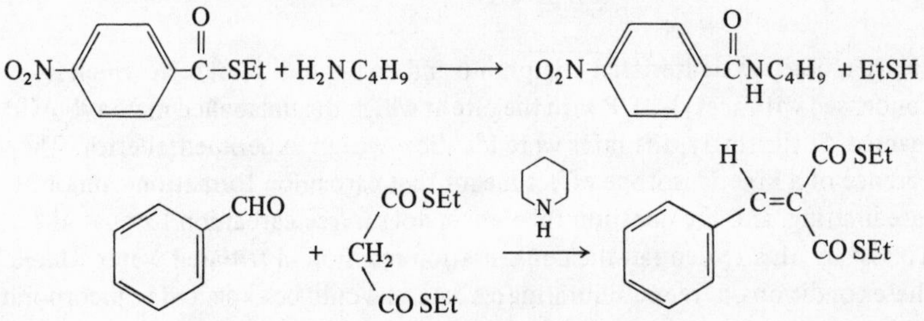

Table 4.2. *Physical properties of esters and thiol esters*

i.r. carbonyl stretching frequencies (cm^{-1})		
Oxygen ester	Thiol ester	Ketone
1725–1740	1670–1715	1690–1725

pK_a	
$CH_3CO.CH_2.CO.SR$	8.5
$CH_3.CO.CH_2.CO.R$	8.2–8.9
$CH_3.CO.CH_2.CO.OR$	10.5

The chief thiol used by nature is coenzyme A, but in some enzyme complexes it is replaced by a protein with a peripheral thiol available for ester formation. This is the case in fatty acid and polyketide biosynthesis, where the thiol protein is known as an acyl carrier protein (ACP).

4.3.1. Chain extending in fatty acid biosynthesis

Much of our knowledge concerning the biosynthesis of fatty acids has come from the work of Feodor Lynen in Munich (Lynen, 1975). He has shown that a multi-enzyme complex is responsible for the biosynthesis of fatty acids; the enzymology is discussed further in Chapter 13. Formally, as we saw, the production of β-dicarbonyl systems in biosynthesis is analogous to the Claisen ester condensation. However, a mere formal analogy tells you little about the mechanism of the reaction, and Lynen was interested to find out whether the enolisation of the thiol ester in fatty acid biosynthesis was the rate-determining step as it is in the Claisen ester condensation.

Lynen observed that the chain-extending molecule, malonyl–ACP, underwent spontaneous hydrogen–deuterium exchange under mildly acidic conditions in deuterium oxide:

$$\text{HO}_2\text{C.CH}_2\text{.CO.SACP} \xrightarrow[\text{pD 6.5–7}]{\text{D}_2\text{O}} \text{HO}_2\text{C.CD}_2\text{.CO.SACP}$$

He then used his deuteriated compound and compared the rate at which it condensed with acetyl–ACP with the rate at which the unlabelled malonyl–ACP reacted. Surprisingly, the rates were identical within experimental error. The absence of a kinetic isotope effect means that carbanion formation cannot be rate limiting, and the question then arises does a free carbanion form at all? To test for this Lynen ran the condensation reaction in tritiated water. Under these conditions, a free equilibrating carbanion would be expected to incorporate tritium, but no tritium was detectable in the product. Lynen concluded that the enzyme causes a rapidly formed carbanion to react at once with acetyl–ACP with concomitant loss of carbon dioxide, a result completely unexpected on the basis of simple chemical analogy. It is noteworthy that if acetyl–ACP were used by nature instead of malonyl–ACP, the additional powerful driving force for completing the condensation reaction, the elimination of carbon dioxide, would be lost.

4.3.2. Hydroxyacid synthases

There are two hydroxyacids in the tricarboxylic acid cycle (also known as the citric acid cycle; Appendix 1) that can be produced by condensation reactions with coenzyme A esters (see Arigoni, Rétey & Lüthy, 1969, 1970; Cornforth *et al.*, 1969). They are citric acid and *S*-malic acid. In each reaction, the eliminated group is coenzyme A (Figure 4.18). Both enzymes follow stereospecific pathways, as is usual with enzyme-catalysed reactions (Battersby & Staunton, 1974; Cornforth, 1976). The mechanism of these two reactions has been especially well studied and makes a sharp contrast with the

Figure 4.18.

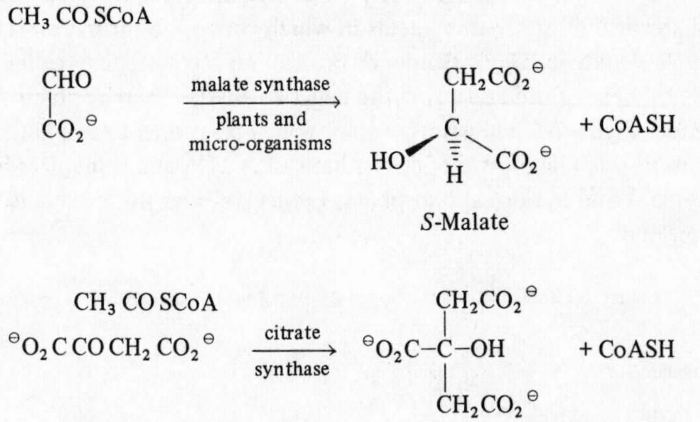

$CH_3 CO SCoA$

$$\underset{CO_2^\ominus}{\overset{CHO}{|}} \xrightarrow[\substack{\text{plants and} \\ \text{micro-organisms}}]{\text{malate synthase}} \underset{H}{\overset{CH_2CO_2^\ominus}{\underset{HO}{|}}} C \overset{|}{\underset{}{}} CO_2^\ominus + CoASH$$

S-Malate

$CH_3 CO SCoA$

$$^\ominus O_2 C\, CO\, CH_2\, CO_2{}^\ominus \xrightarrow[\text{synthase}]{\text{citrate}} \underset{CH_2CO_2^\ominus}{\overset{CH_2CO_2^\ominus}{{}^\ominus O_2C-\underset{|}{\overset{|}{C}}-OH}} + CoASH$$

previous example. Because the two hydroxyacids are so closely related, we
shall concentrate upon the simpler malate synthase. This enzyme is widespread
in micro-organisms but absent in mammals; it requires magnesium cations for
activity. When deuteriated acetyl coenzyme A was used, a small primary isotope
effect, k_H/k_D, of 1.2-1.5 was found. When the reaction was carried out in tritiated
water and the unreacted acetate re-isolated before it was all consumed, tritium
was found to be incorporated into the methyl group. These two results are
the exact opposite of the condensation in fatty acid biosynthesis and clearly
implicate a carbanion in the condensation mechanism. Analogous results have
been obtained for citrate synthase, in this case the isotope effect was greater,
about 3.5. A plausible mechanism is illustrated schematically by Figure 4.19;
the equilibrium (2) is the stage at which tritium exchange takes place.

Figure 4.19.

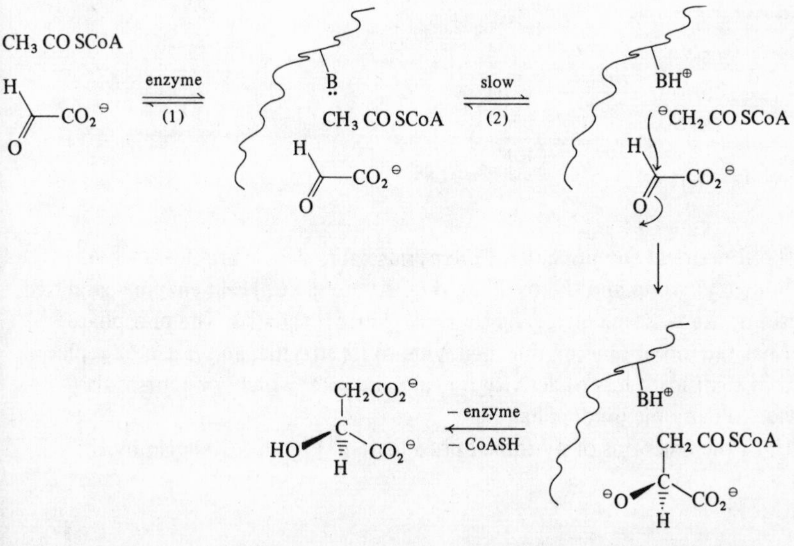

4.4. Pyridoxal phosphate

Pyridoxal phosphate (PLP) is a coenzyme containing a pyridine ring; it mediates reactions of amino acids in which carbanion formation is required (Bruice & Benkovic, 1966; Bender & Brubacher, 1973). The pyridine ring provides a means for delocalising the negative charge, thereby stabilising the carbanion. Figure 4.20 illustrates some typical reactions. Example (*d*) is the condensation sequence in the biosynthesis of δ-ALA and it involves both coenzyme A and pyridoxal phosphate. Let us consider the mechanism of this reaction first.

Figure 4.20. Reactions mediated by pyridoxal phosphate.

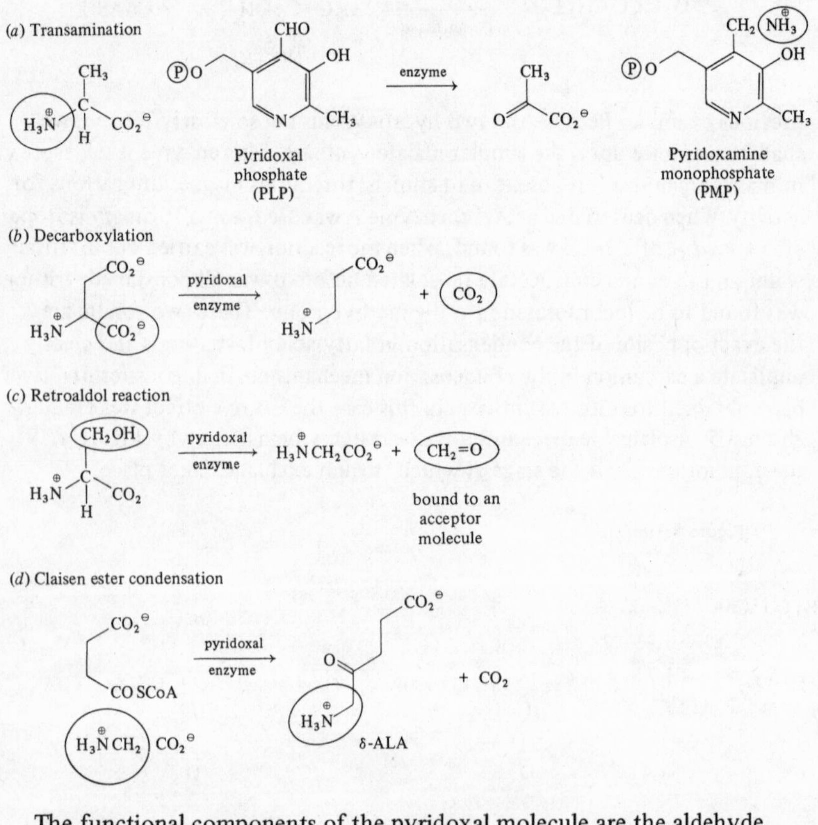

The functional components of the pyridoxal molecule are the aldehyde, the hydroxyl group and the pyridine ring. At ambient pHs in enzyme-catalysed reactions, the pyridine nitrogen atom is largely protonated. The phosphate ester has the job of binding the coenzyme to its enzyme, and it can be replaced without significant loss of activity by sulphate ester which, of course, also provides an anionic binding handle.

All of the reactions of pyridoxal phosphate in Figure 4.20 begin by the condensation of the amino group with the aldehyde to form an imine (Figure 4.21). The imine-forming and hydrolysing reactions are all subject to acid/base catalysis, especially with the help of the neighbouring hydroxyl group at position 3. Hydroxyl groups in the 3-position of pyridine rings are essentially phenolic and acidic. Also the hydroxyl group acts as a stabilising influence on the imines through cyclic hydrogen bonding. In some enzymes, the aldehyde of pyridoxal phosphate is already bound to the protein through imine formation with the side-chain of lysine; the first step of the reaction in such cases is imine exchange.

Figure 4.21. The detailed mechanism of δ-aminolaevulinic acid formation mediated by pyridoxal phosphate.

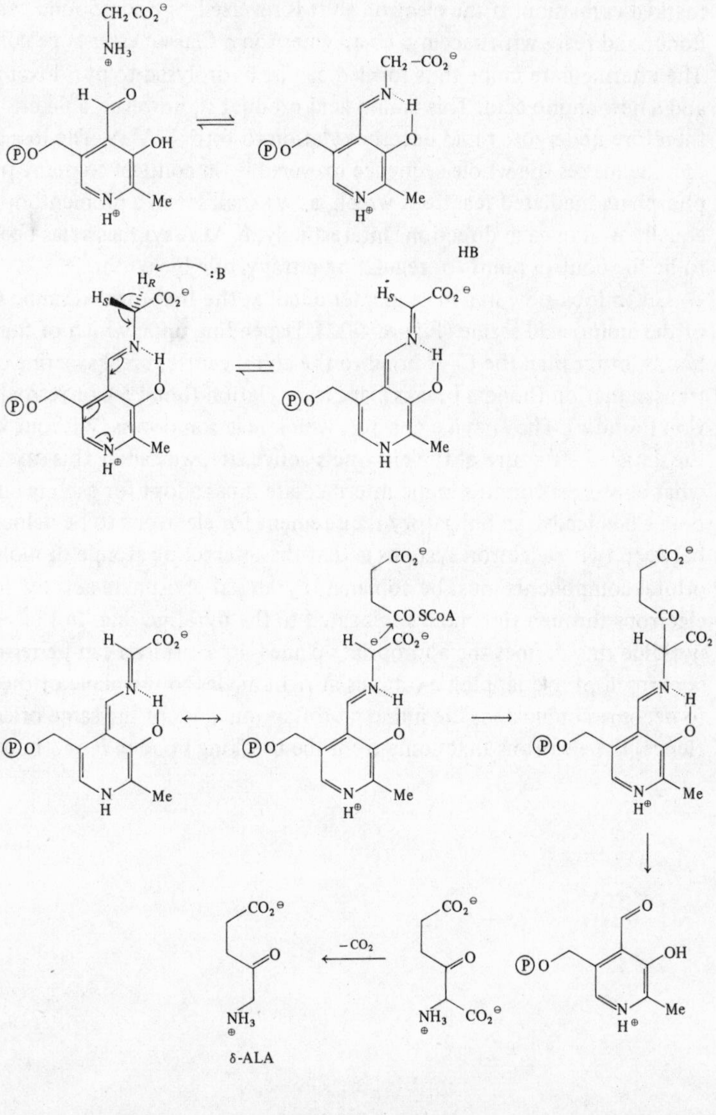

Once the imine is formed, the protons α to the imine nitrogen atom are in an environment similar to protons in a β-keto-ester and their pK_a accordingly decreases by about 10. It is now easy for one of them to be lost to a base on the enzyme surface. The anion formed can be delocalised not just over one carbonyl group but throughout the whole π-electron system down to the positively charged pyridine nitrogen atom. By means of isotopic labelling with deuterium and tritium it has been shown that the proton lost in δ-ALA synthesis is the one occupying the pro-R site (see Eliel (1962) for a definition and explanation of the R and S system of nomenclature for chiral centres). Such stereospecificity is normal in pyridoxal phosphate mediated reactions in the presence of enzymes.

Having removed a proton, we have produced an intermediate that is a masked carbanion. If the electron shift is reversed, the compound can act as a donor and react with succinyl coenzyme A in a Claisen ester type condensation. The intermediate imine thus formed can be hydrolysed to pyridoxal phosphate and a new amino acid. This amino-acid product is, however, a β-keto-acid and therefore undergoes rapid decarboxylation to form δ-ALA. The loss of carbon dioxide makes the whole sequence irreversible, in contrast to many pyridoxal phosphate mediated reactions which, as we shall see in a moment, operate equally well in each direction. Interestingly, δ-ALA synthesis has been shown to be the control point for regulating tetrapyrrole biosynthesis.

Let us look now in a little greater detail at the first-formed imine in the case of the amino acid serine (Figure 4.22). Depending upon which of the three bonds, other than the C—N bond to the chiral centre, breaks, serine can undergo transamination (bond *a* breaks), decarboxylation (bond *b*), or retroaldol reaction (bond *c*). The enzyme controls which reaction occurs. Without knowing the detailed structure of the enzyme's active site, we can in this case deduce what conformation this imine intermediate must adopt for each reaction *a*, *b*, or *c*. Chemically, an obligatory requirement for electrons to be delocalised between two π-electron systems is that the interacting atomic or molecular orbital components must be coplanar. Pyridoxal phosphate acts by delocalising electrons through the imine conjugated to the pyridine ring. In this case, the pyridine ring defines the appropriate plane – its π-orbitals can be regarded as composed of overlapping p-orbitals at right angles to the plane of the ring and to become conjugated, the imine p-orbitals must adopt the same orientation. Hence, the electrons that come from the breaking bond *a*, *b* or *c* must fit into the same plane to be accepted into the π-system of the pyridoxal molecule. Chemists call this type of obligatory orbital alignment **stereoelectronic control**. It is common whenever a molecule has a particular rigid geometry imposed either by its own structure (as in steroids and carbohydrates) or by binding to the active sites of enzymes. Figure 4.22 illustrates the situation in the case of pyridoxal phosphate.

A large number of enzyme-catalysed reactions mediated by pyridoxal phosphate have been replicated using laboratory model systems. There has also been an immense amount of work on the detailed physical organic chemistry of imine formation, isomerisation and hydrolysis (Bruice & Benkovic, 1966).

Figure 4.22. Stereoelectronic control in pyridoxal-mediated reactions.

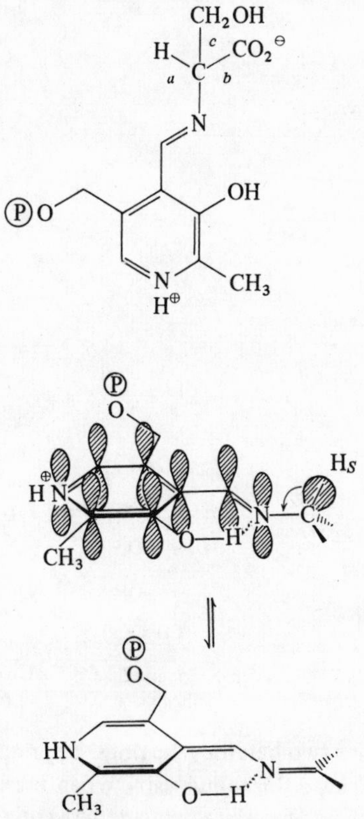

4.5.   Thiamine pyrophosphate

The structure of thiamine pyrophosphate is more complex than pyridoxal, and it is not obvious at first sight what portions of the molecule take part actively in its reactions, especially since the mechanisms of the reactions themselves are difficult to infer from the structures of the reactants. The only obvious common feature (Figure 4.23) is that all the reactants contain carbonyl groups. (See Bruice & Benkovic, 1966; Bender & Brubacher, 1973).

Figure 4.23.

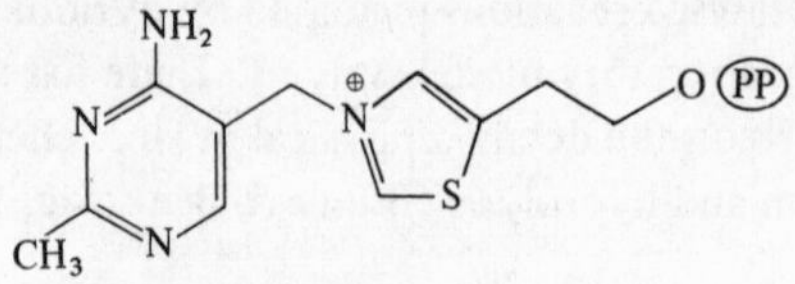

Thiamine pyrophosphate (TPP)

(a) Decarboxylation

$$R\,CO\,CO_2H \xrightarrow[\text{enzyme}]{\text{TPP}} RCOSCoA + CO_2$$

R = Me, pyruvate
dehydrogenase

(b) Acetoin formation

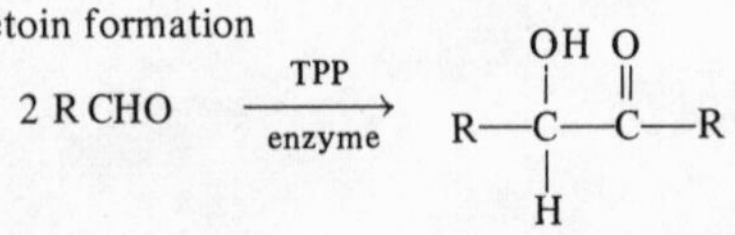

An acetoin

(c) Transketolase reaction

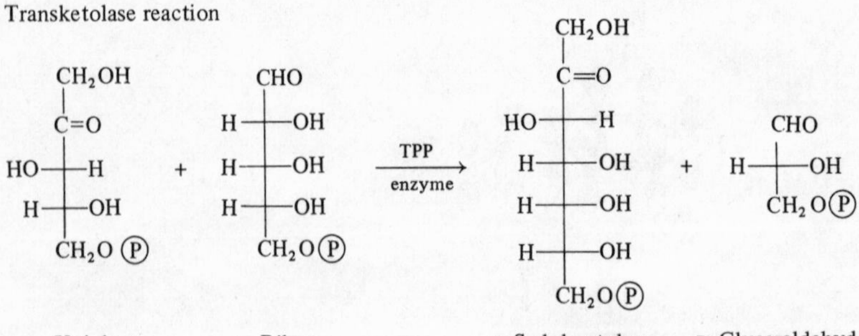

| D-Xylulose 5-phosphate | D-Ribose 5-phosphate | D-Sedoheptulose 7-phosphate | D-Glyceraldehyde 3-phosphate |

Thiamine pyrophosphate contains two heterocyclic rings, a pyrimidine and a thiazole alkylated on nitrogen (giving a thiazolium salt). When Breslow observed that a solution of thiamine in deuterium oxide underwent extremely rapid exchange of the proton at C-2 in the thiazole ring for deuterium, the reactive region of the molecule was discovered. This exchange reaction is not peculiar to thiazolium salts, but is general to all heterocyclic ompounds under the appropriate conditions. However, the reaction is exceptionally facile in the case of thiamine and other thiazolium salts (Figure 4.24). Of the 1,3-azoles

Figure 4.24.

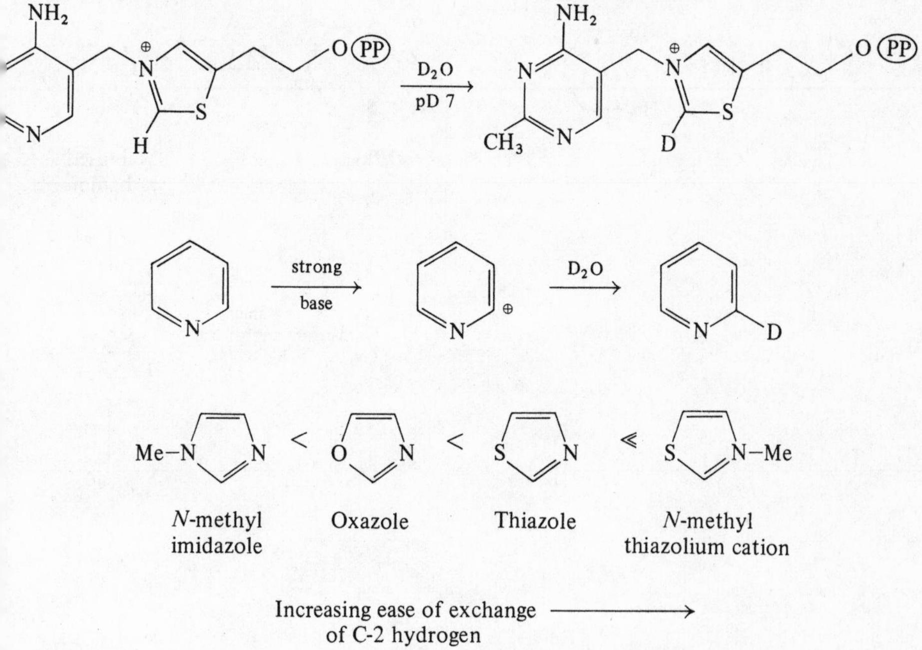

N-methyl imidazole	Oxazole	Thiazole	N-methyl thiazolium cation

Increasing ease of exchange ⟶
of C-2 hydrogen

shown, *N*-methyl imidazole is the least reactive; the greater electronegativity of oxygen makes oxazole more reactive than imidazole. When sulphur is included, its low-lying 3d-orbitals are available to accept electrons and stabilise the anion. Thiazolium salts owe their reactivity to this effect and to the positive charge introduced by quaternising the nitrogen atom; electron withdrawal by the positive charge makes the proton at C-2 much more labile still. Stabilisation of anions by sulphur is unexceptional – dithianes prepared from aldehydes are useful synthetic reagents (Figure 4.25). Anions formed on a carbon atom adjacent to a heteroatom bearing a positive charge are also well-known as intermediates in the Wittig olefin synthesis. Such species are called **ylids**, and in the Wittig reaction the heteroatom is phosphorus. The phosphorus ylid adds to a carbonyl group en route to an alkene; an ylid is thus a stabilised carbanion. Sulphur ylids react with carbonyl groups giving epoxides.

The formation of an ylid from thiamine is the key to understanding its mechanism of action. As with pyridoxal, the ability of a heterocyclic ring to delocalise or accept a negative charge is of the utmost importance. Let us follow through a decarboxylation reaction (Figure 4.26). The first step is the addition of the ylid to the ketone of pyruvate, an α-keto-acid, in the usual way. In the adduct we can recognise the presence of a β-keto-acid type of system (heavy type) if we compare C=N⁺ with C=O, and consequently it is reasonable to

Figure 4.25. Anions and ylids stabilised by phosphorus and sulphur and their reactions.

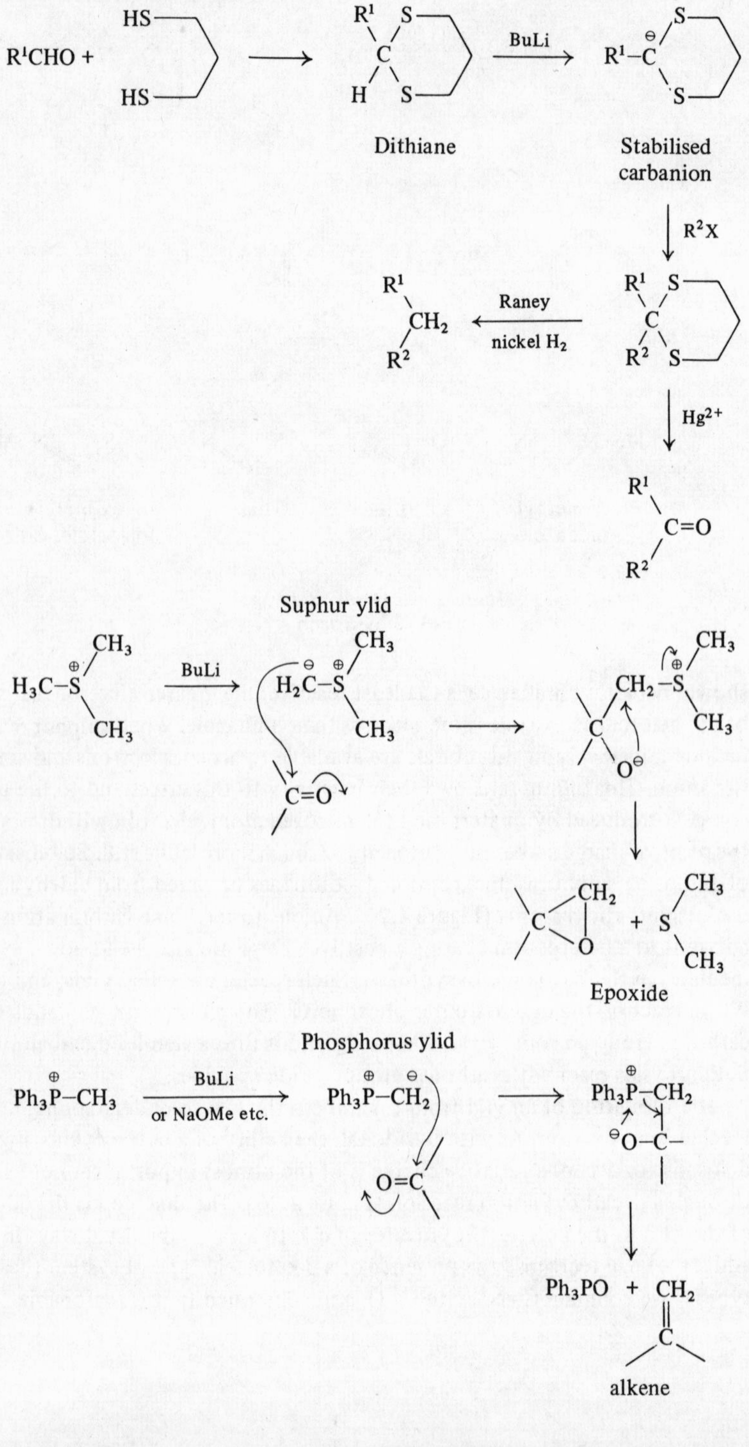
R¹CHO +
HS
HS
BuLi
Dithiane
Stabilised carbanion
R²X
Raney nickel H₂
Hg²⁺
Suphur ylid
BuLi
Epoxide
Phosphorus ylid
BuLi or NaOMe etc.
Ph₃PO + CH₂
alkene

expect this intermediate to decarboxylate. If the electron-accepting thiazolium ring were not present, the highly unstable acetyl anion, $CH_3.CO^-$, would have to form on decarboxylation of pyruvate. The reaction is completed by reversal of the addition to the carbonyl group leading to acetaldehyde and the regeneration of the ylid.

Figure 4.26.

Abbreviated structure of
TPP ylid

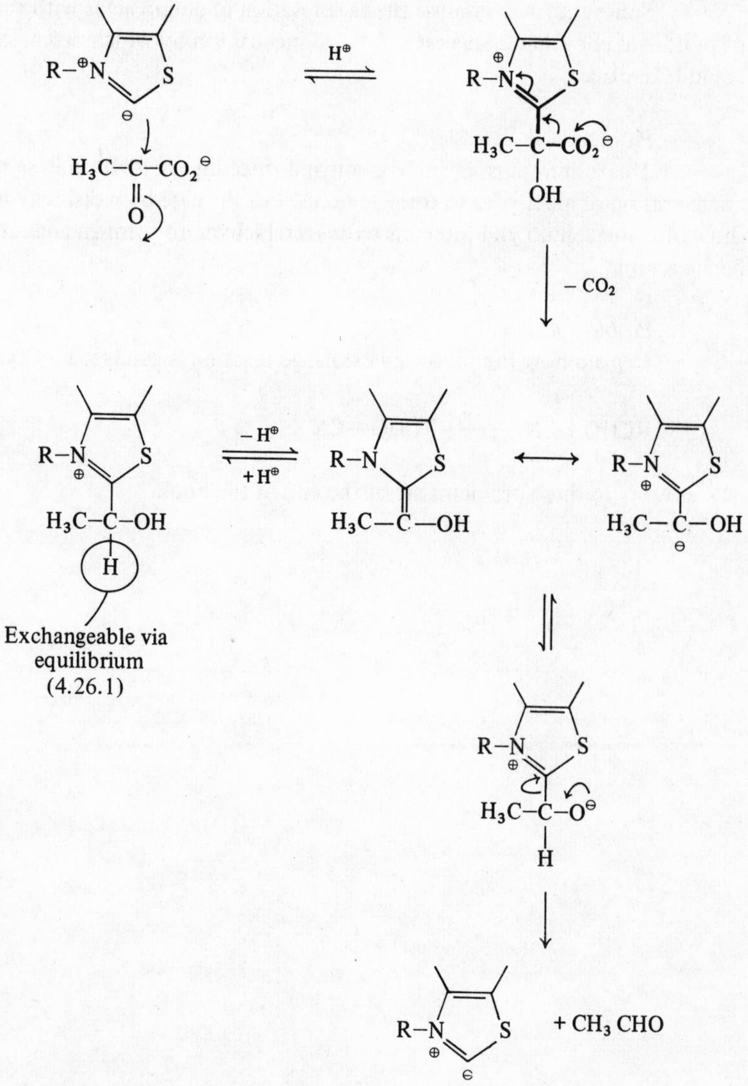

Exchangeable via
equilibrium
(4.26.1)

There is further evidence in favour of this type of mechanism. If the 2-position is blocked by a substituent, the substituted thiamine pyrophosphate is catalytically inactive. Also it is possible to isolate the hydroxyethyl thiamine intermediate (4.26.1) from a reaction mixture containing pyruvate decarboxylase, thiamine and pyruvate. Finally, it can be shown in the absence of enzyme that this same intermediate undergoes deuterium exchange α to the thiazolium ring, as one would expect from the analogy between $C=N^+$ and $C=O$; (the reaction is equivalent to enol formation).

Problem 4.3
Some enzymes catalyse the racemisation of amino acids with the aid of pyridoxal phosphate. Suggest a detailed mechanism by which racemisation could take place.

Problem 4.4
Thiazolium salts, cyanide anion and thiamine itself all catalyse the condensation of aldehydes to form acyloins. For the case of acetaldehyde show how thiazolium ylid interacts with acetaldehyde to form an adduct and hence acetoin.

Problem 4.5
Explain how the following exchange reaction is catalysed by cyanide.

$$RCHO + CN^- \xrightarrow{\ D_2O\ } RCDO + CN^-$$

The answers to these problems are at the end of the book.

5 Some reactions of carbon–carbon double bonds

5.1. Chemical background

When two molecules react to form a product composed of all the atoms of both molecules, an addition reaction occurs. Carbonyl groups, as Chapters 3 and 4 showed, readily undergo nucleophilic addition to the polarised $C=O$ bond. In contrast, carbon–carbon double bonds are not strongly polarised because the component carbon atoms have the same electronegativity. Consequently, nucleophilic addition to carbon–carbon double bonds is rare but electrophilic addition is common. Several biosynthetic reactions involving addition and elimination can be found in Appendix 1. This shows the citric acid cycle, which is at the centre of energy-producing metabolism. In it, there are two reactions catalysed by enzymes that add and eliminate water from their substrates. The enzymes are fumarase and aconitase, and we shall discuss fumarase in more detail later. Further examples are to be found in the biosynthesis of fatty acids (Figure 13.18) and the biosynthesis of isopentenyl pyrophosphate (Figure 5.15). Notice that the reverse reaction of addition, elimination, is a common method of forming carbon–carbon double bonds. Stereochemistry is important in all addition and elimination reactions and, in particular, the concepts of conformation and stereoelectronic control that we met in our discussion of pyridoxal phosphate are significant. If you do not understand these topics, you should consult either a general organic chemistry text or Eliel (1962, p. 139).

5.1.1. Addition reactions

It is obvious from its electronic structure that a carbon–carbon double bond should react readily with electrophiles. The polarisable π-electrons can be donated to an electron acceptor such as a proton, a cation, or a halogen molecule. This is true for pure carbon–carbon double bonds and for those derived by enolisation of carbonyl groups (Figure 5.1).

In examples (*a*), (*b*), (*c*), (*d*) and (*f*), a cationic intermediate is formed with a positive charge located wholly or partly on carbon. Such intermediates in modern nomenclature are known as **carbocations**: if the carbon bearing the

Figure 5.1.

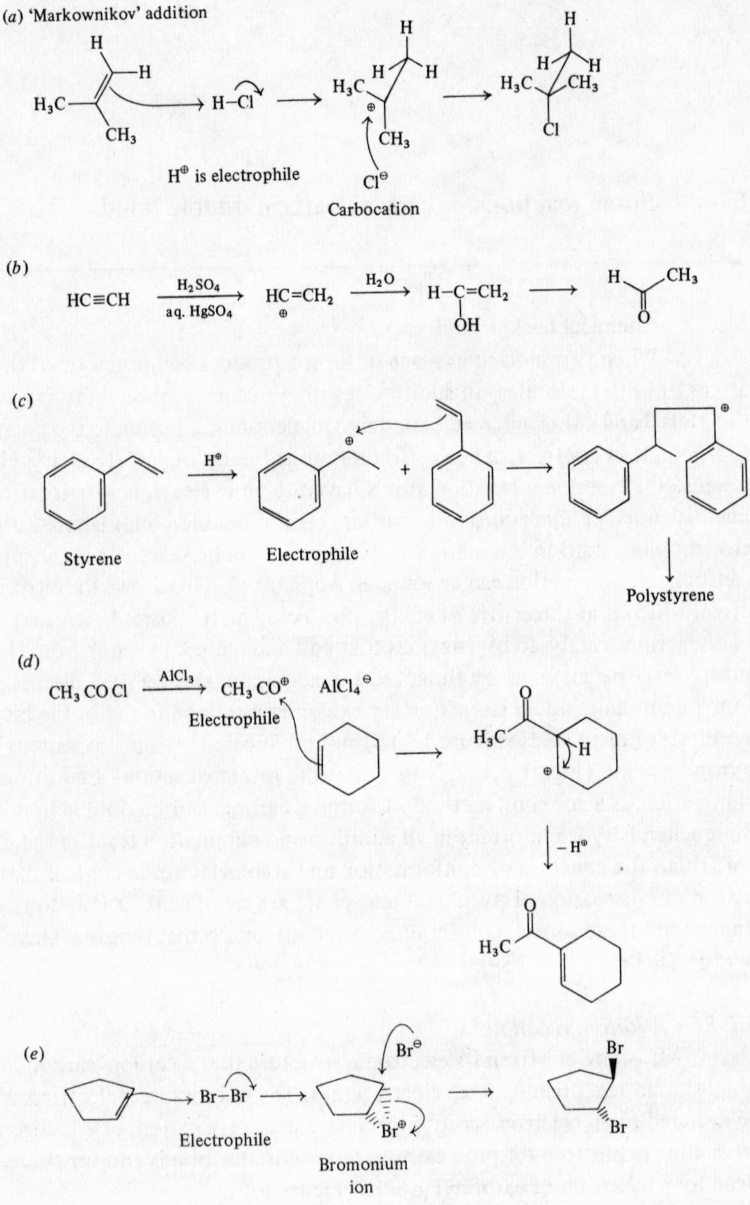
(a) 'Markownikov' addition
H⊕ is electrophile
Carbocation
(b)
H₂SO₄
aq. HgSO₄
H₂O
(c)
Styrene
Electrophile
Polystyrene
(d)
AlCl₃
Electrophile
– H⊕
(e)
Electrophile
Bromonium
ion

Figure 5.1. (continued)

(f)

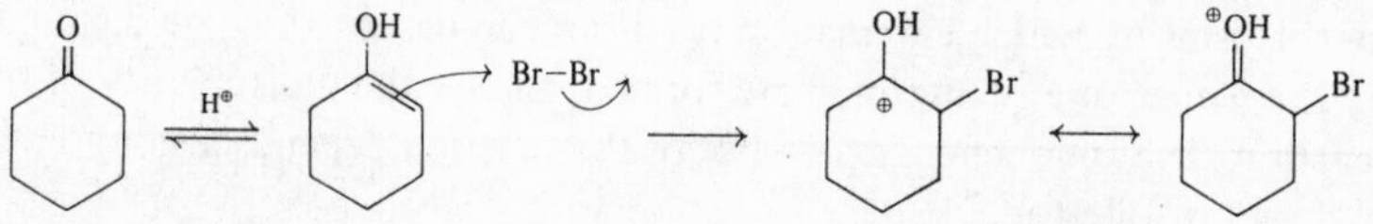

Oxonium ion

positive charge is trivalent, the cation is called a **carbenium ion** (5.1 *a–d* and *f*). Figure 5.1*f* illustrates an intermediate in which part of the positive charge is shared by oxygen, as indicated by the canonical forms. Such cations are known as **oxonium ions**; we met them in Chapters 3 and 4 as intermediates in acid-catalysed addition reactions to carbonyl groups. The stereochemistry of addition reactions is exemplified by Figure 5.1*e*, in which bromine adds to cyclopentene to give *trans*-1,2-dibromocyclopentane. In this case, the stereo-chemical course of the reaction can be understood in terms of an intermediate halonium ion, a three-membered ring species containing a positively charged halogen. This ring is formed on one face of the alkene, the lower face in the figure, and it is most easily opened by a bromide ion attacking from the top face. The two bromine atoms end up *trans* to each other. Attack from the lower face is hindered by the bulk of the halonium ion and consequently no *cis* product forms. Direct evidence for the existence of positively charged inter-mediates such as bromonium ions has been obtained by magnetic resonance spectroscopy (Olah, 1973).

In laboratory reactions, unless there is a structural feature like a ring that will enforce a specific direction upon an addition reaction (e.g. Figure 5.1*e*), a mixture of stereoisomers will usually result. Thus, if water were added to fumaric acid, we would expect to obtain a mixture of *R* and *S* malic acid (Figure 5.2). In contrast, the enzyme-catalysed reaction (fumarase) affords specifically *S*-malic acid. Here, the stereochemical control is supplied by the binding of the substrate fumaric acid to the enzyme at the active site.

Figure 5.2.

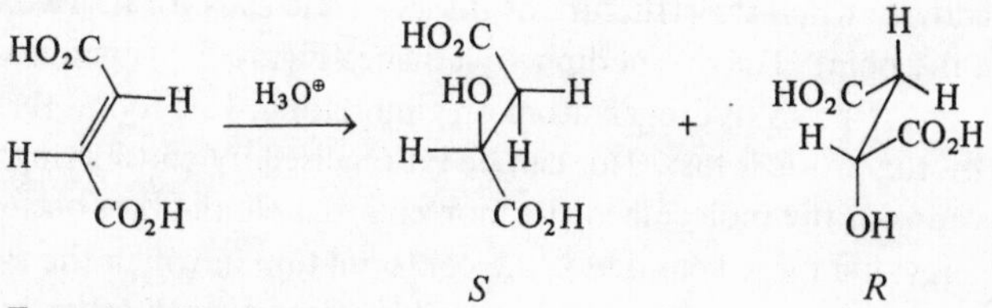

Fumaric acid

Malic acid

5.1.2.   Elimination reactions

Although formally the reverse of addition, there is a wide variety of mechanisms by which elimination reactions can occur. They are distinguished by the nature of any intermediate formed, and by the molecularity of the reaction. The stereochemical course of the reaction depends also upon which pathway is followed.

The most significant mechanism from the biological standpoint is the concerted bimolecular elimination, the E2 mechanism for short. In this case, a base removes the proton to be lost at the same time as the leaving group departs. Because the π-bond of the double bond is formed optimally when two parallel p-orbitals overlap, E2 elimination will be most favoured when the proton that is removed and the leaving groups are *trans* with respect to each other , in what is called the anti-peri-planar conformation (Figure 5.3). This is an example of a stereoelectronic effect on a reaction. E2 eliminations with halides as the leaving group are common.

Figure 5.3.

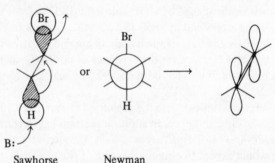

Sawhorse   Newman

Projections of anti-peri-planar conformation

In many cases a molecule has a choice which of two hydrogen atoms should be eliminated along with the leaving group. Enzymes usually direct the reaction and remove only one, as we shall see, but even laboratory reactions can show stereoselectivity when the structure of the substrate calls for it. Two examples will make the point. The bromodiphenylethane (Figure 5.4) could lose either the pro-*R* or the pro-*S* hydrogen atom on elimination. It is found that predominantly the pro-*R* is lost. This can be rationalised by considering the conformations of the molecule at the moment that elimination occurs. The lowest energy and most favoured is the conformation in which the two bulky phenyl groups are kept as far apart as possible, *trans* to each other, and this conformation accordingly affords the bulk of the elimination product.

In the axial bromodecalin (Figure 5.5), elimination affords exclusively the alkene by loss of the pro-*R* hydrogen. This is because the *trans*-decalin ring

Figure 5.4.

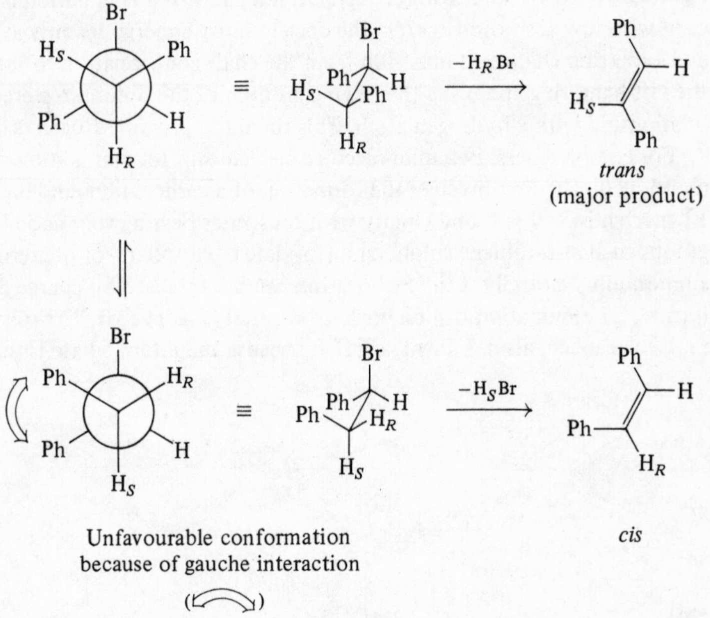

trans
(major product)

Unfavourable conformation
because of gauche interaction

cis

system, which is found as part of the steroid skeleton, is conformationally
rigid. It cannot undergo a change in configuration at the ring junction without
breaking a bond. Consequently there is no choice but to remove the pro-*R*
hydrogen: it is the only one *trans* to the leaving group.

Figure 5.5.

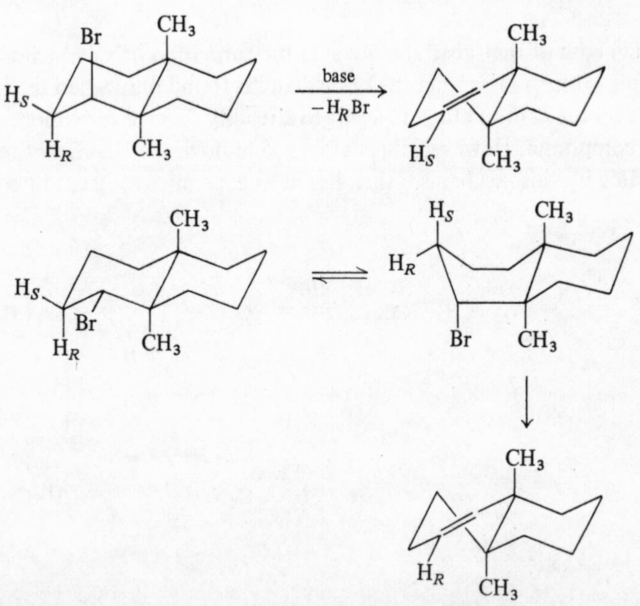

If we had taken the equatorial isomer, a difficulty would arise: there is no hydrogen *trans* to the leaving group and, if a concerted elimination is to proceed with low activation energy, the decalin must undergo its only available conformation change. It must flip from the chair conformation to the boat in the left-hand ring and bring the leaving group into the optimum stereochemical relationship with a hydrogen atom. This time, the pro-*S* hydrogen is lost.

For completeness, two unimolecular mechanisms for elimination should be mentioned. The first involves the formation of a carbocation, and is called the E1 mechanism. It is found chiefly with substrates bearing very good leaving groups such as *p*-toluene sulphonate (tosylate), phosphate, or quaternary ammonium. Naturally if the carbocation can be stabilised by charge delocalisation, E1 elimination will be further favoured (Figure 5.6). The other unimolecular mechanism, known as E1CB because the intermediate is the

Figure 5.6.

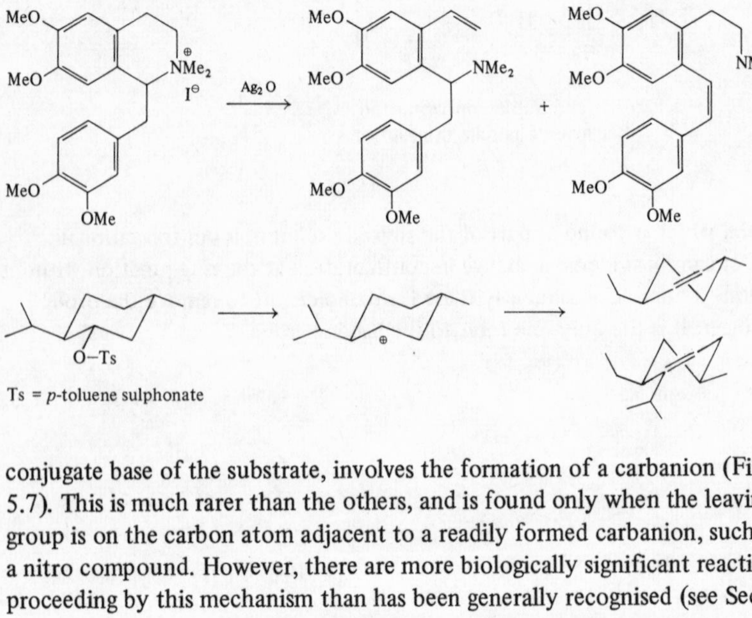

Ts = *p*-toluene sulphonate

conjugate base of the substrate, involves the formation of a carbanion (Figure 5.7). This is much rarer than the others, and is found only when the leaving group is on the carbon atom adjacent to a readily formed carbanion, such as in a nitro compound. However, there are more biologically significant reactions proceeding by this mechanism than has been generally recognised (see Section

Figure 5.7.

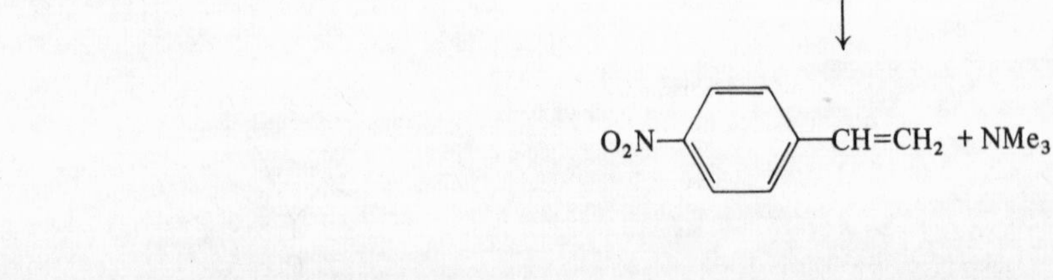

5.2.2.). In contrast to the stereospecificity shown by the concerted E2 elimination, unimolecular elimination reactions via intermediates usually yield a mixture of stereoisomeric products because the intermediate has a long enough lifetime for rotation about the carbon–carbon bond at which elimination is occurring to generate an equilibrium mixture of conformations. Loss of the leaving group or proton from a mixture of conformers of the intermediate affords a mixture of geometrical isomers of the product. As with addition reactions, if there are special structural features in the substrate such as a rigid ring, or if the reaction is controlled by the active site of an enzyme, the intermediate ion may be able to adopt only one conformation. In this case, stereospecificity will be observed.

Addition and elimination reactions are important because of their widespread occurrence both in enzyme-catalysed and in synthetic and degradative reactions in biosynthetic studies. Their stereochemical course is of interest because of its close relationship to the mechanism of the reaction. In order to follow the stereochemical course of many reactions, it is necessary to label one of the hydrogen atoms that could be eliminated with either tritium or deuterium. The reaction path is then apparent from the retention or loss of the label in the product. Conversely, if it is required to establish the position of a label in a molecule, it is useful to be able to employ a reaction of known stereospecificity, such as an E2 elimination. Examples of both situations are featured in this chapter.

5.2. Hydration and dehydration in enzyme-catalysed reactions

The enzyme-catalysed hydration of alkenes is usually a reversible reaction, and both hydrated and dehydrated substrates exist in the equilibrium mixture. In both fatty acid metabolism and the citric acid cycle, nature requires synthetic and degradative reactions. A key member of the sequence is hydration/dehydration.

5.2.1. Fumarase and aconitase

Fumarase, an enzyme of the citric acid cycle, catalyses the reversible stereospecific interconversion of fumarate and *S*-malate. An equilibrium mixture consisting of about 20% fumarate and 80% malate is formed at 37 °C (Figure 5.8). Fumarase is an allosteric enzyme, an unusual feature for an enzyme not at a key beginning, end, or branch-point of a metabolic sequence (Chapter 13). Fumarase catalyses the *trans*-addition–elimination of water. The stereochemical course of the reaction was proved by the stereospecific chemical synthesis of [3-*R*-^{2}H] malate and correlation with the enzymic product.

Figure 5.8.

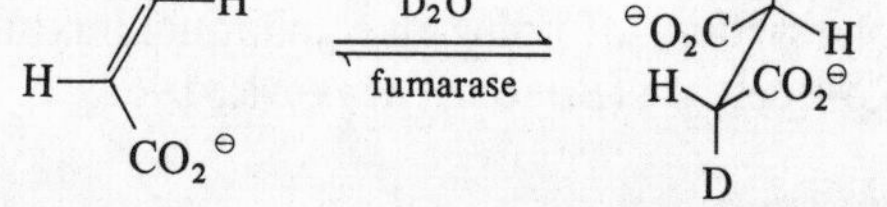

2-*S*-[3-*R*-^{2}H]malate

The pH/rate profile for fumarase suggests that groups of pK_a 6.2 and 6.8 are involved in the catalytic step (Bruice & Benkovic, 1966). The imidazole ring of histidine is one obvious candidate (pK_a 6.8) and the other group is probably a carboxylate anion (pK_a 6.2). From our précis of the chemistry of addition and elimination reactions, we realise that an acidic group will be necessary to donate a proton to the double bond of fumarate and, in the dehydration step, a base will be required to accept a proton from malate. The acid/base catalysis implied by the pH/rate observations is therefore to be expected.

Let us take the argument a little further. The addition of acids to alkenes usually involves the intermediacy of carbocations (Figure 5.1). Is there any evidence for the intermediacy of a carbocation in the reaction catalysed by fumarase?

Fumarase is able to hydrate a range of acids that are close structural analogues of fumarate such as halofumarates, hydroxyfumarates, and acetylene dicarboxylate (Figure 5.9). In every case, *trans*-addition of the hydroxyl group gives the S-stereoisomer at C-2. Because the hydroxyl group is added at the carbon atom not bearing an electronegative oxygen or halogen atom, it has been argued that a positively charged intermediate is formed; this is consistent with the 'Markownikoff' sense of addition (Nigh & Richards, 1969; Hill & Teipel, 1971). If this suggestion is correct, fumarase must hold a carbocation firmly in one configuration and permit the addition of water from one side only, otherwise a racemic product would result.

Figure 5.9.

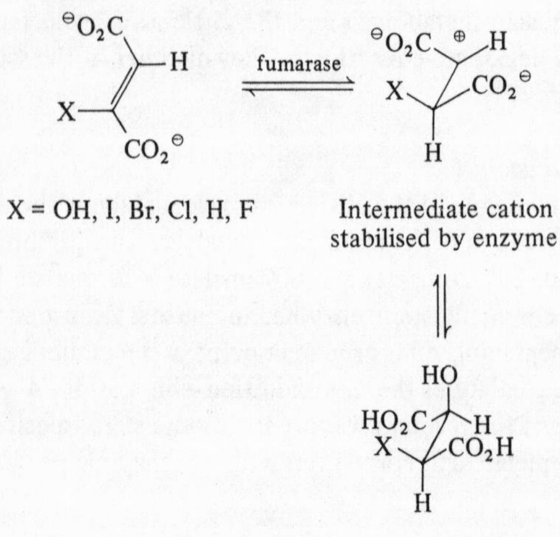

X = OH, I, Br, Cl, H, F   Intermediate cation
                               stabilised by enzyme

Aconitase is another hydrating/dehydrating enzyme of the citric acid cycle (see Appendix 1). It has many similar properties to fumarase, and catalyses a stereospecific *trans*-addition reaction, affording an equilibrium mixture of citrate, *cis*-aconitate and S,S-isocitrate (Bruice & Benkovic, 1966).

5.2.2. Enoylhydrases in fatty acid biosynthesis

The condensation of malonyl–ACP with a fatty acyl–ACP yields a β-hydroxyester, as we saw in Chapter 4. The β-hydroxyester undergoes dehydration as the next step of fatty acid biosynthesis (Figure 3.18, p. 366) and, like the enzyme-catalysed hydration of fumarate, hydroxy ester dehydration is reversible. The mechanism of the reaction, however, is quite different. The stereochemical course of enzyme-catalysed hydration of alkenoyl coenzyme A esters is *syn*, not *trans* as was the case with fumarate (Willadsen & Eggerer, 1975). Tritium-labelled substrates were used to prove the stereochemical course (Figure 5.10). Notice that the results are complementary: *syn* elimination of the [2-*R*-³H]hydroxyacid substrate affords the *trans*-α,β-unsaturated

Figure 5.10.

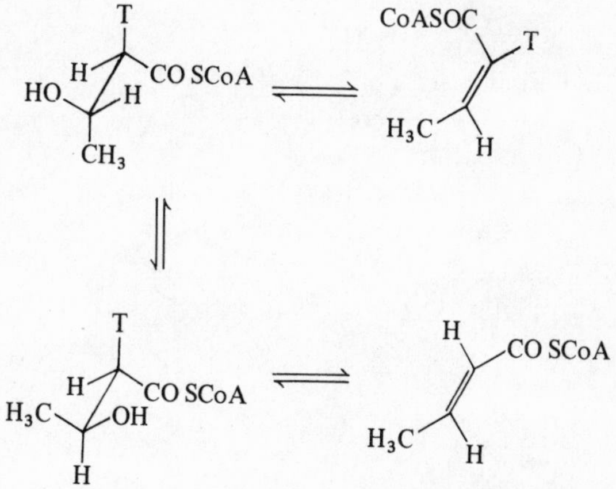

ester in which tritium has been lost; but the opposite (tritium retention) is found in the case of the *cis*-α,β-unsaturated ester. Experiments producing complementary results such as these are good evidence upon which mechanistic discussions can be based. We know (Section 5.1.2) that the favoured stereochemical course for a concerted elimination is *trans*. Here we have the opposite, a *syn*-elimination, and it is therefore probable that an intermediate is involved. This means that either a cation or an anion must be formed α to the thiol ester of the substrate in the enzyme-catalysed reaction. Since anion formation α to thiol esters is facile, as was discussed in Chapter 4, it seems probable that enoylhydrases operate by an E1CB type mechanism (Figure 5.10). If this is the case, it is another example of nature capitalising upon the inherent reactivity of a substrate in order to bring about the required transformation.

5.3.   Ammonia lyases

By removing the elements of ammonia from α-amino acids, ammonia lyase enzymes link amino-acid metabolism, energy-producing metabolism and the biosynthesis of a number of complex natural products in plants. Transaminases (Chapter 4) and glutamate dehydrogenase perform a similar function.

Aspartate ammonia lyase, for instance, equilibrates asparate and fumarate by the addition/elimination of ammonia. This is a common degradation of many amino acids in plants. Elimination of ammonia from phenylalanine and tyrosine yields cinnamic acids, which are precursors for the biosynthesis of many phenolic compounds via oxidative coupling reactions (see Chapter 7) and also some alkaloids (Figure 5.11).

Although aspartate ammonia lyase catalyses the equilibration of fumarate

Figure 5.11.

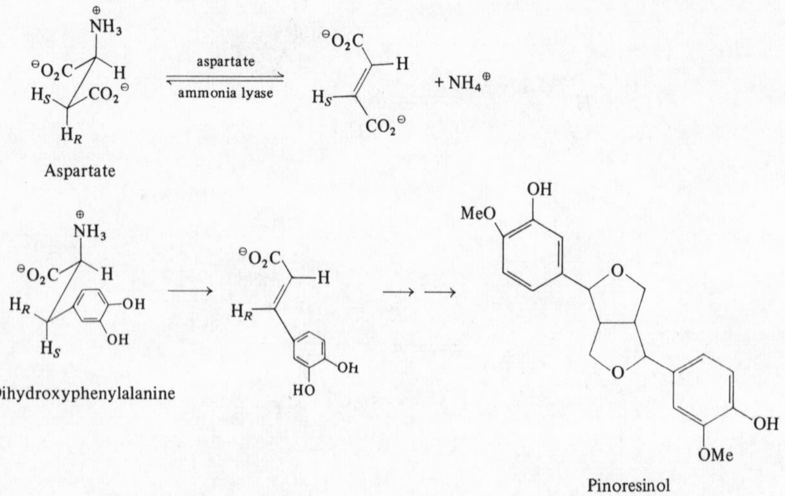

Aspartate

Dihydroxyphenylalanine

Pinoresinol

and aspartate, most other ammonia lyases are essentially irreversible under normal operating conditions in the plant or animal, but they can be made to catalyse the addition of ammonia to the appropriate alkene in the laboratory if a large excess of ammonia (usually ammonium ions) is provided. The stereochemical course and mechanism of action of these enzymes is of interest, especially in comparison with the enzymes that catalyse hydration and dehydration.

5.3.1. Phenylalanine and tyrosine ammonia lyases

Battersby and others have demonstrated that phenylalanine and tyrosine ammonia lyases catalyse the *trans*-elimination of ammonia from their substrates (Battersby *et al.*, 1972; Battersby & Staunton, 1974). One approach involved the synthesis of the stereospecifically labelled substrate amino acids (Figure 5.12). Stereospecific reduction of deuteriated benzaldehyde with liver alcohol dehydrogenase (Chapter 8) yielded the *S*-alcohol, which was converted into its *p*-toluene sulphonate. The sulphonate was treated with the carbanion derived from malonic acid under conditions which gave clean inversion of configuration at the benzylic centre via an S_N2 mechanism. Decarboxylation and bromination led to the 2-*RS*-[3-*S*-^{2}H] bromophenyl propionate, which on treatment with ammonia afforded the required labelled phenylalanine. The sample produced is racemic at C-2 and only the enantiomer for which the enzyme is specific, the naturally occurring *S*-isomer will undergo enzyme-catalysed elimination. This leaves *R*-phenylalanine and a labelled cinnamic acid as products. All the deuterium label was found to be in the cinnamic acid as shown, and consequently it could be concluded that a *trans*-elimination, in which proton and amino group are in the anti-peri-planar conformation, occurs at the active site of phenylalanine ammonia lyase. To reinforce this conclusion, the complementary result was obtained with [3-*R*-^{2}H]phenylalanine as substrate.

Figure 5.12.

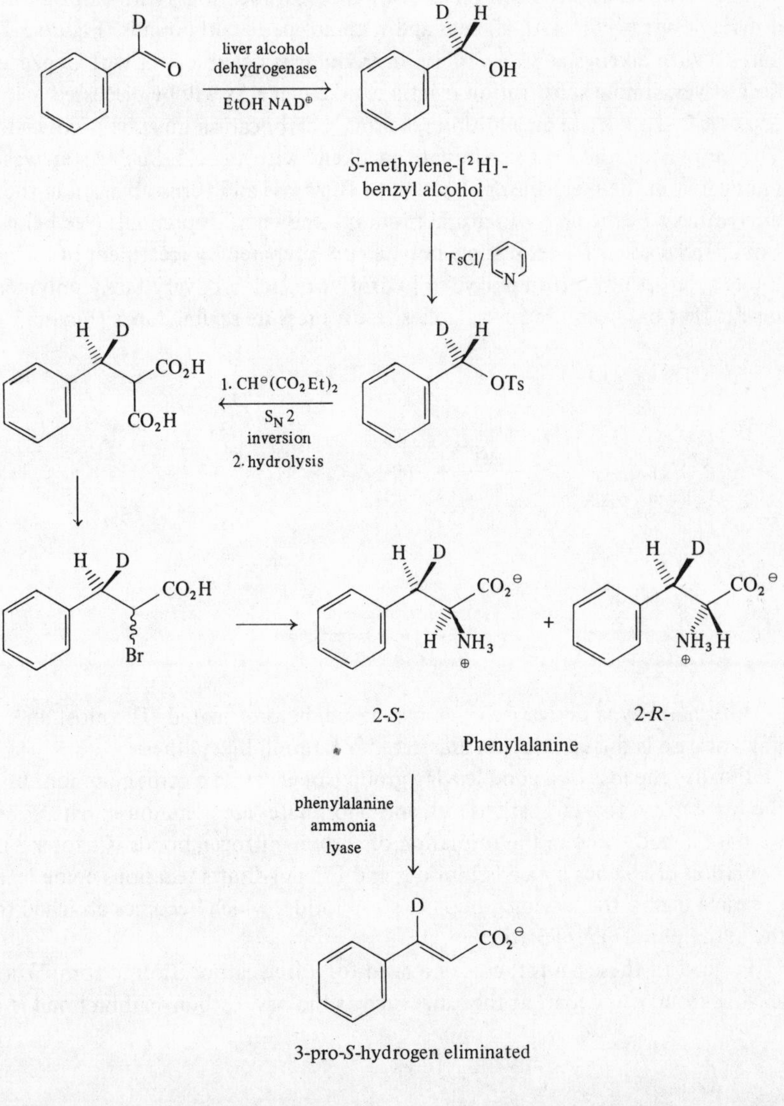
liver alcohol
dehydrogenase
EtOH NAD⊕
S-methylene-[²H]-
benzyl alcohol
TsCl/
1. CH⊖(CO₂Et)₂
S_N2
inversion
2. hydrolysis
2-S-
2-R-
Phenylalanine
phenylalanine
ammonia
lyase
3-pro-S-hydrogen eliminated

With these examples of the ammonia lyases, we can begin to see how the net of correlation of enzymic stereospecificities is established. In order to determine the stereochemical course of the lyase-catalysed reaction, it was necessary to use the known stereochemical properties of the dehydrogenase in addition to chemical reactions that do not affect the chirality at the important centres. Thus modern advances in the stereochemistry of enzyme-catalysed reactions follow the same logical sequences of correlation that were used by the chemists of the nineteenth and earlier twentieth century in their fundamental work on the configurations of organic compounds and natural products (Robinson, 1974).

5.4. Formation of carbon–carbon bonds by addition to alkenes

The reaction of carbon electrophiles (carbocations) with carbon-carbon double bonds occurs with alkenes and with aromatic compounds (Figure 5.1c, d and e). With alkenes, a wide variety of products is possible, but with benzene derivatives, simple substitution reactions take place, as will be discussed in Chapter 7. To initiate an addition reaction, a carbocation must be produced. The simplest method is to protonate an alkene with an acid. Such a step was implicated in the mechanism of action of fumarase and turns up again in the biosynthesis of the class of natural products known as isoprenoids (see below). Commercial use is made of some polyalkenes prepared by treatment of 2-methylpropene with sulphuric or hydrofluoric acids. A very tacky polymer results that has been used as an adhesive for pressure sealing tapes (Figure 5.13).

Figure 5.13.

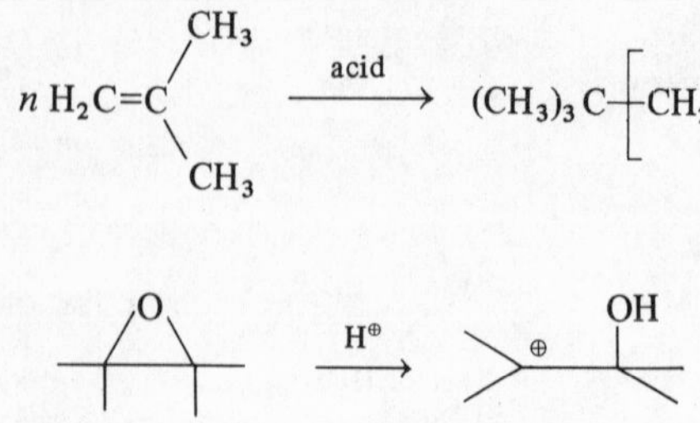

Alternatively, a derivative of an alkene can be protonated. The most important case is the epoxide, as illustrated by steroid biosynthesis.

Finally, the loss of a good leaving group can generate a carbenium ion. In the laboratory, tosylates are useful, and phosphates are common in nature, as we have already seen in the formation of carbon–nitrogen bonds (Chapter 3). Acylation of alkenes by acid chlorides and Friedel–Crafts reactions come into the same class – the leaving group here is chloride, which becomes attached to the aluminium (Figure 5.1d).

In most of these cases there is no need for a free carbocation to form. The leaving group can depart at the same time as the new carbon–carbon bond is formed so that the reaction becomes concerted. Although there is no evidence for the existence of free cations in enzyme-catalysed reactions, the chemistry of the transformations that take place is so easily understandable in terms of the reactivity of carbocations that we shall discuss cations even when no firm justification exists. It is worth emphasising that much of what follows in this section is a plausible rationalisation of the complex behaviour observed. However, all of the biological reactions have laboratory counterparts in which the mechanism is better understood, and within the limits of such analogies, it is helpful to unify the vast number of biological transformations by means of mechanistic arguments.

Once a cation has formed, it can react in a number of distinct ways (Figure 5.14). The multiplicity of pathways available is in part responsible for the wide variety of natural products formed via addition of carbon electrophiles to alkenes. Firstly, simple loss of a proton regenerates an alkene. Secondly, the cation can add to an alkene in the same or in another molecule. Thirdly, an external nucleophile such as a molecule of water can quench the cation, affording an alcohol. Finally, rearrangement to another cation can occur.

Figure 5.14.

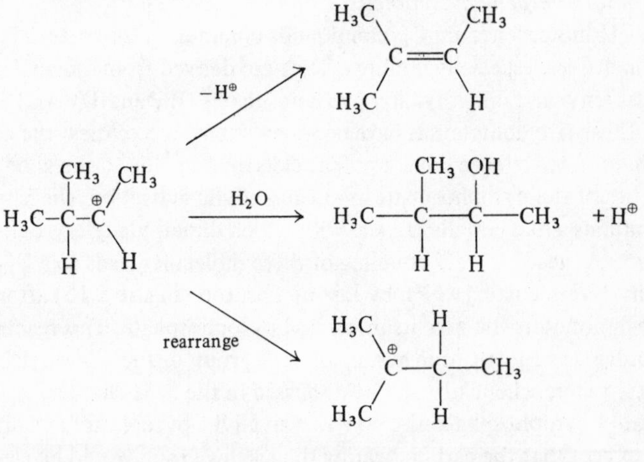

Let us illustrate these points by examining the biosynthesis of some monoterpenes from dimethylallyl pyrophosphate and isopentenyl pyrophosphate. These two compounds are formed from mevalonate by an elimination reaction coupled to a decarboxylation (Figure 5.15). The elucidation of the structures of the compounds biosynthesised via these two C_5 compounds has occupied the lives of a great many chemists, and has led to the discovery of many important chemical principles, some of which we use in these pages. Both chemical and physical methods have been extensively applied to structure elucidation, and the interested reader will find useful discussions in Tedder, Nechvatal, Murray & Carnduff (1972).

Figure 5.15.

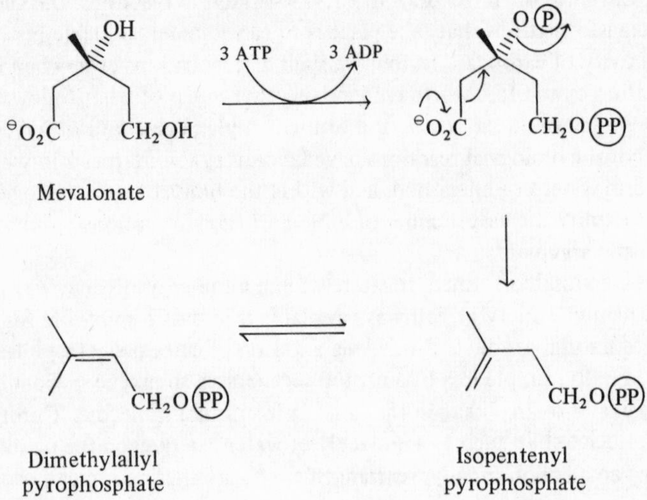

Mevalonate

Dimethylallyl
pyrophosphate

Isopentenyl
pyrophosphate

5.4.1. Monoterpene biosynthesis

Monoterpenes are C_{10} compounds common in the essential oils of plants and trees, especially conifers. They are derived from one molecule each of isopentenyl and dimethylallyl pyrophosphates (IPP and DMAPP; Figure 5.16). These two compounds have complementary reactivities: the allyl pyrophosphate is, as we have seen, a potent electrophile, due to the good leaving properties of the pyrophosphate anion and allylic activation; the isopentenyl unit contains a nucleophilic alkene with which dimethylallyl pyrophosphate can react. A 'head' to 'tail' coupling of these molecules leads to a C_{10} cation (a formal intermediate) which by loss of a proton (Figure 5.16) affords the principal monoterpene precursor, geranyl pyrophosphate. This reaction can also be regarded as a substitution at a methylene group of the dimethylallyl component; its stereochemistry will be discussed in the next chapter.

Geranyl pyrophosphate also contains an allylic pyrophosphate, and we would expect that the carbon bearing the leaving group would be electrophilic. It could, therefore, react with water to give geraniol, or, after isomerisation to neryl pyrophosphate, undergo cyclisation, as shown in Figure 5.17, to yield a further cation. This cation can in turn undergo trapping, loss of a proton, or rearrangement, thereby opening the way to a large number of bicyclic monoterpenes. Notice that the ring-forming steps are additions of carbon electrophiles to alkenes.

Carbocations are high-energy intermediates, and they are most unlikely to have a long lifetime unless stabilised by an enzyme (compare carbanions,

Chapter 4). There is some stereochemical evidence to suggest that a nucleophilic group, such as SMe of methionine, on the enzyme bonds to the cation to form an intermediate, a sulphonium salt of similar reactivity to the cation (Figure 5.16). Sulphonium salts are also involved in biological methylation (Chapter 6).

Figure 5.16.

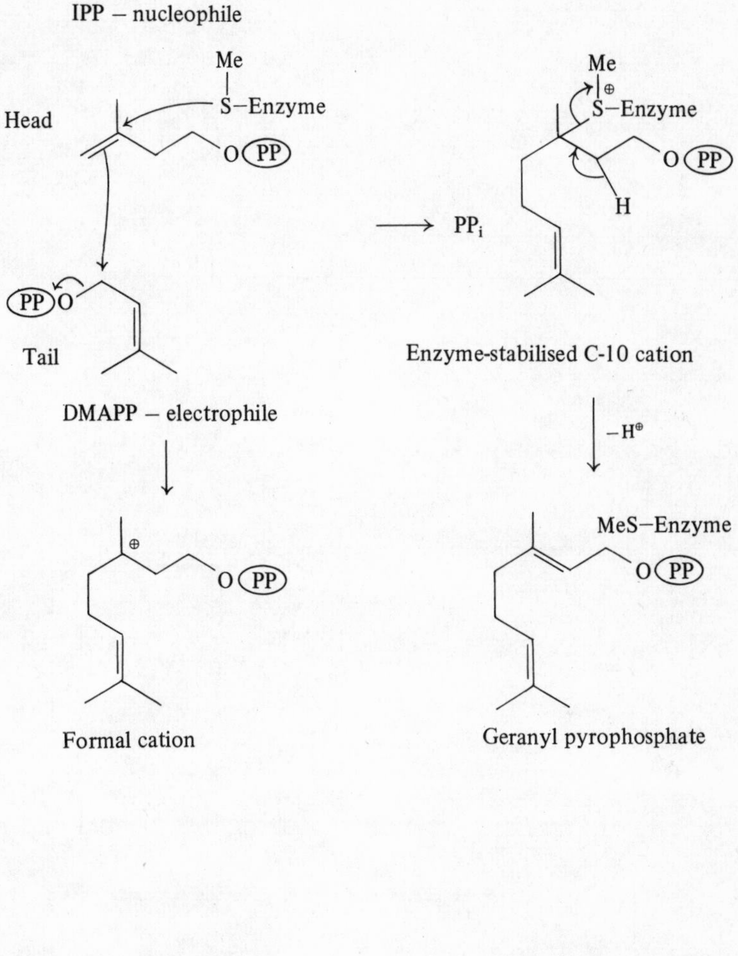
IPP – nucleophile
Me
S–Enzyme
Head
O PP
PP O
Tail
DMAPP – electrophile
⊕
O PP
Formal cation
PPi
Me
⊕
S–Enzyme
O PP
H
Enzyme-stabilised C-10 cation
–H⊕
MeS–Enzyme
O PP
Geranyl pyrophosphate

Figure 5.17.

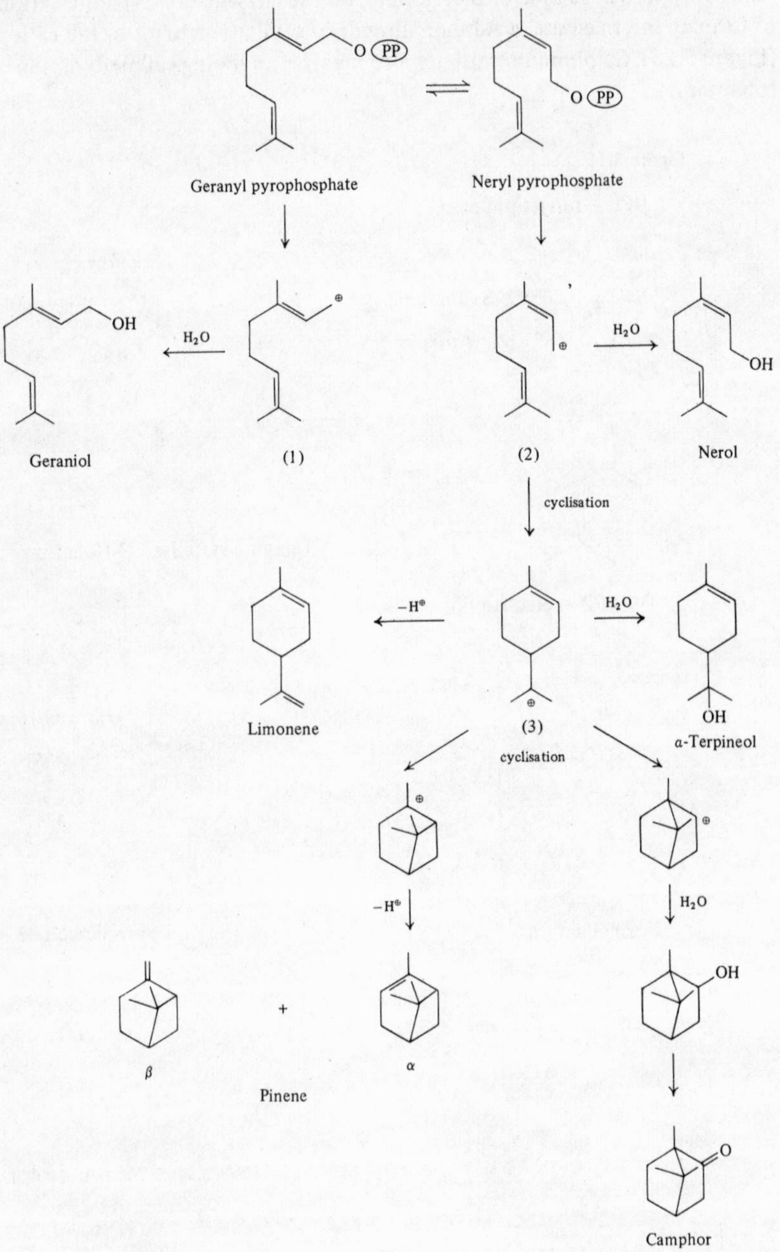
O PP
O PP
Geranyl pyrophosphate
Neryl pyrophosphate
OH
H_2O
⊕
H_2O
OH
Geraniol
(1)
(2)
Nerol
cyclisation
$-H^⊕$
H_2O
Limonene
OH
(3)
α-Terpineol
cyclisation
$-H^⊕$
H_2O
OH
+
β
α
Pinene
O
Camphor

Figure 5.18.

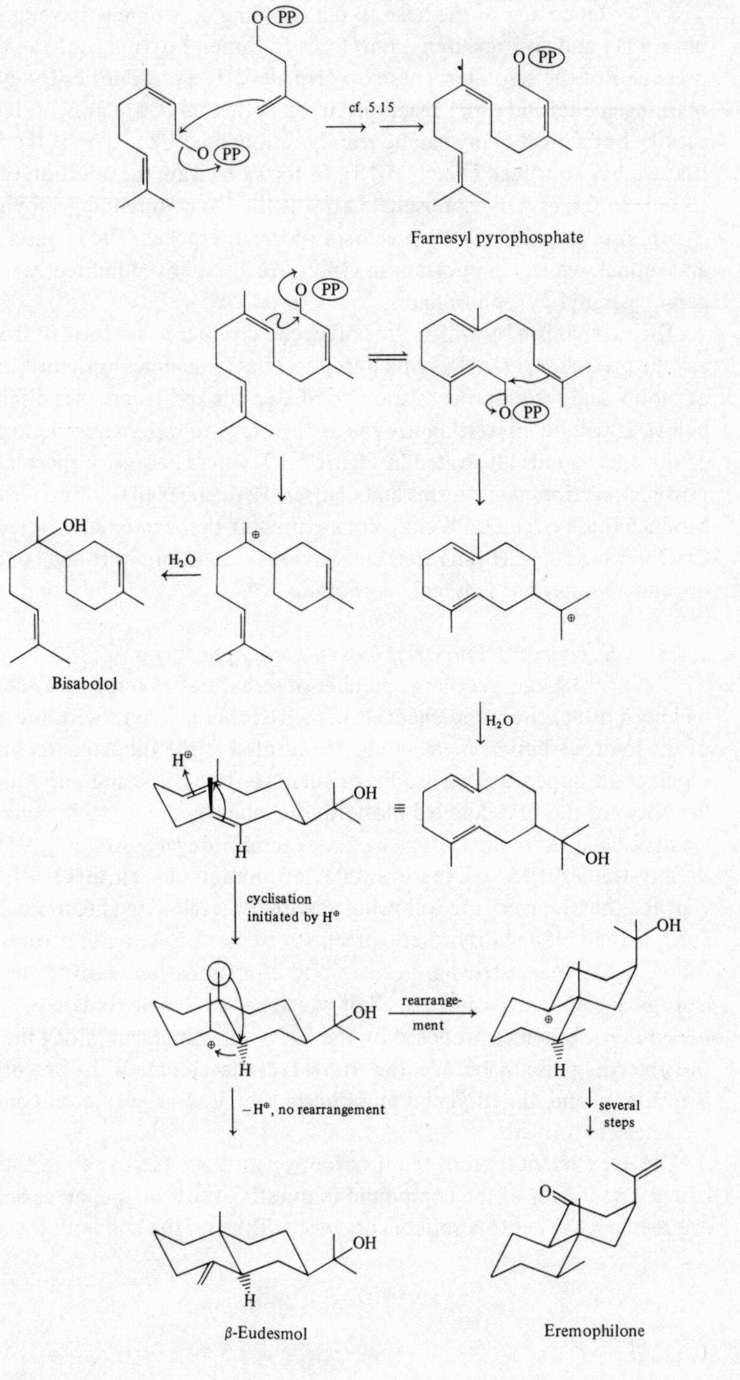

O PP
cf. 5.15
O PP
Farnesyl pyrophosphate
O PP
O PP
OH
H₂O
Bisabolol
H₂O
H⊕
OH
H
OH
cyclisation
initiated by H⊕
OH
rearrange-
ment
OH
H
H
–H⊕, no rearrangement
several
steps
O
OH
H
β-Eudesmol
Eremophilone

5.4.2. *Biosynthesis of sesquiterpenes and diterpenes*

Repetition of the head to tail coupling, this time with geranyl pyrophosphate and an isopentenyl unit leads to farnesyl pyrophosphate, the precursor of the sesquiterpenes (C_{15} terpenes). The possibilities for cyclisations, rearrangements and other reactions in this series are bewildering in their complexity but a great many can be readily rationalised by means of the reactions that we have outlined (Figure 5.18). In the same way, the addition of another C_5 unit to farnesyl pyrophosphate affords the C_{20} compound geranylgeranyl pyrophosphate,which is the precursor of the diterpenes. The vitamin A group and retinal, which is important in vision, are also derived indirectly from geranylgeranyl pyrophosphate.

The mechanism by which the polycyclic diterpenes are formed from their acyclic precursor geranylgeranyl pyrophosphate resembles both the biosyntheses of mono- and sesquiterpenes and also of steroids and triterpenes discussed below. It will be interesting for you at this stage to rationalise the formation of the compounds illustrated in Figure 5.19, where necessary speculating upon possible reaction mechanisms and courses. Remember that in the plants that produce these compounds enzymes are present that can oxidise or reduce C=C, C=O, −OH and −CH functions, and you may therefore introduce oxidative or reductive steps as you feel appropriate.

5.4.3. *Experimental proof of biosynthetic hypotheses*

By 1950, a very large number of structures of terpenoid derivatives had been established, and chemists perceived that a C_5 unit with the structure of the hydrocarbon isoprene could be regarded as the fundamental building block of all terpenes. Two groups of chemists, one Swiss and one American, put forward the first detailed biosynthetic schemes similar to the ones that we have been examining (Ruzicka, Eschenmoser, Jeger & Arigoni, 1955; Stork & Burgstrahler, 1955). Experimental confirmation of their theories had to wait for the discovery the following year that mevalonate (Figures 5.15 and 5.20) is the unique biosynthetic precursor of all steroids and terpenes. In the following 25 years there has been a flood of information, derived chiefly from isotopic labelling experiments, all of which can be accommodated by the biosynthetic schemes proposed by the Swiss and Americans along the lines of the preceding discussion. We now know that mevalonate is the precursor of isopentenyl and dimethylallyl pyrophosphates, whence all steroids and terpenes are formed.

The experimental proof requires four operations. Firstly, an isotopically labelled precursor of the compound in question must be prepared. Secondly, the precursor is fed to a suitable organism. Thirdly, the compound in question

Figure 5.19. Examples of sesquiterpene biosynthesis.

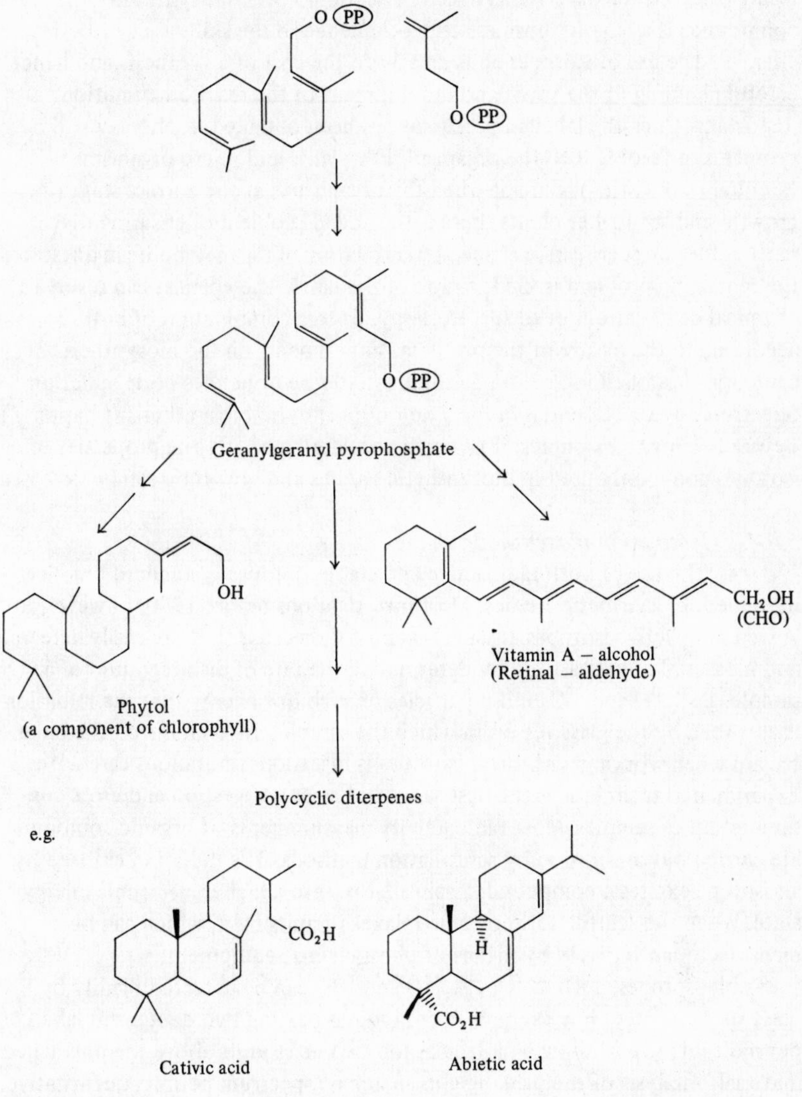
O PP
O PP
O PP
Geranylgeranyl pyrophosphate
OH
Phytol
(a component of chlorophyll)
CH_2OH
(CHO)
Vitamin A – alcohol
(Retinal – aldehyde)
Polycyclic diterpenes
e.g.
CO_2H
Cativic acid
H
CO_2H
Abietic acid

is isolated from the organism and purified, and, finally, the labels are located in the product. We have already seen in Section 5.3.1 that synthesis of labelled compounds is a lengthy business and a challenge to the skill of a synthetic chemist. The use of isotopes adds greatly to the cost of a synthesis and hence careful planning of the route and development of the reaction conditions is important. Once the labelled precursor has been obtained, there may still be problems in feeding it to the organism. For plants and micro-organisms it is essential to feed the precursor when the organism is at the correct stage of growth, and for higher plants there is the added problem of ensuring that it reaches the correct organ or tissue. After isolation of the metabolite in question, the remaining problem is the location of the label. The chemist can resort to chemical degradation, or to spectroscopy, or to a combination of both according to the nature of the problem. Experiments on the biosynthesis of camphor, described in Section 5.4.5, illustrate the principles of degradation. Spectroscopy was essential in the study of porphyrin biosynthesis (Chapter 7). Before turning to examples, it is worth considering briefly the properties of isotopes commonly used in biosynthetic studies and how they can be detected.

5.4.4. *Isotopes in biosynthetic studies*

The use of both radioactive and stable isotopes is standard practice in modern biosynthetic studies. Most investigations before 1970, however, involved radioactive isotopes, usually ^3H and ^{14}C, because they are easily detected and measured quantitatively by determining the rate of disintegration of a sample. Both ^3H and ^{14}C emit β-particles of such low energy that the radiation is absorbed by the glass vessels in which the samples are contained. The chief hazard when working with these isotopes is ingestion; scrupulous care with experimental technique is the best safeguard against ingestion and cross-contamination of samples. Most radioactivity measurements of organic compounds are carried out in solution by scintillation methods. The radiation emitted by an isotope excites a compound, a scintillator, into a higher electronic energy state. When the scintillator molecule relaxes it emits light, which can be measured quantitatively by means of photoelectric equipment.

Stable isotopes, such as ^2H, ^{13}C, ^{18}O and ^{15}N, can be detected readily by mass spectrometry. For example, a compound bearing two deuterium labels per molecule would show a molecular ion two mass units above the unlabelled material. Analysis of the peak heights in a mass spectrum permits quantitative determination of the label to be accomplished. However, because molecules often undergo rearrrangement in the mass spectrometer, it can be difficult to define the location of a label in a sample. Magnetic resonance spectroscopy provides the best solution to the problem of locating stable isotopes. ^{13}C, ^{15}N and ^{31}P are nuclei with a spin quantum number of $\frac{1}{2}$, and with modern instrumentation it is possible to measure magnetic resonance spectra of these nuclei, even at the low natural abundance (*c.* 1 %) of ^{13}C and ^{15}N (Simpson, 1975;

Knowles, Marsh & Rattle, 1976). [13]C spectra of both enriched and natural abundance samples are now routinely available and form the basis of modern biosynthetic studies, as we shall see in Chapter 7. The use of ^{13}C instead of ^{14}C is especially recommended whenever chemical degradation is an impractical method of locating labels. [2]H n.m.r. spectroscopy is also possible, and deuterium is now becoming more common in biosynthetic studies. Magnetic resonance experiments are also valuable in studying enzyme mechanisms and membrane phenomena in larger biological systems. These topics are discussed later.

5.4.5. Camphor

The postulated biosynthesis of camphor is shown in Figure 5.17. It includes two cyclisation reactions in which carbenium ions add to carbon–carbon double bonds. Oxidative steps follow. The veracity of this sequence has been tested by both [3]H- and ^{14}C-labelling experiments (Banthorpe & Baxendale, 1970). Let us consider the tritium experiments first (Figure 5.20).

Chemically synthesised [2-^{3}H]mevalonate, both 3-*R* and 3-*S* isomers, was fed to chrysanthemums (as well as other flowering plants). All of the enzymes that have mevalonate as substrate accept only the 3-*R* isomer, and this is converted via dimethylallyl and isopentenyl pyrophosphates into camphor with the label in the positions marked. The isolated camphor was found to be radioactive. It remained to be proved that the tritium was present at the predicted positions and the proof made use of specific degradation reactions. For example, the protons on the carbon α to a carbonyl group are acidic and labile to exchange under either acidic or basic catalysis (see Chapter 4). Therefore, if the isolated camphor has a tritium label at C-3, and is treated with a mild base, the tritium atoms should exchange with the medium and be washed out by the vast excess of solvent protons. A decrease in radioactivity of the re-isolated camphor should be observed. In this experiment there was no such decrease, and accordingly it was concluded that no label was present at C-3. However the biosynthetic scheme suggests that one tritium atom should be present at C-6. This was tested by submitting camphor to much more drastic basic conditions. Bicyclic ketones like camphor react with bases under vigorous conditions to form homoenolates, that is three-membered ring anions with oxygen at one apex. Just as exchange occurs at an enolate ion, so under these conditions (OH[-] at 220 °C) exchange of the protons at C-6 will occur, and any label present will be washed out. This is what was observed with the camphor isolated from the feeding experiment. Since no label is present at C-3, one tritium label is therefore located at C-6, as predicted. However, the other should be located at the gem-dimethyl group on the one-carbon bridge; unfortunately, there is no simple chemical reaction that will allow these positions to be examined. Therefore it was necessary to use ^{14}C labels to demonstrate whether the C-2 of mevalonic acid ends up at the one-carbon bridge of camphor.

The first step in locating the two ^{14}C atoms was to brominate camphor (an

Figure 5.20.

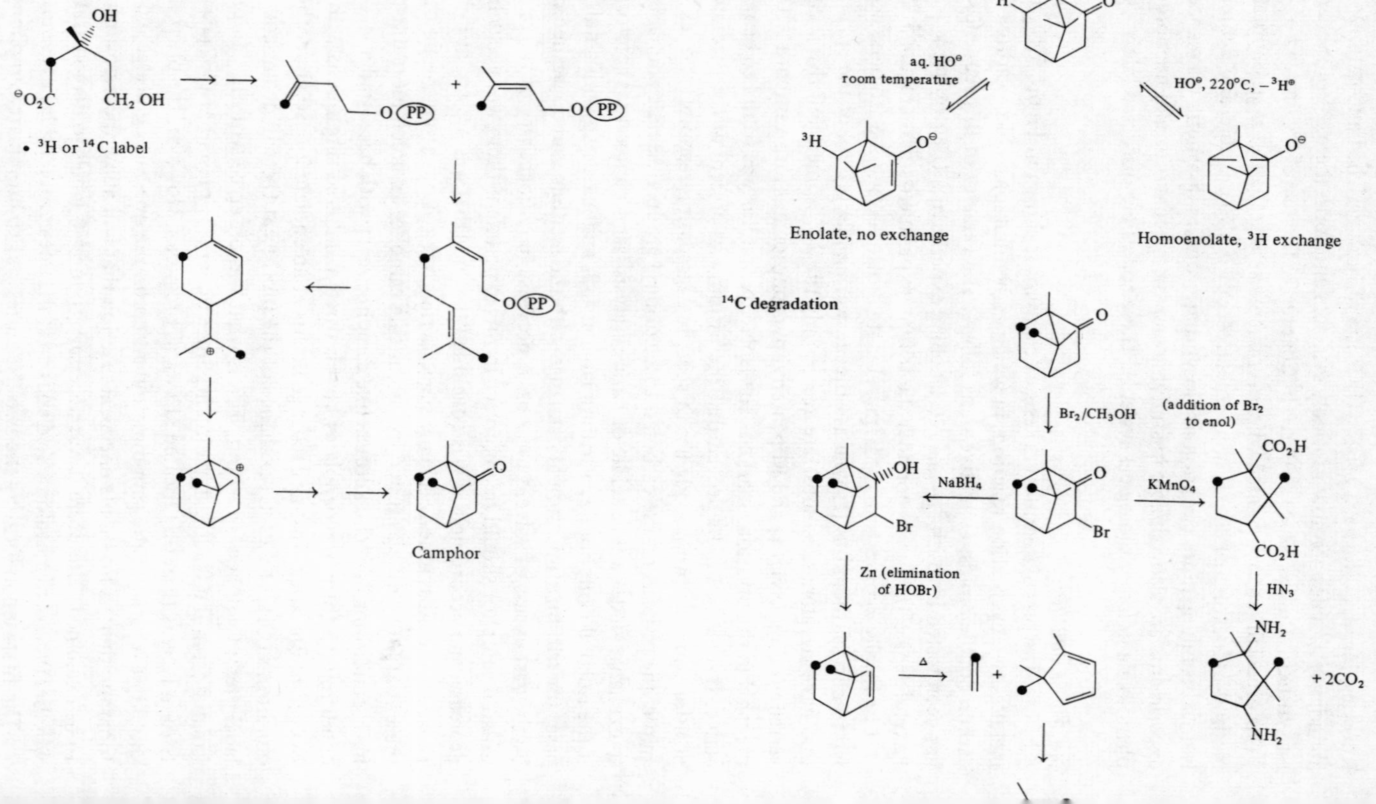
OH
$^{\ominus}O_2C$
CH_2OH
• ^3H or ^{14}C label
O PP
Camphor
^3H degradation
aq. $HO^{\ominus}$
room temperature
$HO^{\ominus}$, 220°C, $-^3H^{\oplus}$
Enolate, no exchange
Homoenolate, ^3H exchange
^{14}C degradation
Br_2/CH_3OH
(addition of Br_2 to enol)
NaBH$_4$
KMnO$_4$
OH
Br
CO_2H
Zn (elimination of HOBr)
HN_3
NH_2
$+ 2CO_2$

Figure 5.21.

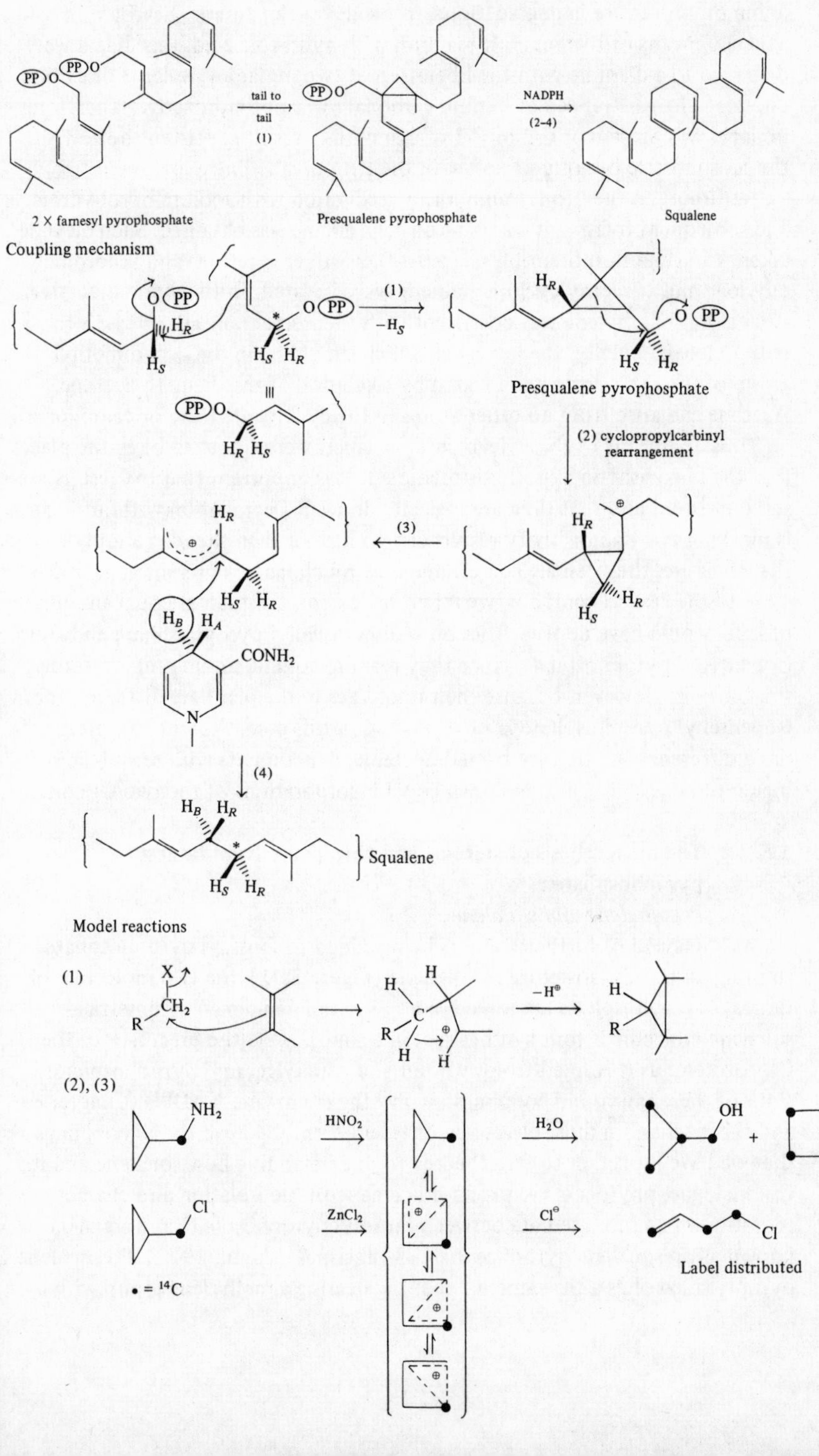
PP
tail to
tail
(1)
NADPH
(2–4)
2 × farnesyl pyrophosphate
Presqualene pyrophosphate
Squalene
Coupling mechanism
O PP
H_R
H_S
H_S H_R
(1)
$-H_S$
Presqualene pyrophosphate
(2) cyclopropylcarbinyl
rearrangement
(3)
H_B H_A
$CONH_2$
N
(4)
Squalene
Model reactions
(1)
X
CH_2
R
$-H^{\oplus}$
(2), (3)
NH_2
HNO_2
H_2O
OH
Cl
$ZnCl_2$
$Cl^{\ominus}$
Label distributed
• = ^{14}C

addition to C=C of the enol). The bromo-compound was then oxidised, a very common procedure in degradations, to yield a cyclopentane dicarboxylic acid. By means of the Schmidt reaction with hydrazoic acid, this diacid was degraded to a diamine with the liberation of two molar equivalents of carbon dioxide. The gas trapped as barium carbonate was not radioactive. Therefore no label was present at C-2 or C-3 of camphor, a result consistent both with the biosynthetic hypothesis and with the tritium-labelling experiment.

Returning to the bromocamphor, by reduction with sodium borohydride and subsequent treatment with zinc dust, an alkene was obtained. Such bicyclic alkenes fragment on thermolysis (retro-Diels-Alder reaction) and accordingly ethylene and trimethylcyclopentadiene were isolated. Both were radioactive. A label in the ethylene was consistent with the homoenolisation results for tritium labels. Finally, the last label, which should be in the gem-dimethyl group of the cyclodiene was located by oxidation of the diene to acetone. Acetone can arise from no other atoms in the cyclopentadiene or camphor.

Thus, by means of these degradations, labels were found to be in the places that the biosynthetic hypothesis predicted. It is important that the results are self consistent and that they are logically derived. Degradation without logic is meaningless. Radioactivity results can go further than showing simply where the labels are, they can also determine how much label is present at each position. In the case of camphor, we might have expected that an equal amount of label would have derived from both dimethylallyl pyrophosphate and from isopentenyl pyrophosphate, since they react in equimolar amounts to form the C_{10} unit. However, because their pool sizes in the plant are different, more isopentenyl pyrophosphate is incorporated in this case. Only in monoterpenes has a difference in pool size been detectable; experiments with sesqui- and higher terpenes have always shown equal incorporations of the two C_5 units.

5.5.   The biosynthesis of steroids and triterpenes from farnesyl pyrophosphate.

5.5.1.   *Presqualene and squalene*

Instead of further C_5 units being added to farnesyl pyrophosphate to bring up the C_{30} structure of squalene (Figure 5.21), the two molecules of farnesyl pyrophosphate are joined in a tail to tail manner. A symmetrical squalene molecule is thus synthesised; in a similar way, the precursor of the C_{40} carotenoids is formed from two units of geranylgeranyl pyrophosphate. Both of these tail to tail couplings involve the coenzyme NADPH (Chapter 8) and the mechanism must clearly be different from the head to tail couplings that we have looked at so far. The key to understanding how squalene and its C_{40} analogue, phytoene, are produced came with the isolation and characterisation of an intermediate between farnesyl pyrophosphate and squalene known as presqualene pyrophosphate (Mulheirn & Ramm, 1972). Presqualene pyrophosphate has a three-membered ring bearing a methylene group with a leaving group (OPP) attached to it (Figure 5.21). The current understanding of the formation of squalene is illustrated in this figure. Reaction (1) is simply an alternative mode of addition of an electrophile to a carbon–carbon double bond. Instead of losing a proton to afford another alkene, a three-membered ring results. There are many parallels between cyclopropane and alkene chemistry because both readily give rise to carbocations. It is this property that is important in the rearrangement and reduction of presqualene to give squalene. Reactions (2) and (3) are examples of a group of rearrangements known as cyclopropylcarbinyl rearrangments. They were first recognised in experiments of the type illustrated in the model reactions (1) and (2) of Figure 5.21; the properties of the cyclopropylcarbinyl cation allow us to understand the detail of the biosynthesis of squalene, as the scheme illustrates. In the biosynthesis, instead of being trapped by water or chloride ion, the cation produced is reduced by hydride from the coenzyme NADPH. The stereospecificity of all of these transformations has been established and overall inversion of configuration occurs at the carbon atom marked * during the sequence.

5.5.2. The biosynthesis of steroids and triterpenes from squalene

For many years it had been recognised that the vast majority of naturally occurring steroids and triterpenes possess an equatorial hydroxyl group at C-3. Obviously this oxygen atom must be inserted into squalene or a squalene derivative, and it is now well established that an oxidase specifically converts squalene into its 2,3-epoxide (Figure 5.22). As we saw at the beginning of the discussion of terpene chemistry, protonation of an epoxide is a useful way of generating a carbenium ion with a hydroxyl group on the neighbouring carbon atom. If we look at the surroundings of the cation derived from squalene-2,3-oxide, we find that it is very close to a carbon–carbon double bond further up the squalene chain, and ideally set up for an addition reaction to form a six-membered ring. If this process is repeated with all subsequent double bonds, a pentacyclic system is obtained, the skeleton of the pentacyclic triterpenes. Alternatively, if the enzymes stop cyclisation part way and direct a five-membered ring to be formed, the steroid skeleton is obtained. These cyclisations by additions of cations to double bonds are thought to be concerted, and their direction is controlled by the conformation that the enzymes enforce squalene to adopt. Thus a chair–boat–chair type of conformation will lead to steroids and an all-chair conformation to pentacyclic triterpenes, as Figure 5.22 shows.

Having followed through the concerted sequence of electrophilic addition reactions, let us concentrate upon the cation $S^{\oplus}$. This cation is the precursor of all steroids. For example, if further oxidation steps occur after loss of a proton, the skeleton of the antibiotic fusidic acid can be reached. This is not the normal stereochemistry of the steroid skeleton. More usually a backbone

Figure 5.22.

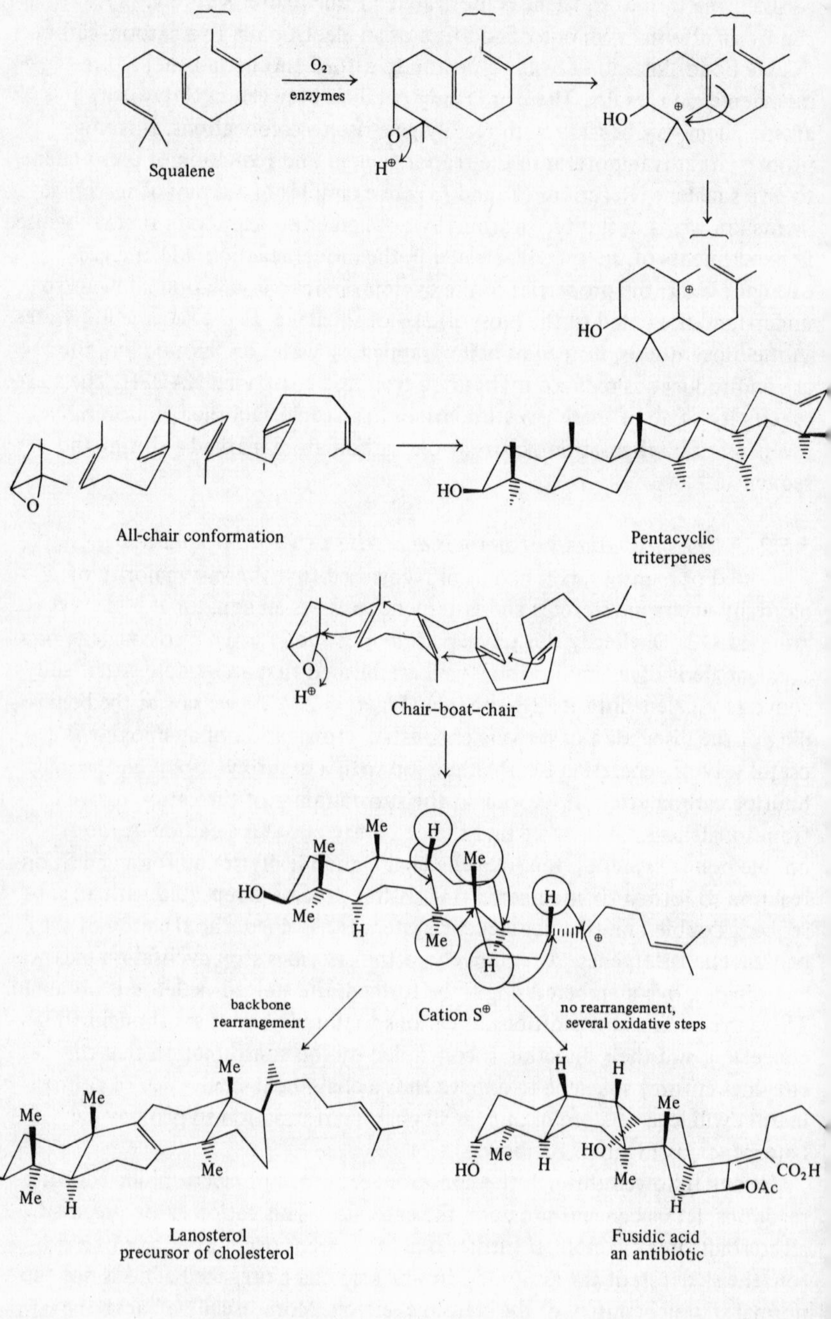

Squalene

O₂ enzymes

All-chair conformation

Pentacyclic triterpenes

Chair–boat–chair

backbone rearrangement

Cation S⊕

no rearrangement, several oxidative steps

Lanosterol precursor of cholesterol

Fusidic acid an antibiotic

rearrangement of cation $S^{\oplus}$ takes place, affording lanosterol in animals. Inspection of the conformational arrangement of the migrating groups in $S^{\oplus}$ shows that every shift is a *trans*-1,2-rearrangement: stereoelectronically, this is the ideal orientation for reacting groups because it maximises orbital overlap in the bond-forming process (recall pyridoxal and see Figure 5.18). The backbone rearrangement is also thought to be concerted like the cyclisation step.

The efficiency of reactions in which the orbitals involved are correctly aligned is very significant, both to develop a new synthetic reaction and to understand enzyme catalysis. The addition reactions discussed earlier in this chapter illustrate how the conformation of a substrate controls the stereochemistry of the reaction and what products are obtained. The backbone rearrangement provides another example of a ring system controlling stereochemistry. If we liken the groups on the enzyme's active site that bind the substrate in the correct position for reaction to the atoms of a ring that do not take part directly in the reaction, then we can regard the stereochemical control that enzymes exert as a special case of stereoelectronic control. For this reason, it is very important for the chemist or biochemist interested in the behaviour of enzymes to understand the fundamental stereochemical and electronic properties of the molecules with which he is concerned.

5.5.3. *Experimental verification of the biosynthetic sequence*

A very substantial research effort has shown even the detailed migrations of single hydrogen atoms through this biosynthetic sequence – virtually every point of stereochemistry of every centre in the steroid molecule has been examined and found to be in accord with the mechanistic scheme outlined above. Once again, the use of radioactive isotopic labels of both carbon and hydrogen has provided the information; this is discussed in detail in Mulheirn & Ramm (1972).

5.6. Biogenetic-type synthesis

The specificity and efficiency with which nature constructs molecules of the complexity of steroids has long aroused the envy of synthetic chemists. Traditional steroid synthesis involved multiple applications of condensation reactions, as we mentioned in Chapter 4; side-reactions were always a potential hazard and yields in many steps were poor. When it was realised that the biosynthesis of steroids was controlled to a high degree by the conformation that the polyene squalene adopted during cyclisation, chemists wondered if synthetic polyenes would also cyclise stereoselectively. To investigate this with a view to improving the methods available for steroid syntheses, a large research effort was expended into the synthesis of suitable alkenes, and several new methods were developed (Sammes, 1971). Initial results were encouraging and, within a few years, steroid synthesis itself was attempted (Johnson, Gravestock & McCarry 1971; Figure 5.23). A precursor polyene was designed so that when treated with acid, cyclisation would occur in a similar manner to squalene. The intermediate, formed in good yield, was stereochemically pure, and easily converted intro progesterone.

A synthesis such as this that is consciously designed following the outlines of a biosynthetic pathway is known as a biogenetic-type synthesis (Scott, 1976). It is based upon the premise that if an enzyme can catalyse and control a reaction, it should be possible to devise conditions to carry out an analogous laboratory synthesis. Biogenetic-type syntheses have been applied to many fields, including phenolic metabolites derived from acetate and alkaloids. However, only in isolated cases such as progesterone have the yields been reasonably good. Consequently, biogenetic-type syntheses are more of interest in their relationship to the enzyme-catalysed reaction and the insight into the biosynthesis that can be obtained than in their efficiency as syntheses of complex natural products. Enzymes are still more efficient than chemists, but then they have had many millions of years more practice.

Figure 5.23.

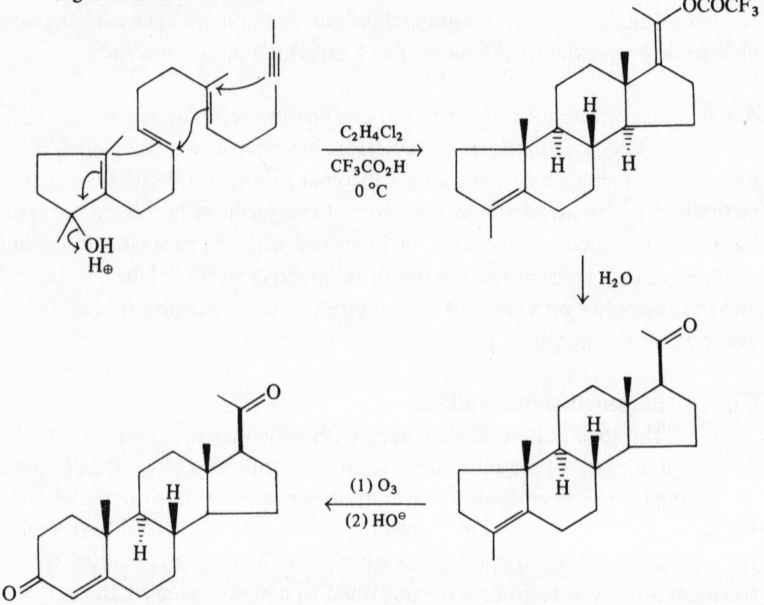

Problem 5.1

The biosynthesis of protoporphyrin-IX from coproporphyrinogen-III involves the formation of two vinyl groups from two of the propionate sidechains of the precursor (Figure 5.24). It is thought that this reaction proceeds in two steps. Firstly, a hydroxylation reaction α to the pyrrole ring takes place (see Chapter 6); secondly, elimination to give the vinyl group occurs. Suggest a plausible mechanism for the elimination, giving due consideration to the stereochemistry. How might it be possible to test your mechanistic proposals? (Battersby *et al.*, 1975.)

Figure 5.24.

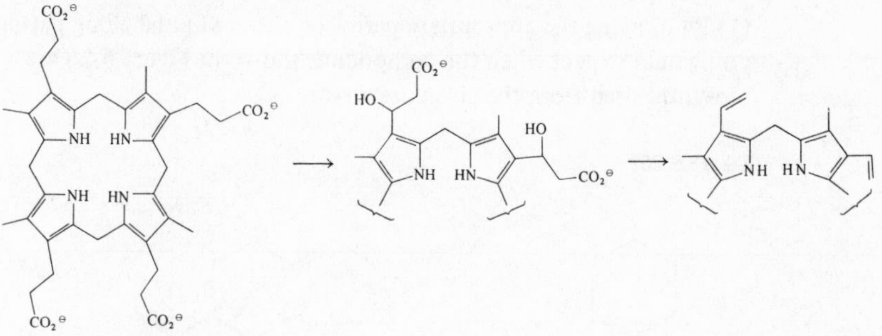

Coproporphyrinogen-III

Protoporphyrinogen-IX

Problem 5.2

Protocatechuic acid arises in nature by the enzyme-catalysed dehydrogenation (= oxidation) and dehydration of shikimic acid (Figure 5.25). Stereospecifically labelled shikimic acid was prepared, with tritium occupying the 6-pro-*S* position. On incubation with the enzyme system, the label was retained in the isolated protocatechuic acid. In contrast, the non-enzymic reaction was not stereospecific. (Scharf *et al.*, 1971).

Figure 5.25.

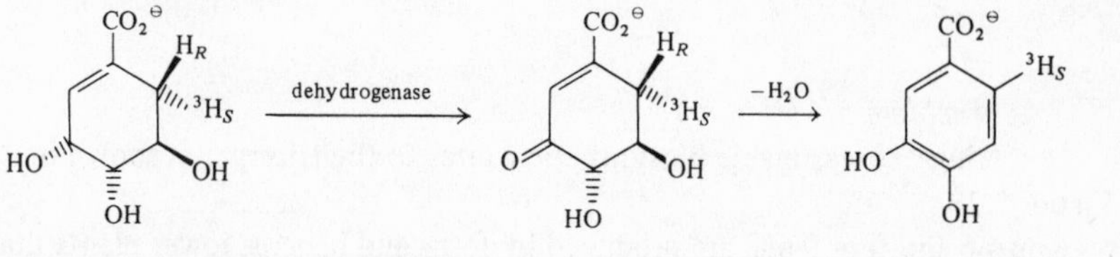

Shikimic acid

Protocatechuic acid

(1) How can the stereochemical course of the enzyme-catalysed reaction be described in a few words?

(2) Bearing in mind that the position from which the hydrogen atom is lost is adjacent to an α,β-unsaturated ketone, suggest a mechanism for the enzyme-catalysed reaction.

(3) Is the non-enzymic result consistent with the results of the enzyme-catalysed reaction and the expected chemistry of the substrate?

The following problems provide opportunities for you to propose some plausible biosynthetic routes to terpenes and also to suggest simple degradations of labelled compounds, using the chemistry that you have seen in this chapter.

Problem 5.3

(1) By drawing the appropriate pathway, show what labelling pattern you would expect when the compounds shown in Figure 5.26 are biosynthesised from the given precursor.

Figure 5.26.

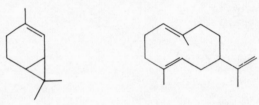

from [4-¹⁴C] mevalonate   from [2-¹⁴C] mevalonate

(2) Suggest degradations that would be suitable for locating the label in the positions indicated on Figure 5.27.

Figure 5.27.

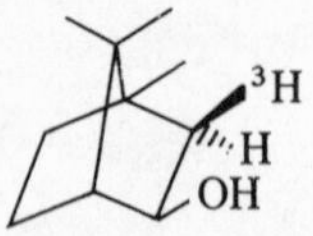

Problem 5.4 (Harder)

Suggest reasonable biosynthetic routes to the triterpenes shown in Figure 5.28.

Hopene and fern-9-ene are produced by ferns and lichens, lower plants that have not evolved the ability to epoxidise squalene (Barton, Mellows & Widdowson, 1971).

Figure 5.28.

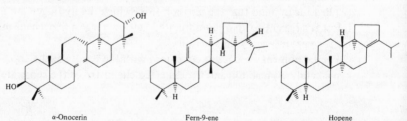

α-Onocerin   Fern-9-ene   Hopene

6   Substitution reactions at aliphatic carbon atoms

6.1.   The chemical background of substitution reactions

A substitution reaction occurs whenever a group or atom bonded to carbon is replaced by another. Both aliphatic and aromatic carbon atoms can be substrates for substitution reactions, but the characters of the two classes of substrate differ so much that it is necessary to consider each in turn. In this chapter, we shall concentrate upon aliphatic substitution, and we shall turn our attention to aromatic compounds in Chapter 7.

There are three classes of aliphatic substitution reaction, depending upon the **leaving group**. If an electronegative group or ion is displaced, a **nucleophilic** substitution takes place (Figure 6.1); alternatively, replacement of hydrogen usually occurs via **radical** or sometimes **electrophilic** mechanisms.

Figure 6.1.

Nucleophilic aliphatic substitution

$$\overset{\delta+}{R\,CH_2}-\overset{\delta-}{I} \longrightarrow R\,CH_2\,OH + I^{\ominus}$$

$HO^{\ominus}$
Nucleophile

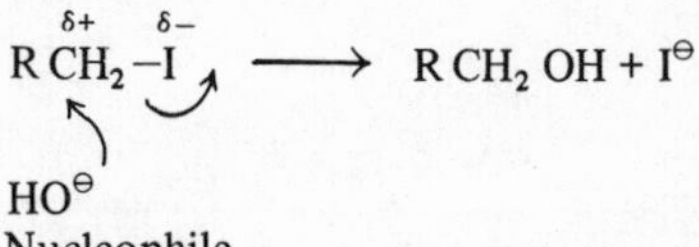

Free-radical aliphatic substitution

Alkyl radical

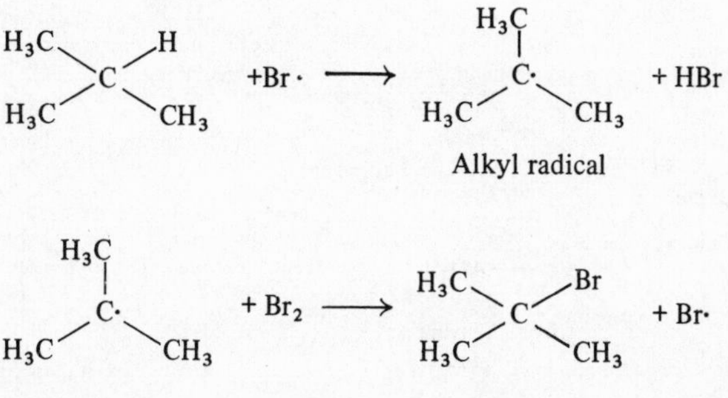

Nucleophilic substitution reactions are common in nature and in the laboratory (Figure 6.2). Notice that the leaving groups are either anions of acids (OAc⁻, $H_2PO_3^-$, p-$CH_3.C_6H_4.SO_3^-$, X⁻) or derived from positively charged groups (H_3O^+, R_4N^+, R_3S^+). All of these are groups that withdraw electrons inductively from their neighbouring carbon atom, leaving it somewhat positively charged and hence susceptible to nucleophilic attack. Nature uses such reactions to construct chains of aliphatic carbon atoms in the biosynthesis of isoprenoids. A great deal is known about the mechanism of these reactions from stereochemical studies as we shall see. On the other hand, it is still to be established how alkylation of heteroatom nucleophiles (O, S, and N) takes place in nature. For example, the stereochemical course of reactions of the biological methylating agent, *S*-adenosylmethionine, is unknown. A different situation pertains to reactions at glycosidic atoms in carbohydrates such as the synthesis of polysaccharides and nucleotides. Here the intrinsic stereochemical requirements of the sugar largely determine the course of the reactions.

Radical substitutions at aliphatic carbon atoms are easiest whenever the radical intermediate can be stabilised by delocalisation of the odd electron into a π-system. This generalisation applies both to biosynthetic and laboratory reactions (Figure 6.3). Although it has been shown that the laboratory reactions illustrated are radical processes, work is still in progress on the detailed mechanism of the biological reactions. We shall see that is is possible to demonstrate the stereochemical course of such reactions, but it is difficult to decide whether the reactions involve radicals or other reactive intermediates. Most arguments on this point centre around hydroxylation via the highly electrophilic oxygen

Figure 6.2.

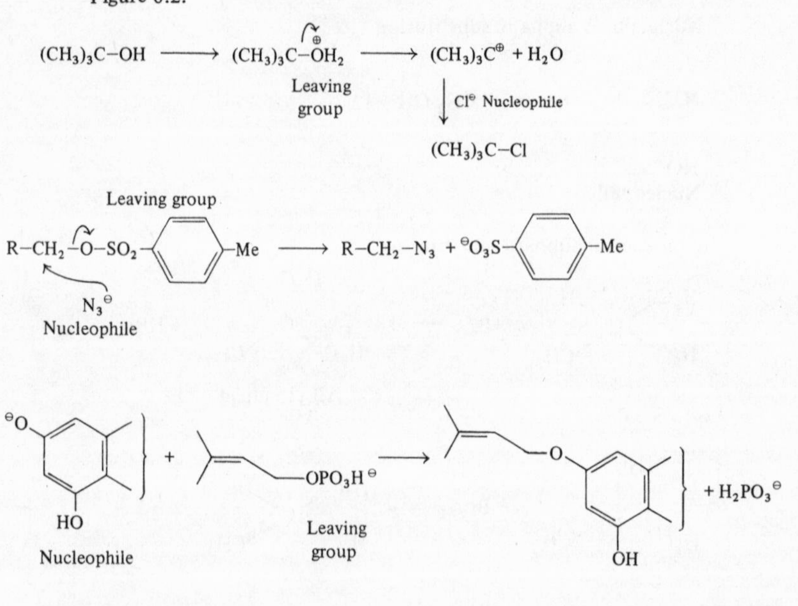

Figure 6.3.

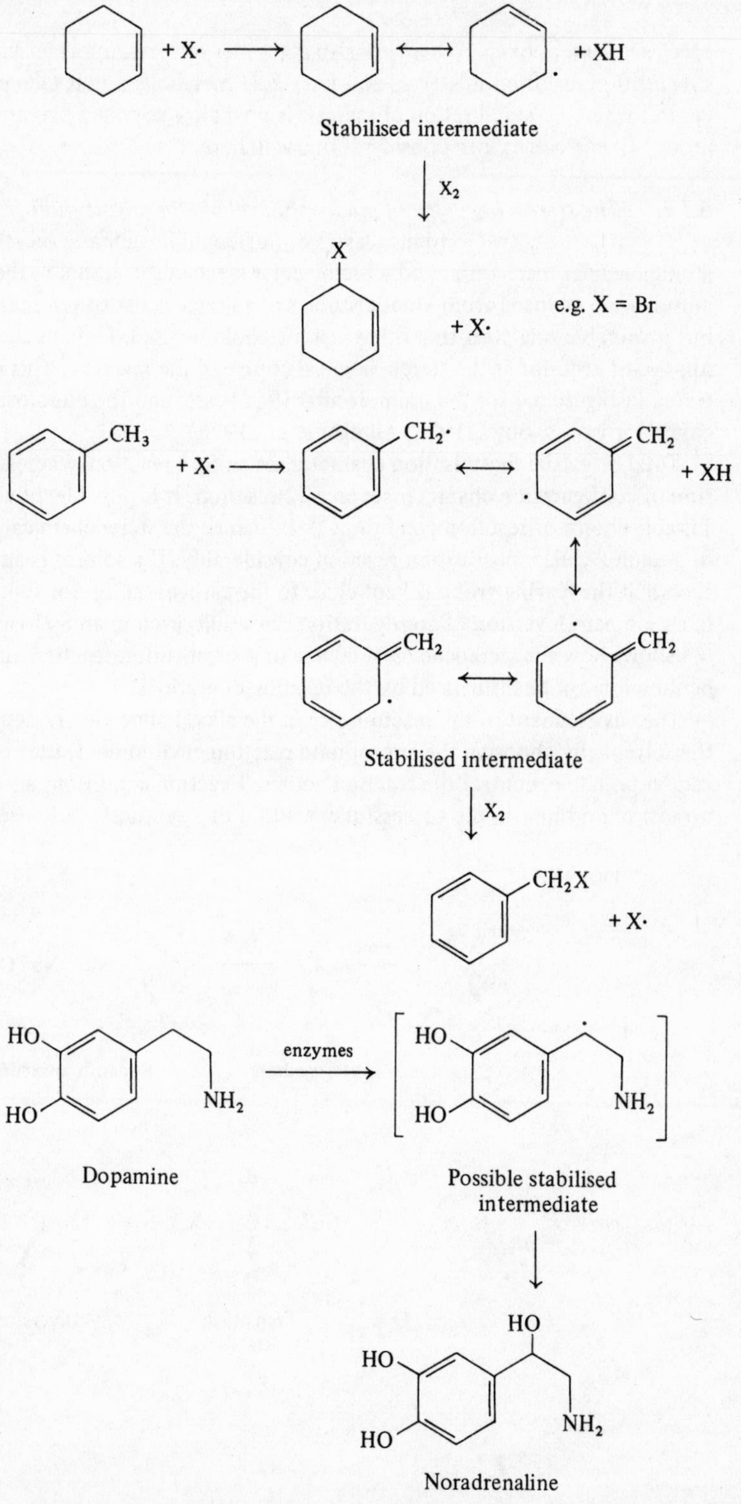

species known as oxene. There are also a number of very important biological substitution reactions in steroid and fatty acid metabolism that take place at centres where no stabilisation of radicals is possible – so-called non-activated atoms. Oxene comes into consideration again here.

6.1.1. *The stereochemistry of nucleophilic aliphatic substitution*

There are two extreme classes of nucleophilic aliphatic substitution, a unimolecular mechanism and a bimolecular mechanism. Initially, these mechanisms were deduced from kinetic studies of suitable substitution reactions, but it soon became clear that other criteria could be used. Perhaps the most important criterion is the stereochemical course of the reaction. This is illustrated in Figure 6.4 for the unimolecular (S_N1) case, and the bimolecular (S_N2) case (Morrison & Boyd, 1973; Allinger *et al.*, 1976).

Total or partial racemisation characterises an S_N1 reaction, whereas inversion of configuration characterises an S_N2 reaction. It is possible, however, by suitable choice of reaction conditions to influence the stereochemical course of a unimolecular substitution reaction considerably. If a solvent is chosen such that the leaving group is kept close to the carbenium ion for some time in an ion pair, inversion of configuration can result, even in an S_N1 case. Figure 6.5 shows how the stereochemical course of a substitution reaction via a carbenium ion can be influenced by the reaction conditions.

The environment of the reaction site in the alkyl halide clearly depends upon the solvent. By choosing the appropriate reaction environment, the chemist can, in principle, control the reaction course. Reaction conditions are of the utmost importance in the successful execution of a synthesis. It is also possible

Figure 6.4.

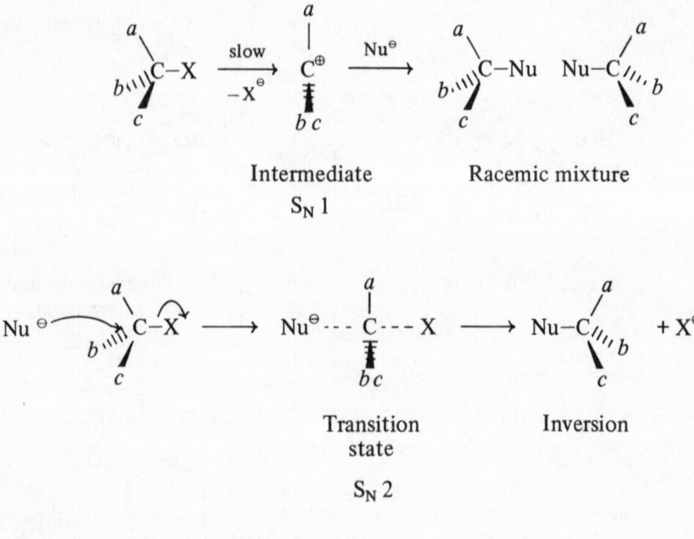

Intermediate
$S_N 1$

Racemic mixture

Transition state

Inversion

$S_N 2$

to use solvent effects as a diagnostic tool for the reaction mechanism. Look once again at Figure 6.4. The full positive charge on the intermediate carbenium ion of the S_N1 case compared with the dispersed charge in the S_N2 reaction suggests that the rate-controlling transition state for the former will be more highly charged than the transition state for the latter. Consequently, if we increase the polarity of the solvent, we would expect an S_N1 reaction to be accelerated more than an S_N2.

So far, we have seen how the chemist can produce inversion or racemisation at will in a subsitution reaction, but how can retention of configuration be accomplished? The answer lies in an intramolecular substitution known as an S_Ni mechanism (Figure 6.5). Here the structure of the reacting groups and solvent chosen impose the illustrated geometry of the reaction – the leaving group departs from the same side attacked by the intramolecular nucleophile. Such intramolecular control of a reaction course can be likened to the stereochemical control of an enzyme's active site.

Figure 6.5.

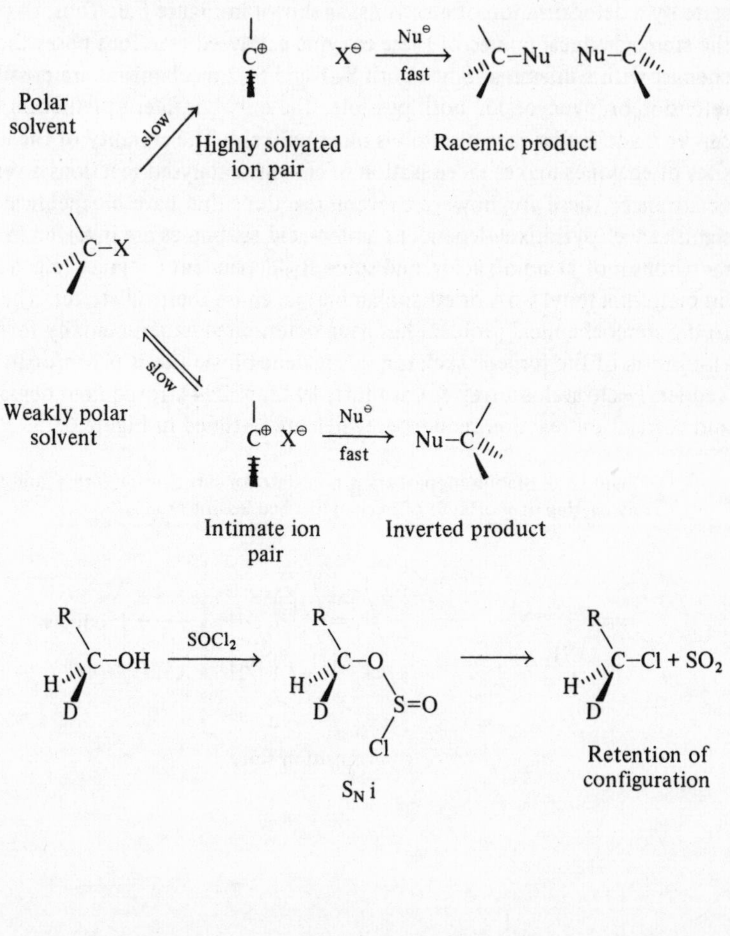

If we are interested in the mechanism of substitution reactions catalysed by enzymes, we cannot use simple molecularity as a criterion of mechanism and we must, therefore, look to other methods. Since the enzyme active site provides the environment for the reaction, solvent effects are also inapplicable. We are left with the powerful tool of the stereochemical course of the reaction.

6.2. Biosynthetic nucleophilic substitution reactions

6.2.1. Reactions of active phosphate esters

Several biosynthetic substitutions in which carbon–carbon bonds are formed from allylic pyrophosphates are found in the biosynthesis of terpenoid compounds (Chapter 5). There are a number of others that we have met in passing in which carbon–heteroatom bonds are formed, for example pteroic acid (Section 3.1.3.) and thiamine pyrophosphate (Section 4.5). All of these reactions are interesting from the stereochemical viewpoint, to see whether enzymes choose the charge-delocalised allylic and benzylic cations as intermediates via an S_N1 mechanism or whether an S_N2 or concerted pathway is followed. In the latter case we could envisage stabilisation of the transition state by a delocalisation of electrons, as shown in Figure 6.6. Thus, to predict the stereochemical course of these enzyme-catalysed reactions poses the chemist with a dilemma. Since both S_N1 and S_N2 mechanisms are possible, retention or inversion are both possible. The only confident prediction that can be made is that racemisation is most unlikely. The chirality of the active sites of enzymes makes racemisation in enzyme-catalysed reactions a very rare occurrence. There are, however, several reactions that have biochemical significance: pyridoxal-dependent amino-acid racemases are involved in the metabolism of D-amino acids, and some B_{12}-dependent enzymes (e.g. acting on methylmalonyl CoA or ethanolamine) racemise their substrates. The answer to the stereochemical problem has been determined experimentally for the elongation of the terpene skeleton in squalene biosynthesis (Cornforth, 1969; Tedder, Nechvatal, Murray & Carnduff, 1972, p. 224). It required degradation and correlation reaction sequences which are outlined in Figure 6.7.

Figure 6.6. Stabilisation of transition state for substitution α to a double bond by overlap of π-orbitals of nucleophile and leaving group.

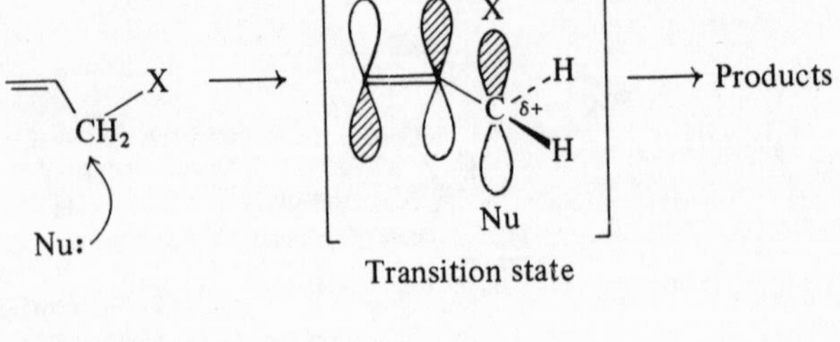

Transition state

Figure 6.7. Squalene biosynthesis. The stereochemical structures shown in (a) were confirmed by comparison of the degradation product with the compound prepared as shown in (b).

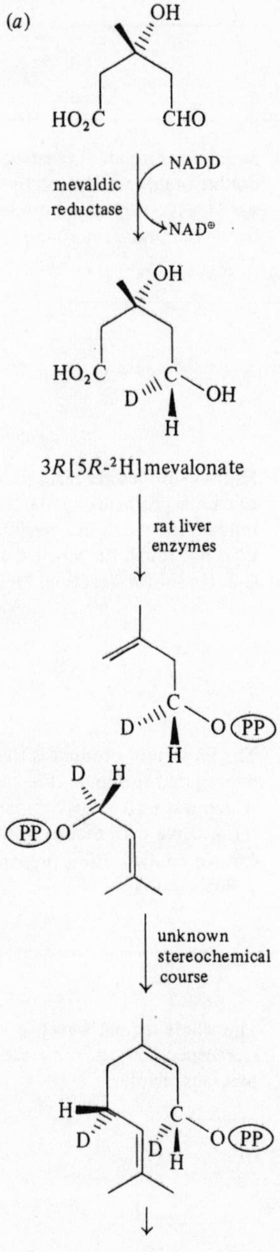

(a)

3-R-mevaldic aldehyde is reduced sterospecifically with [4-^{2}H]NADH (NADD, see Chapter 8) and mevaldic reductase to give chiral deuteriated mevalonate which is then incubated with rat liver homogenates.

$3R[5R-^2H]$mevalonate

The rat liver enzymes transform the labelled mevalonate into the IPP/DMAPP pool and then convert them into squalene in a reaction of unknown stereochemical course.

The structures drawn here assume the answer that will be derived from the following degradation and correlation.

Figure 6.7 (continued).

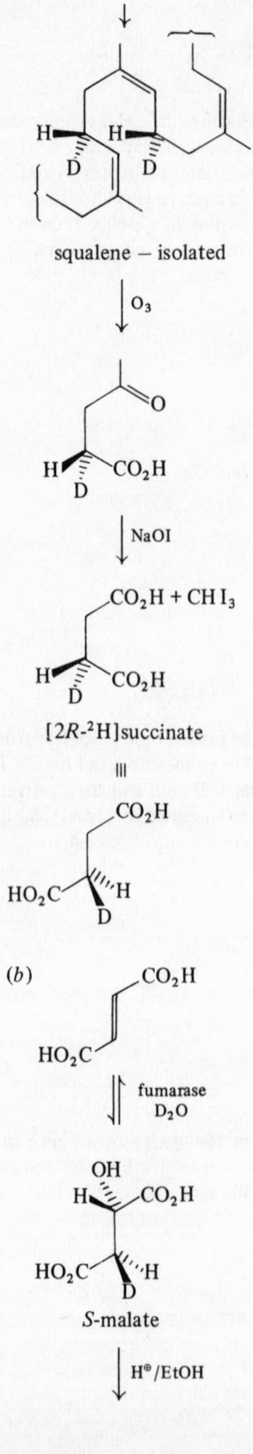

squalene — isolated

Squalene is isolated, purified and the double bonds are cleaved by ozonolysis. This is a very common degradation for compounds containing C=C bonds.

| O₃

Methyl ketones are readily degraded to carboxylic acids by alkali and iodine. The reaction is very mild and does not touch the labelled atom at C-2. (Iodoform reaction. Mechanism?)

| NaOI

$CO_2H + CHI_3$

[2*R*-²H]succinate

|||

The important product is the chiral deuteriated succinate, the chirality of which was proved by comparing its o.r.d. curve with that of a standard of known configuration prepared as follows in (*b*).

(*b*)

The whole sequence rests upon the stereospecificity of fumarase (see previous chapter).

| fumarase
D₂O

S-malate

| H⊕/EtOH

Figure 6.7. (continued)

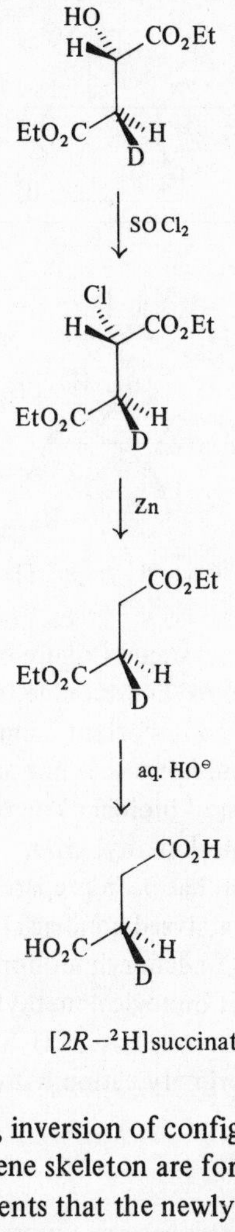

HO
H·····CO$_2$Et

EtO$_2$C········H
D

All that needs to be done is to remove the hydroxyl group and insert a hydrogen atom. Thionyl chloride converts the alcohol into the corresponding chloride with retention of configuration (S_Ni). The acids have been protected as esters to prevent conversion into acid chlorides.

| SO Cl$_2$

Cl
H·····CO$_2$Et

EtO$_2$C········H
D

| Zn

CO$_2$Et

EtO$_2$C········H
D

Care is required to choose reducing conditions that neither reduce the esters nor cause elimination of HCl.

| aq. HO$^\ominus$

CO$_2$H

HO$_2$C········H
D

Mild base hydrolysis cleaves the protecting groups without affecting the chiral centre.

[$2R-^2$H]succinate

Thus, inversion of configuration takes place when carbon–carbon bonds of the terpene skeleton are formed. Cornforth has also shown by labelling experiments that the newly formed double bond is *trans* with respect to the growing chain. A mechanism consistent with both of these results is shown in Figure 6.8. Notice the *trans*-elimination that gives rise to the double bond. The conformation from which elimination occurs is the most stable available for such a reaction because the bulkiest groups are as far away as possible from each other ($-CH_2.OPP$ and $-CH_2.CH_2-$). Once again, we see how an enzyme-catalysed reaction makes use of the optimum intrinsic reactivity of its substrates.

6.2.2. Biological methylation reactions

Although the number of biosynthetic reactions involving methyl group transfer is large, direct evidence concerning the stereochemistry and mechanism is not available to date. *S*-adenosylmethionine is the principal methylating agent and it is reactive, as we saw earlier, because the sulphonium

Figure 6.8.

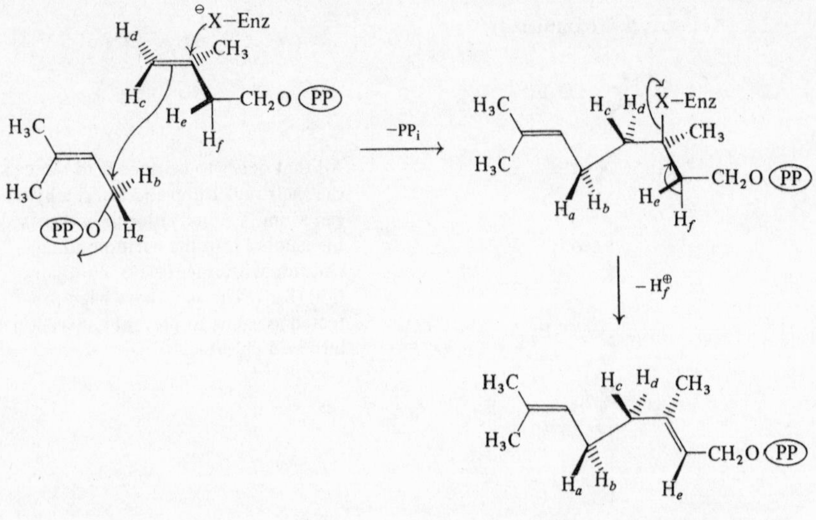

Geranyl pyrophosphate

salt contains a good leaving group, *S*-adenosylhomocysteine (Figure 6.9). Through the medium of methyl transferases, *S*-adenosylmethionine takes part in the biosynthesis of alkaloids (Chapter 8), and is an important compound in the deactivation of excitatory neurotransmitters and hormones like adrenaline (Chapter 9). However, to determine the mechanism of biological methylation by a stereochemical approach, we would need a chiral methyl group. A chiral methyl group bears all three hydrogen isotopes, and has been prepared in the case of acetic acid and used in studies of enzyme-catalysed condensation reactions (Section 4.3.2.). So far, no-one has prepared *S*-adenosylmethionine with a chiral methyl group. However it is very likely that biological methylation proceeds via S_N2 type processes because the S_N1 intermediate, CH_3^+, is one of the least stable carbocations imaginable; it is a primary cation with no way of delocalising the positive charge.

Figure 6.9.

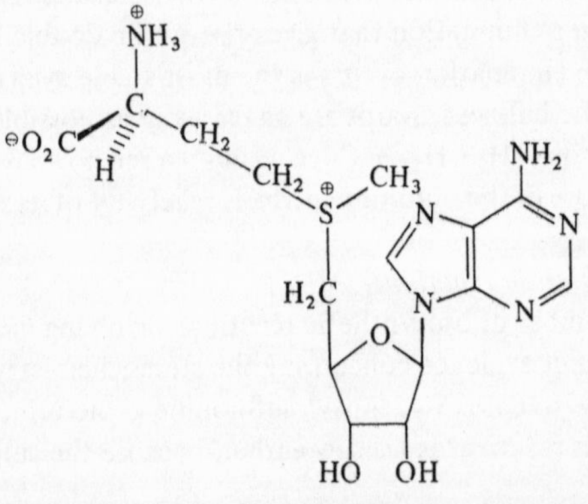

S-adenosylmethionine

6.2.3.   Substitution at glycosidic carbon atoms

The hydrolysis of a polysaccharide by lysozyme (Chapter 1) is an example of nucleophilic substitution at a glycosidic carbon atom. All biochemical reactions in which glycosides are formed and hydrolysed belong to this class. The feature that distinguishes glycoside reactions from other aliphatic centres is the presence of nucleophilic groups close to the centre at which substitution occurs. Such neighbouring groups can act as stabilising influences upon intermediates, and either the enzyme or the substrate can provide one. Unimolecular and bimolecular substitution reactions are again possible (Figure 6.10). In all reactions at glycosidic centres, the cationic intermediate is resonance stabilised via the oxonium ion. However, the stereochemical course

Figure 6.10.

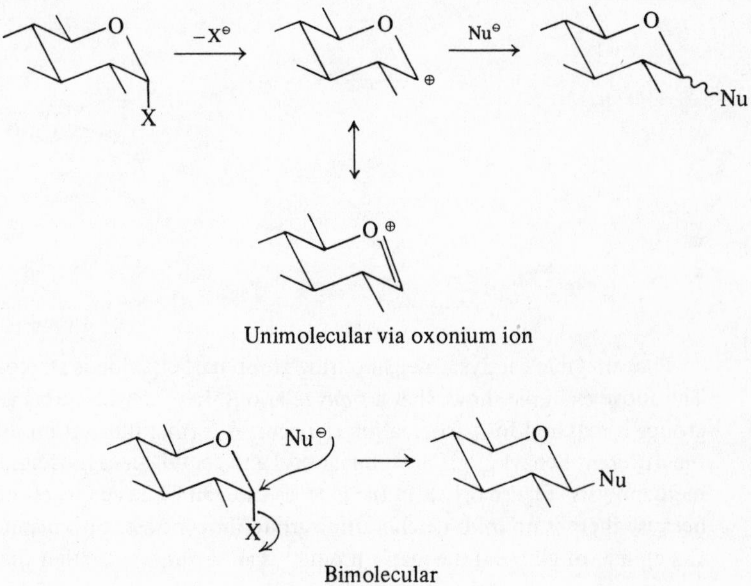

Unimolecular via oxonium ion

Bimolecular

depends greatly upon the nature of substituents at C-2 (Tedder *et al.*, 1972; Figure 6.11). This is well illustrated by the reaction of 1-chloro-2-acetoxy- and 1-chloro-2-hydroxyglucose derivatives. Treatment of 3,4,6-tri-*O*-acetoxy-β-D-glucopyranosyl chloride (6.11.1) with sodium acetate gives the β-tetraacetoxy compound with **inversion** of configuration. No neighbouring group participation occurs. On the other hand, if the 2-substituent is acetoxy (6.11.2), the product is formed with **retention** of configuration. We can rationalise this if the ionisation of the chloride is assisted by the 2-acetoxy group yielding an oxonium ion that blocks the lower or α-face of the ring. The attacking acetate anion can only approach from the top or β-face and thus substitutes with retention of configuration.

Figure 6.11.

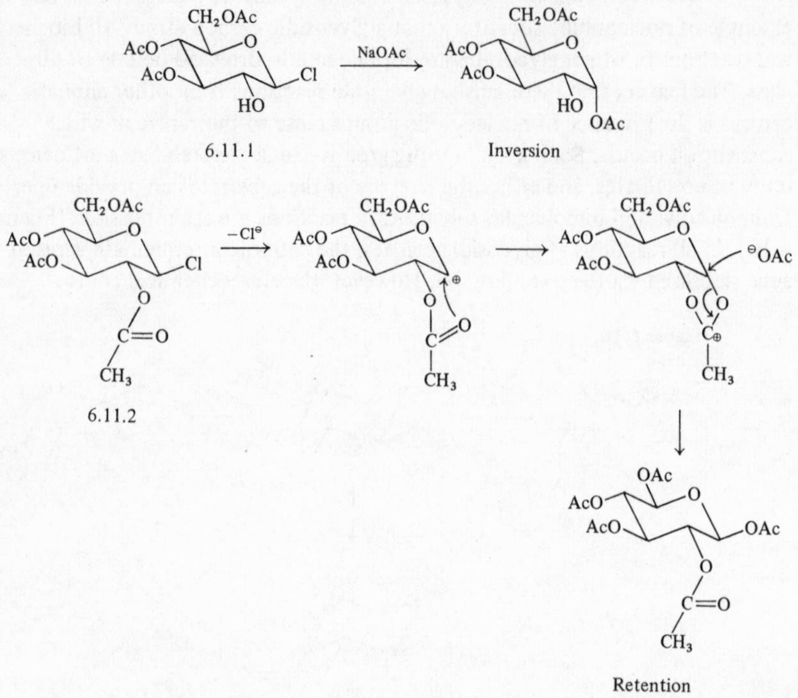

6.11.1   Inversion

6.11.2   Retention

Like enzymic catalysis, neighbouring group participation is stereospecific. The above example shows that a *trans* relationship of neighbouring and leaving groups is required for participation to occur. A further illustration of this is the different behaviour of acetylmannosyl and acetylglucosyl bromides to methanolysis (Figure 6.12). In the glucose case, an S_N2 type reaction occurs because there is no *trans* neighbouring group. In contrast, with mannose (the C-2 epimer of glucose) the major product is an *ortho* ester, rather than the

Figure 6.12.

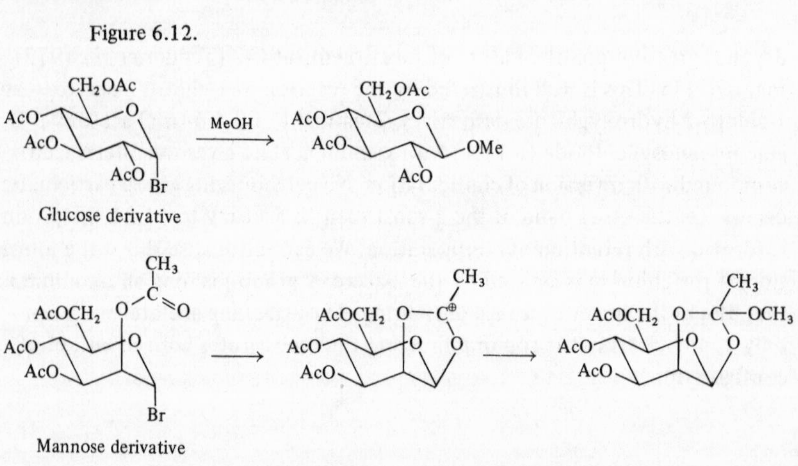

Glucose derivative

Mannose derivative

normal substitution product. The *ortho* ester is formed when methanol captures the cation that was generated by neighbouring group participation of the C-2 acetoxy group.

Bearing in mind that reactive groups on the active sites of enzymes show properties of neighbouring groups, we can transfer these ideas to the mechanism of sucrose formation and hydrolysis (Gray, 1971, p. 246; Figure 6.13).

Figure 6.13.

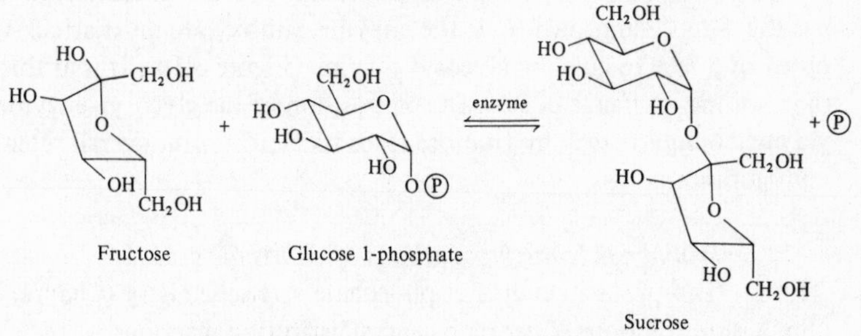

Fructose   Glucose 1-phosphate

Sucrose

$$\text{Glucose 1-}(P) + (^{32}P) \xrightleftharpoons{\text{enzyme}} \text{Glucose 1-}(^{32}P) + (P)$$

$$\text{Glucose 1-}(P) + \text{Enz-XH} \rightleftharpoons \text{Glucose 1-XEnz} + (P)$$

Figure 6.14.

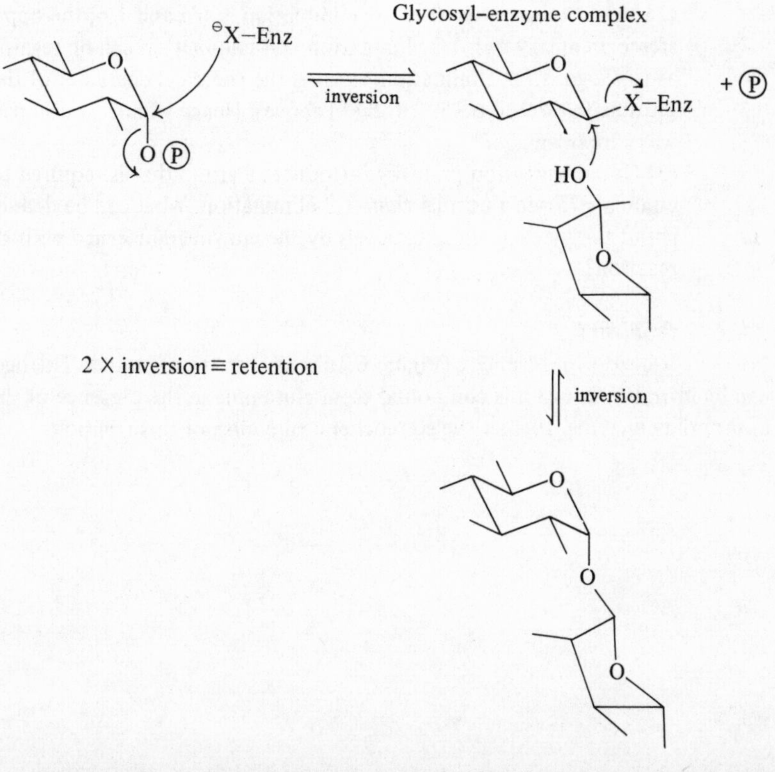

Glycosyl–enzyme complex

inversion

$2 \times$ inversion $\equiv$ retention

inversion

The equation given shows that overall retention of configuration occurs. From what we have just seen, the formation of a glycosyl–enzyme complex is probable in which a neighbouring group on the enzyme surface attacks the glycosidic carbon atom. This idea is supported by the observation that in the absence of fructose, ^{32}P phosphate in the reaction medium exchanges with D-glucose α-1-phosphate. The pH dependence of the rate of reaction implicates a carboxylate anion in catalysis, and it is probable that this is the neighbouring group that forms the glycosyl–enzyme complex. Since glucose 1-phosphate has the α-configuration at C-1, the enzyme carboxylate must attack from the upper or β-face to give the glycosyl enzyme (Figure 6.14). It is at this stage that isotope exchange occurs. Decomposition of the glycosyl–enzyme complex via nucleophilic attack by fructose leads to sucrose with overall retention of configuration.

Problem 6.1; the biosynthesis of chorismic acid
This problem embraces phosphate ester chemistry (Chapter 3), elimination reactions (Chapter 5) and substitution reactions.

(1) Allylic halides and phosphate can undergo substitution reactions by two bimolecular routes (Figure 6.15). Attack by X⁻ is described as S_N2', or as vinylogous S_N2 substitution. Y⁻ carries out pure S_N2 substitution. Look at the stereochemical course of the elimination reaction leading to chorismic acid. What is unusual about it?
(2) You should notice that the elimination is *syn* and 1,4, the opposite stereochemistry to a 1,2-elimination. 1,4-elimination can be regarded as vinylogous 1,2-elimination. What is the chemical character of the phosphate ester in 6.15.1? (cf. (1) above). Hence what reaction pathways are open to it?
(3) One substitution pathway introduces a group that is required to eliminate H⁺ via a normal *trans*-1,2-elimination. What can be deduced about the mechanism of catalysis by the enzyme concerned with this reaction?

Problem 6.2
Ribose 1-diphosphate (Figure 6.16) exists as the α-anomer. Nitrogen can be introduced into this compound from glutamine in the presence of the appropriate enzyme. Predict the stereochemical course of this reaction.

Figure 6.15.

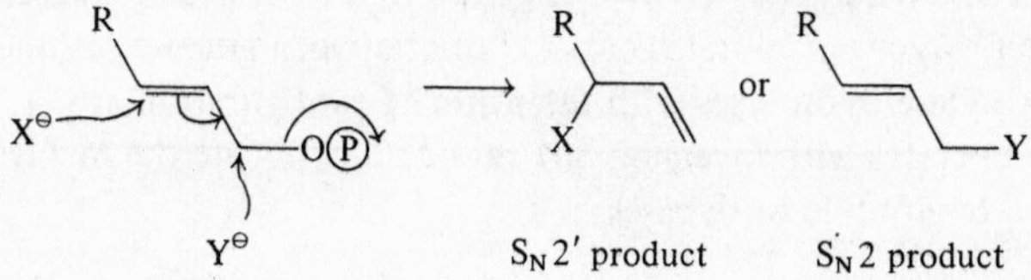

$S_N 2'$ product $S_N 2$ product

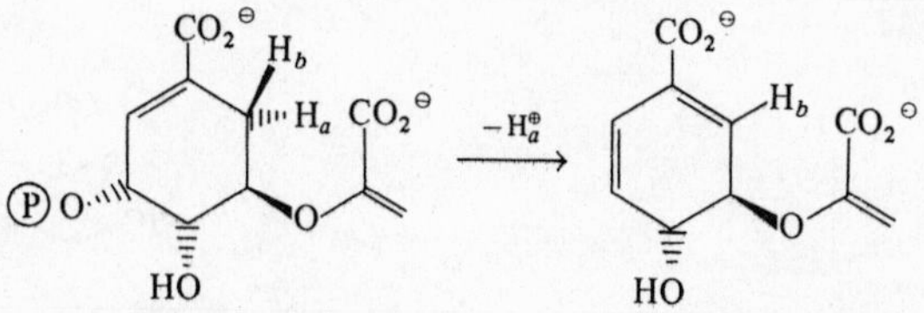

6.15.1   6.15.2
Chorismic acid

Figure 6.16.

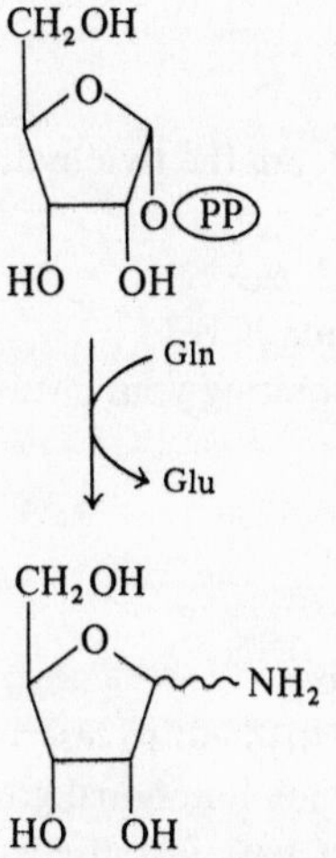

Problem 6.3

α- and β-amylases are two enzymes that hydrolyse polyglucosides such as glycogen. Glycogen contains α(1–4) links between glucose residues (Figure 6.17). α-Amylase hydrolyses with retention of configuration at C-1, whereas β-amylase operates with inversion. pH/rate data implicate the involvement of a carboxylate group in both cases.

Figure 6.17.

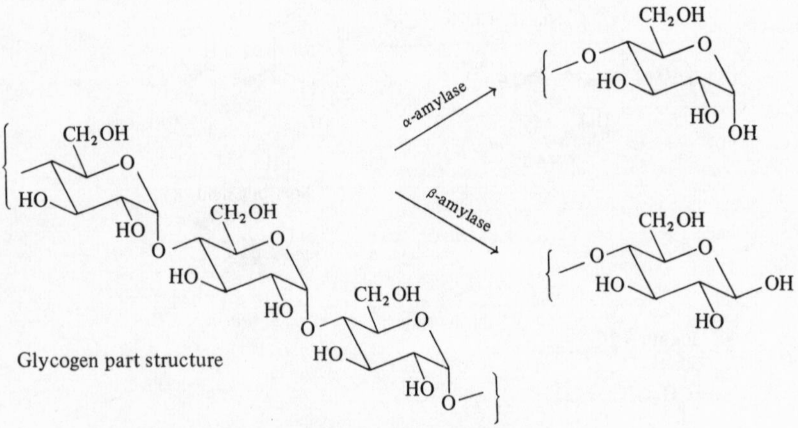

Glycogen part structure

(1) On the basis of the above information, are the two hydrolysis mechanisms likely to be the same?
(2) What intermediates are possible in each case?
(3) What function could the carboxylate ion have?
(4) Suggest a mechanism of action incorporating your answers and the above information.

Answers are at the end of the book.

6.3. Hydroxylation and free-radical substitutions

In laboratory chemistry, it is usually very difficult to insert a new substituent on a carbon atom that does not bear some functional group already. For example, the nucleophilic substitutions that we have just discussed require a leaving group on the attacked carbon. Also the bromination of a ketone requires the presence of the carbonyl group to provide the necessary activation, this time towards an electrophilic reagent, bromine. A further possibility exists – radical substitution (Nonhebel & Walton, 1974; Nonhebel, Tedder & Walton, 1979). A large number of reagents react by the homolytic cleavage of a bond to produce atoms or radicals which can then interact with an organic substrate. For example, halogen molecules dissociate on irradiation with u.v. light. The atoms then can abstract hydrogen atoms from an alkane to yield an alkyl radical and the hydrogen halide. Finally, pairing of the alkyl radical with a halogen atom completes a radical substitution.

$$X\text{-}X \rightarrow 2X\cdot$$

$$X\cdot + RH \rightarrow R\cdot + HX$$

$$R\cdot + X_2 \rightarrow RX + X\cdot$$

$$R\cdot + X\cdot \rightarrow RX$$

Other reagents such as sulphuryl chloride (SO_2Cl_2) undergo similar reactions, and the proportions of products obtained from alkanes in all cases depends upon the relative stabilities of the alkyl radicals that are intermediates. Tertiary alkyl radicals are more stable than secondary, which are more stable than primary. Consequently, the highest degree of substitution occurs at tertiary positions when account is taken for the statistical probability of attack at each atom. It has been estimated that bromine atoms react 1600 times more readily at a tertiary, and 80 times more readily at a secondary than at a primary hydrogen atom.

Like ions, radicals can also be stabilised by delocalisation. Allylic and benzylic radicals are much more stable than simple alkyl radicals and consequently a compound containing an allylic or benzylic hydrogen will undergo selective radical substitution at that position. A wide range of reagents is available to form alkyl halides, and radicals are intermediates in many oxidative synthetic methods (House, 1972; Figure 6.18).

6.3.1. Benzylic hydroxylation in biosynthesis

In both laboratory synthesis and in nature it is often impossible to

Figure 6.18. Radical substitution at benzyl-type positions.

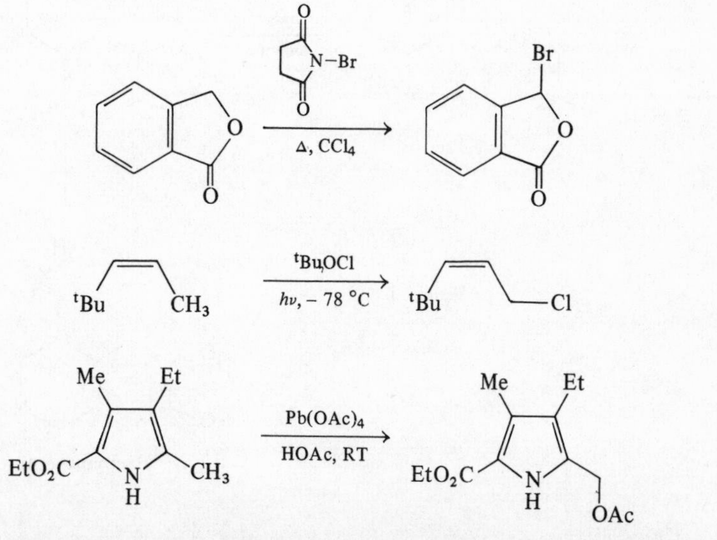

build the required reactivity into a given carbon atom by simple addition or condensation reactions. It is then that reactions that attack C—H bonds directly come into play and, as we have just seen, radical substitution reactions are very suitable. A large number of natural products are formed from amino acids, including the hormone adrenaline, and several alkaloids are formed by hydroxylation at benzylic carbon atoms (Battersby, Kelsey, Staunton & Suckling, 1973; Gunsalus, Pedersen & Sligar, 1975; Figure 6.19). We noted also the example of vinyl group formation in protoporphyrin IX biosynthesis as a problem in Chapter 5.

In most of these hydroxylation reactions, the enzymes involved have proved difficult to characterise. The best studied is dopamine-β-hydroxylase, a copper-containing enzyme that can be isolated from bovine adrenal medulla: it catalyses the hydroxylation of dopamine to noradrenaline (Figure 6.19) using molecular oxygen. The key questions to be answered concerning the mechanism of this oxidation are: how is the oxidising species generated by the enzyme and what type of reaction occurs on the substrate? Chapters 9 and 15 look at some aspects of the former question; here we consider things from the substrate's point of view.

Figure 6.19.

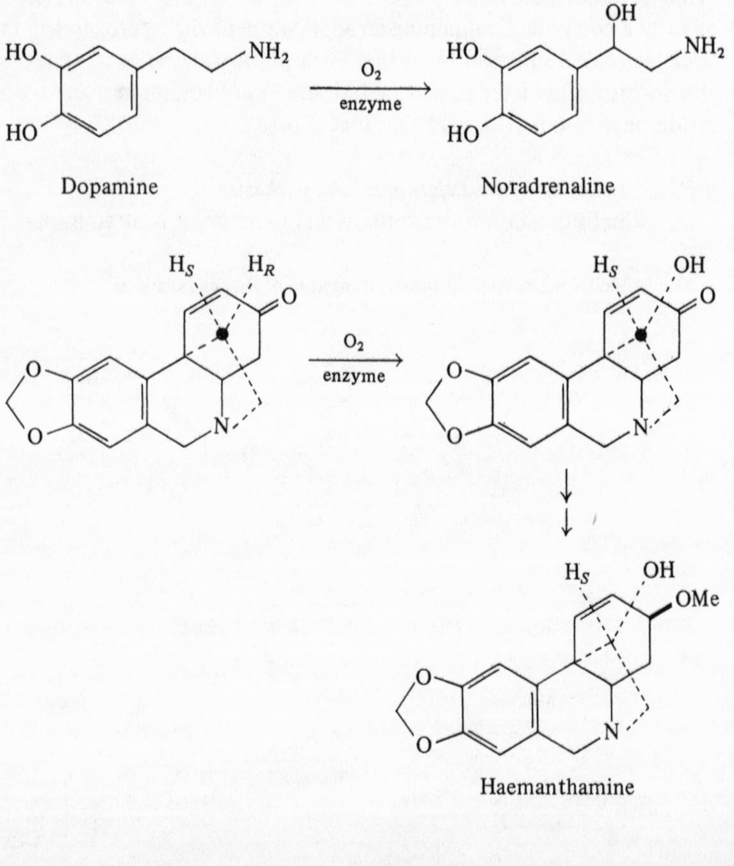

Dopamine   Noradrenaline

Haemanthamine

Two extreme oxidising species may be envisaged; either an oxygen radical could be formed, or an electron-deficient oxygen species known as an oxene is a possibility. Oxene is the oxygen analogue of the well established reactive intermediates carbenes and nitrenes (Morrison & Boyd, 1973; Allinger *et al.*, 1976; Figure 6.20). These intermediates have six valence electrons and can exist in two important electronic states, singlet and triplet. In the singlet state all electrons are paired, whereas the triplet has two unpaired electrons – it is essentially a diradical. The reactivity observed depends upon the electronic structure. A singlet carbene has an empty sp^2-orbital and is highly electrophilic. Typically, it inserts into a carbon–hydrogen or a carbon–carbon bond in a concerted reaction taking over the electrons that previously made the bond. On the other hand, a triplet carbene has one unpaired electron in a p- and one in an sp^2-orbital, and its first move is to abstract a hydrogen atom to yield a pair of radicals. Recombination of the radicals leads to products (Figure 6.20). Nitrenes and oxenes may be considered to react similarly. The question of the mechanism of hydroxylation therefore becomes one of distinguishing between singlet and triplet oxene and other potential oxidising species such as $OH\cdot$ and $O_2H\cdot$; this is no easy task and has not yet been accomplished.

Figure 6.20.

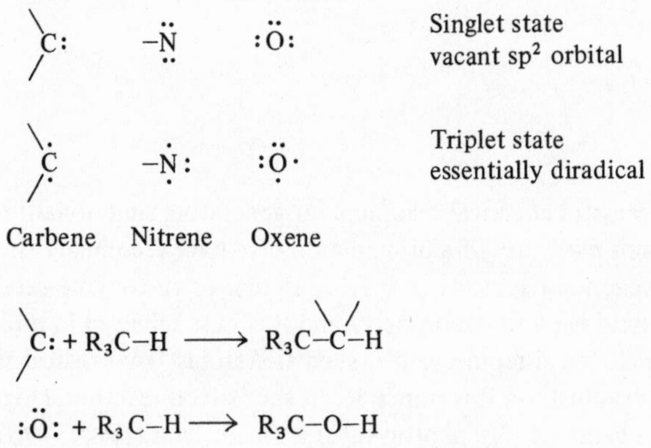

$\diagup\!\!\!\diagdown C:$	$-\ddot{\underset{\displaystyle \cdot\cdot}{N}}$	$:\ddot{O}:$	Singlet state vacant sp^2 orbital
$\underset{\displaystyle \cdot}{\overset{\displaystyle \cdot}{\diagup\!\!\!\diagdown C}}$	$-\dot{N}:$	$:\dot{\ddot{O}}\cdot$	Triplet state essentially diradical
Carbene	Nitrene	Oxene	

$$\diagup\!\!\!\diagdown C: + R_3C\text{–}H \longrightarrow R_3C\text{–}\overset{\diagdown\ \diagup}{C}\text{–}H$$

$$:\ddot{O}: + R_3C\text{–}H \longrightarrow R_3C\text{–}O\text{–}H$$

6.3.2. *Substitution at non-activated positions*

Radical substitution reactions at benzylic or allylic carbon atoms are facile because the intermediate radicals can be stabilised by electron delocalisation. Carbon atoms α to a functional group can be regarded as 'activated' atoms because the functional group confers some chemical reactivity on neighbouring positions, as in the benzyl case. However, there are cases when substitution is required at an atom remote from any functional group. It is a difficult chemical problem to develop reactions with specificity for a given non-activated position. Nature solves the problem by means of the specific binding of the substrate to the active site. In this way, important hydroxylating enzymes synthesise bile acids, unsaturated fatty acids, ω-hydroxyacids and primary alcohols (Figure 6.21). The enzymology has similarities with drug-metabolising enzyme systems, and a discussion of this topic will be found in Chapter 15.

Figure 6.21.

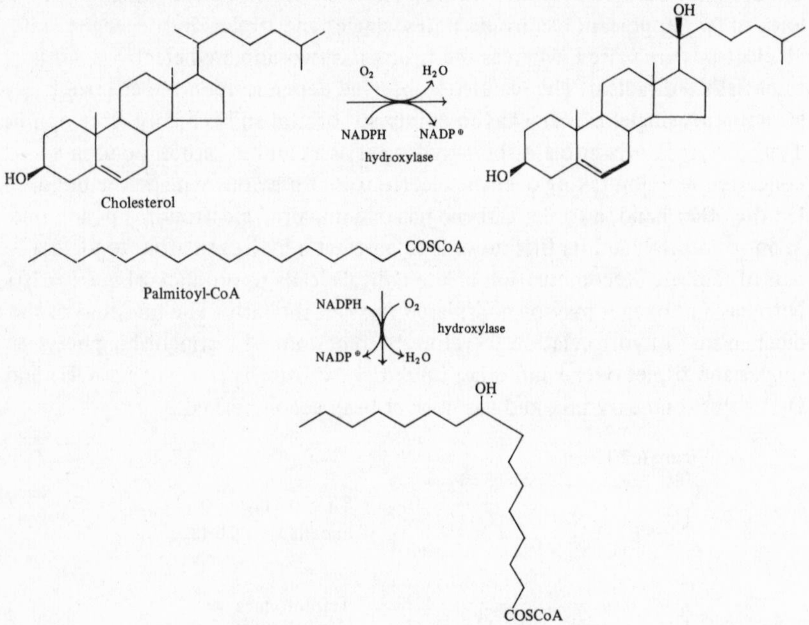

The most successful chemical reactions for generating functionality at non-activated positions make use of a principle that we have seen many times: an intramolecular reaction has many properties similar to an enzyme-catalysed reaction. Selectivity is one such property and it is best achieved in molecules with rigid stereochemical requirements, such as steroids. Two radical substitution reactions illustrate this principle. In the Barton reaction, (Figure 6.22) a radical is generated by photolysis of a nitrite ester (Hesse, 1969); Breslow's system uses photolysis of sulphonyl chloride or benzophenone derivatives (Breslow 1972; Breslow, Wife & Prezant, 1976). In both reactions the first-formed radicals then abstract a hydrogen atom from a suitably placed carbon atom, depending upon the geometry of the compound.

In the case of the Barton reaction, the alkoxy radical effects a 1,5-hydrogen abstraction. It is thought that this is a favourable reaction because a six-membered ring transition state of low energy is likely in hydrogen transfer; indeed 1,5-hydrogen abstraction is a very common general reaction of radicals. Recombination of the nitrosyl radical with the rearranged alkyl radical leads to a nitroso compound which tautomerises to the corresponding oxime. Thus functionalisation at C-20 in the steroid skeleton can be accomplished, and similar reactions generate functional groups at C-19.

Selectivity in Breslow's system is determined by the length of the side chain

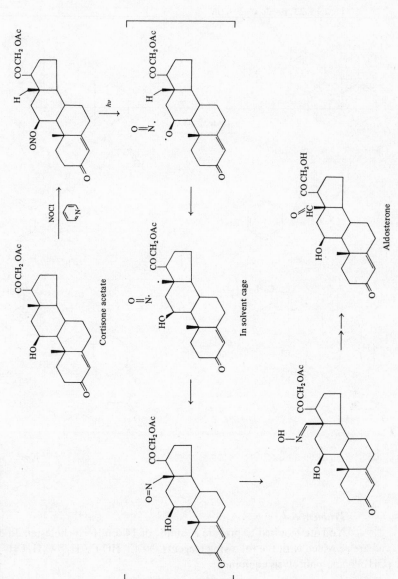

Figure 6.22. Barton reaction.

attached to the α-face at C-3. Clearly, hydrogen abstraction by a side-chain radical, which is the first-formed intermediate, must take place on the α-face, and the hydrogen at C-14 is in a favourable position. The example in Figure 6.23 shows how chlorination may be achieved selectively via an intermediate halosulphenyl radical that abstracts hydrogen from C-14. The reaction chain in this case is initiated by photolysis of sulphuryl chloride.

Figure 6.23. Breslow's system.

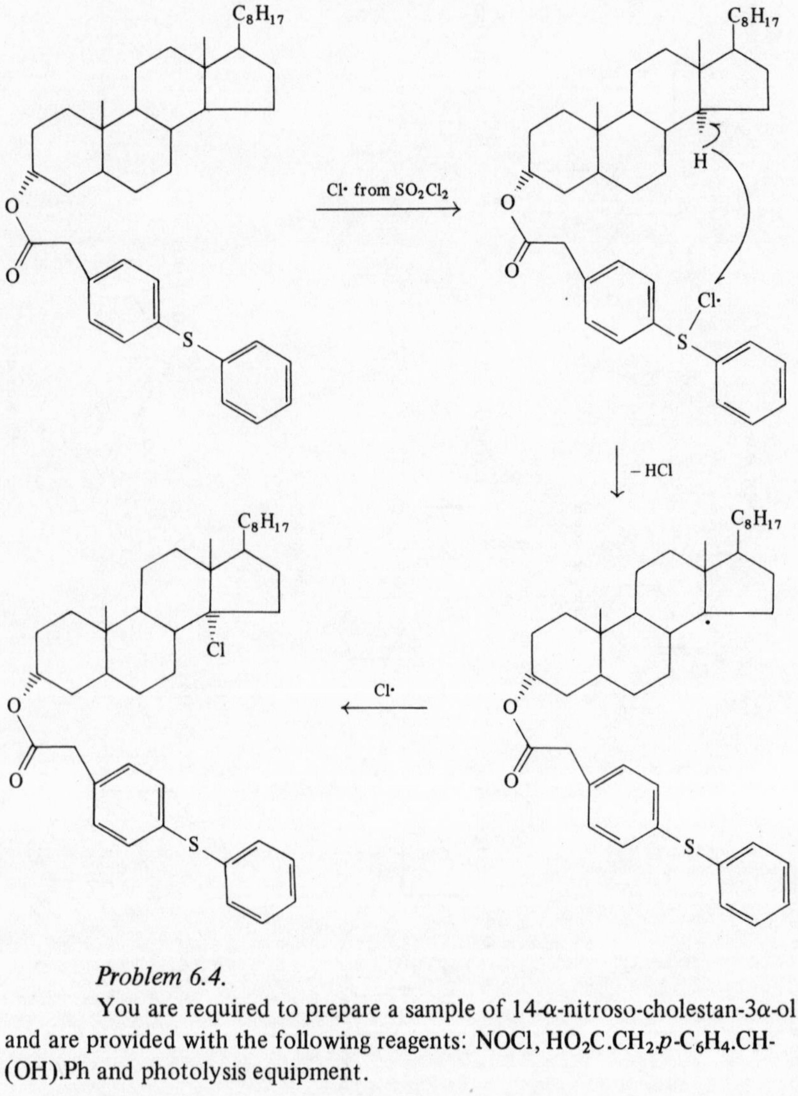

Problem 6.4.

You are required to prepare a sample of 14-α-nitroso-cholestan-3α-ol, and are provided with the following reagents: NOCl, HO$_2$C.CH$_2$.p-C$_6$H$_4$.CH-(OH).Ph and photolysis equipment.

(1) What reaction sequence would you suggest?
(2) How can you justify the selectivity of your scheme?

7   Substitution reactions at aromatic carbon

7.1.   Introduction

It seems appropriate to follow a survey of aliphatic substitution reactions with a discussion of the chemistry and biochemistry of their relatives in aromatic chemistry. As we shall see, the scope of aromatic substitutions in biological reactions is wider than aliphatic ones. As usual, let us first précis the background chemistry.

You should be familiar with electrophilic aromatic substitution in benzene and its derivatives: this is the reaction in which an electrophile (E in Figure 7.1) such as a nitronium or halonium ion or carbocation initially adds to the relatively electron-rich benzene ring and, after elimination of a proton, forms a substituted benzene. Consult your basic organic chemistry textbook if you are unfamiliar with the mechanism of this reaction, especially as applied to derivatives of benzene. It is particularly important for you to be able to draw canonical forms for the intermediates fluently (Figure 7.1).

Figure 7.1.

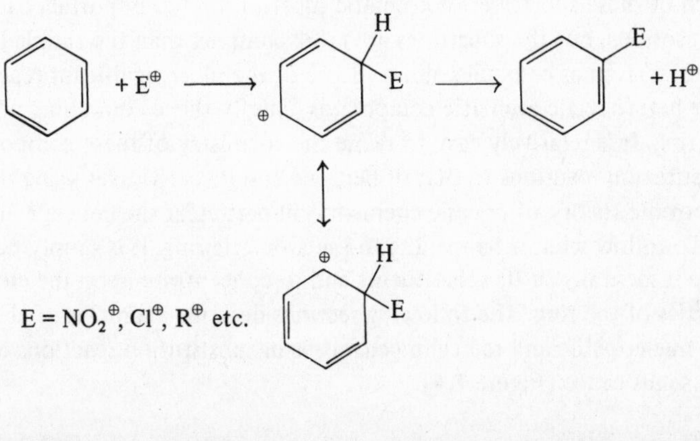

$E = NO_2^{\oplus}, Cl^{\oplus}, R^{\oplus}$ etc.

In aromatic compounds in which the ring is deficient of π-electrons, substitution can take place by another mechanism (Figure 7.2). A decrease in electron density relative to benzene can permit a nucleophile to attack the ring. Unless a leaving group like chloride is involved, hydride ion must be removed to complete the substitution, but because hydride is an ion of relatively high energy, nucleophilic substitutions are only facile reactions when a good leaving group is available. This is of significance in the reactions by which the bases of nucleotides and nucleosides are transformed one into another, as we shall see.

Figure 7.2.

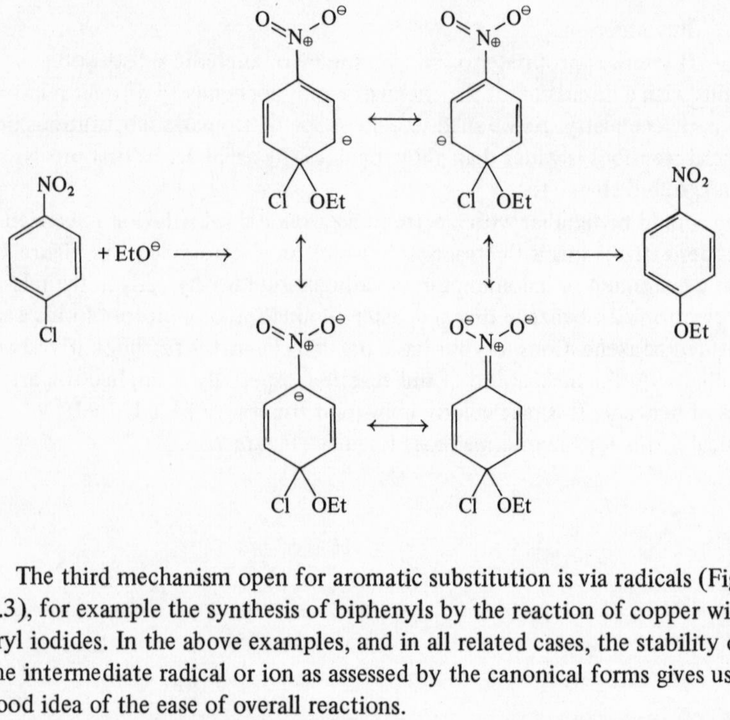

The third mechanism open for aromatic substitution is via radicals (Figure 7.3), for example the synthesis of biphenyls by the reaction of copper with aryl iodides. In the above examples, and in all related cases, the stability of the intermediate radical or ion as assessed by the canonical forms gives us a good idea of the ease of overall reactions.

Each of these subclasses of aromatic substitution has importance in metabolic reactions, but the substrates are more complex than the simple benzene derivatives given as examples above. Some of the most significant reactions involve heterocyclic aromatic compounds, chiefly those containing nitrogen in the ring. It is relatively easy to relate the chemistry of these compounds in substitution reactions to that of benzene and its derivatives using the rules of electronic theory of organic chemistry, in particular the concept of resonance. To follow what is happening to the aromatic ring, it is simply necessary to strip it mentally of its substituents and to concentrate upon the electronic properties of the ring. The following sections describe examples of electrophilic, nucleophilic and radical mechanisms in substitution reactions of biological significance (Figure 7.4).

Figure 7.3.

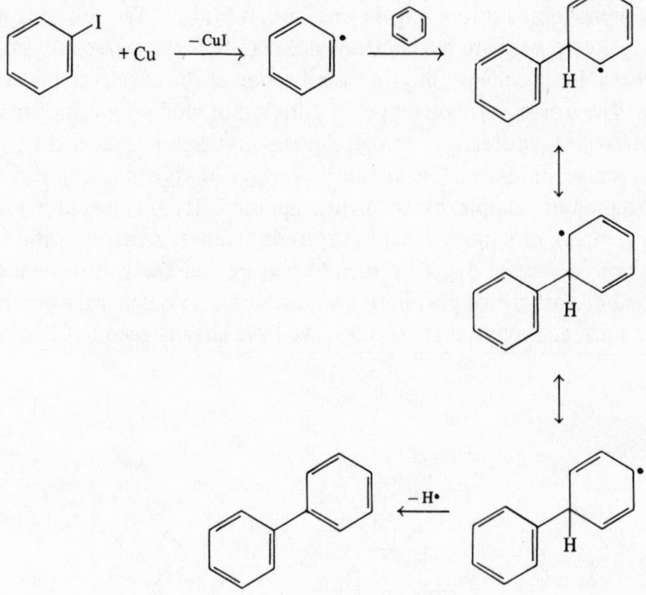

Figure 7.4. Enzyme-catalysed substitution reactions.

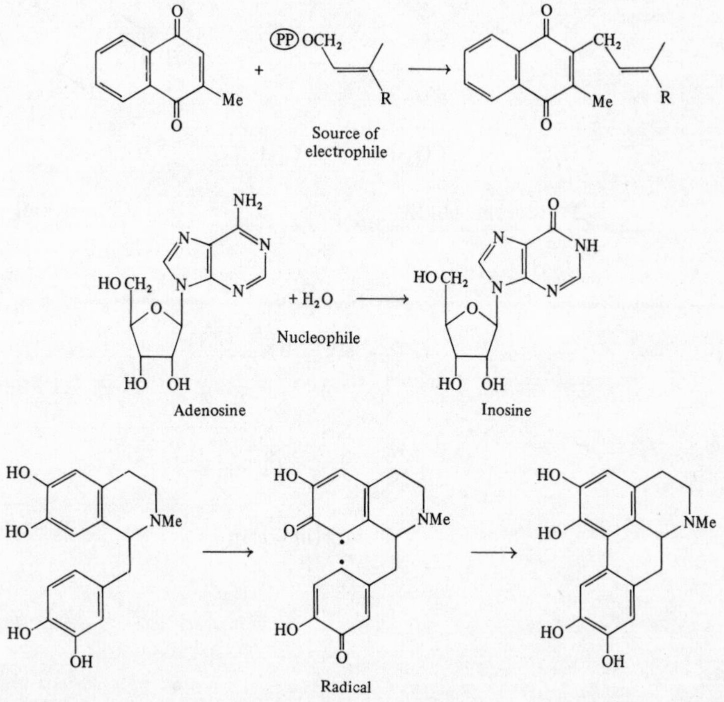

7.2.   The biosynthesis of the porphyrin ring system

Porphyrins (Figure 7.5) are macrocyclic molecules built up of four pyrrole units joined into a ring by one-carbon bridges. The four nitrogen atoms make up a square planar array and readily form a wide range of metal complexes. In fact the porphyrin ligand is one of the strongest chelating agents known. The iron complexes of porphyrins are of vital importance in the oxygen-carrying proteins, haemoglobin and myoglobin: it is at the iron that the oxygen atom binds. Derivatives of porphyrins known as chlorins are, as their magnesium complexes, the light-trapping units of chlorophylls, and the cobalt complex of a more distant porphyrin relative, a corrin, is the functional part of the coenzyme B_{12}. With such important functions, it is small wonder that the biosynthesis of porphyrins has attracted so much attention both from the chemical and medical standpoint. We have already seen in Chapter 4 how

Figure 7.5.

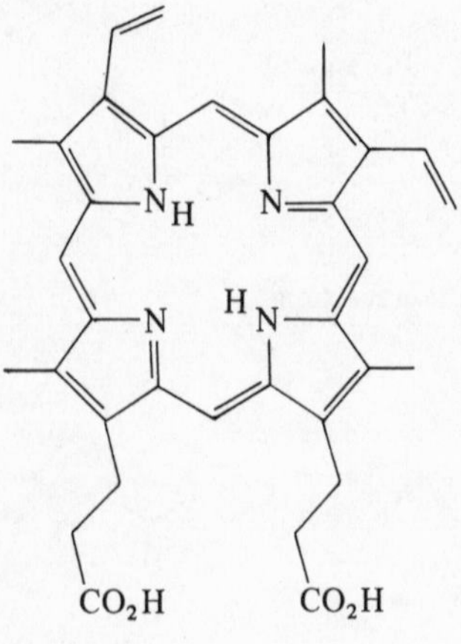

Protoporphyrin IX

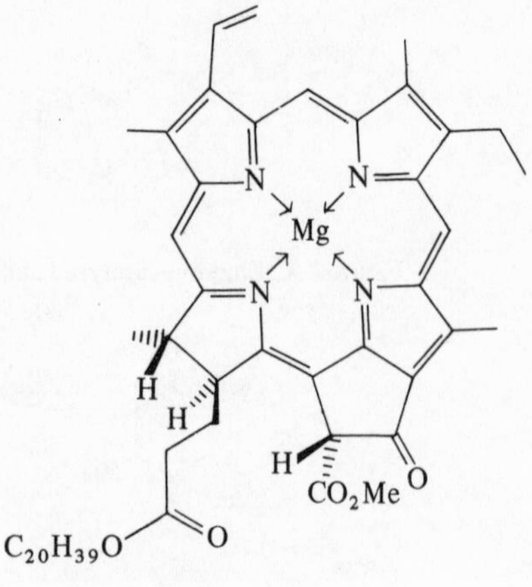

Chlorophyll a

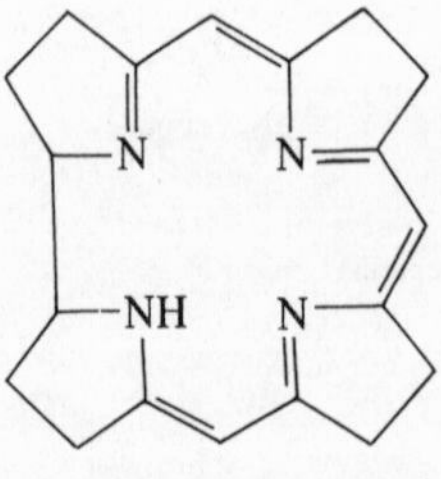

The corrin skeleton

the pyrrole precursor of porphyrins, porphobilinogen (PBG), is synthesised
by means of condensation reactions and we can now study the building up of
the porphyrin macrocycle itself. Essentially, the mechanism of biosynthesis is
a series of electrophilic substitutions, and we shall begin by taking a quick look
at electrophilic substitution in pyrroles in general (Joule & Smith, 1978, p. 194).

7.2.1. *The reactivity of pyrrole*

If pyrrole is to be an aromatic molecule, then by Hückel's rule it
must have six π-electrons in the ring. Two pairs of these come from the double
bonds and the third pair is donated by non-bonded electron pair of the nitrogen atom, which in pyrrole can be considered to be sp^2-hybridised. Intuitively
this suggests to us that the pyrrole ring should be electron rich. Both physical
and chemical measurements bear this out. The dipole moment of pyrrole is
larger than its saturated analogue pyrrolidine, and is directed towards the
carbon atoms of the ring, not towards the nitrogen as it is in pyrrolidine. This

Figure 7.6.

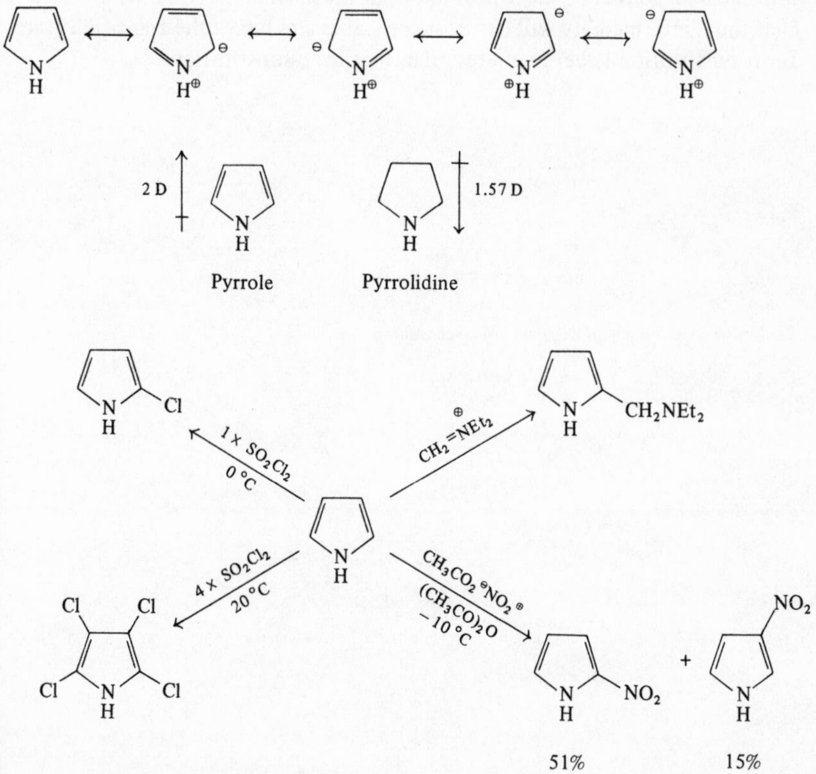

indicates that the carbon atoms must bear a net negative charge, which can be rationalised by the contributions of resonance structures (Figure 7.6 *top*). A high electron density in the ring suggests that electrophilic aromatic substitution would be very easy in pyrrole. Notice the mild conditions required in the examples in Figure 7.6: pyrrole is at least 10^5 times more reactive than benzene.

By selecting pyrrole derivatives as a building block for porphyrins, nature has chosen substrates ideal for electrophilic substitution, and porphobilinogen is the optimum porphyrin precursor. In each porphobilinogen molecule, there is an unsubstituted position available for attack by an electrophile and also an aminomethyl group which, on loss of an NH_2 group, affords a carbocation, an electrophile (7.7.1). If the coupling enzymes remove or modify the NH_2 group, it is easy to formulate the joining of two molecules of porphobilinogen to give a dipyrrylmethane (7.7.2). This reaction is directly analogous to the processes used in synthetic chemistry to prepare dipyrrylmethanes (Figure 7.7). In both of these reactions, the pyrrole undergoes substitution at the carbon atom α to nitrogen, and it is easy to show that this position is more reactive towards electrophiles than the β-carbon. We can draw more canonical forms for the intermediate adduct at the α-position than the β, and consequently we believe that the α-intermediate will be the more stable and hence the transition state for α-substitution lower in energy than that for β-substitution.

Figure 7.7.

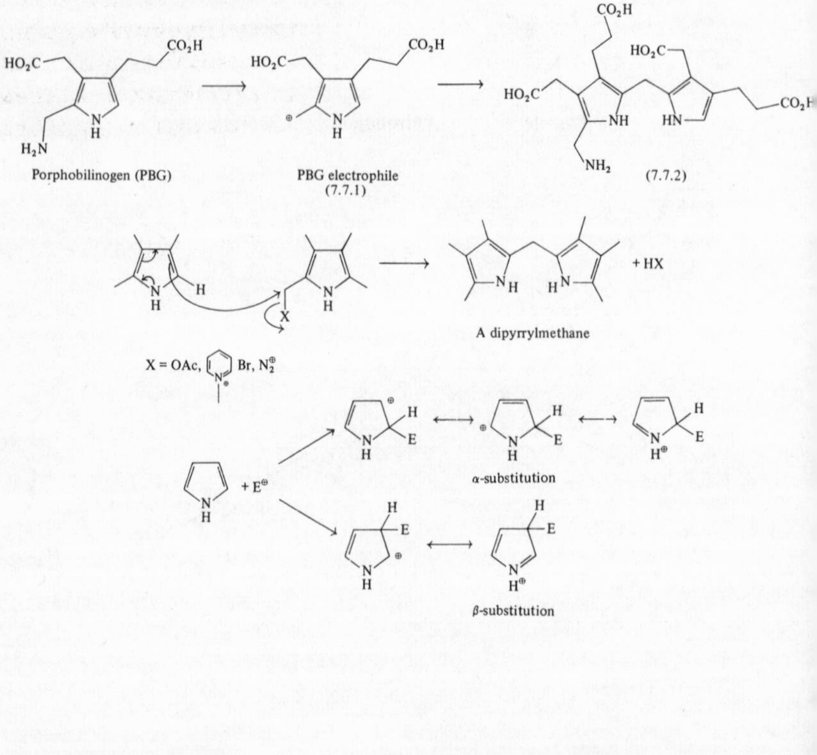

Porphobilinogen (PBG)

PBG electrophile (7.7.1)

(7.7.2)

X = OAc, [pyridine] Br, $N_2^{\oplus}$

A dipyrrylmethane

$+ HX$

α-substitution

β-substitution

7.2.2. The biosynthesis of porphyrins

If you join four porphobilinogen units together and then cyclise, you obtain the octahydro derivatives of porphyrins, which are called **porphyrinogens**. The porphyrinogens are biological precursors of porphyrins, and the immediate product of cyclisation is known as uroporphyrinogen. However, with two side-chain substituents, an acetate and a propionate, there are several isomeric uroporphyrinogens possible. Two of these are shown in Figure 7.8. The type I isomer is only found naturally in certain pathological conditions. All functional naturally occurring porphyrins are derived from the type III isomer of uroporphyrinogen by decarboxylation and oxidation. The formation

Figure 7.8.

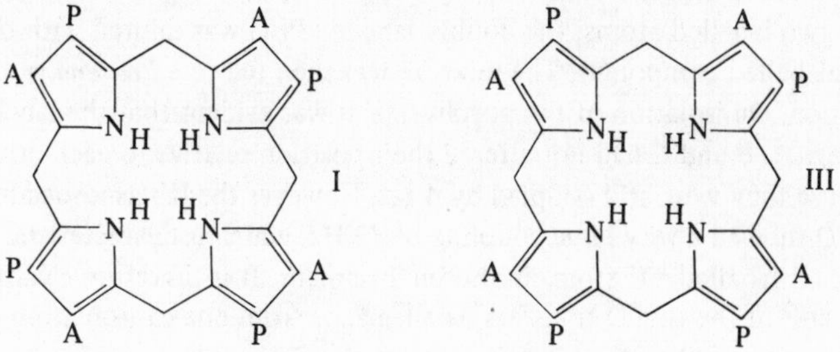

$A = CH_2 CO_2H \quad P = CH_2 CH_2 CO_2H$

Uroporphyrinogen isomers

of only one isomer poses a problem. The type III pattern is not the product of a simple head to tail cyclisation of four molecules – one ring has been turned so that its substituents are the other way round from usual. A vast number of hypotheses were advanced to explain the chemical mechanism of this transformation and progress has required the subtle probe of ^{13}C magnetic resonance spectroscopy and the labours of several research groups, but Battersby's work has been the most revealing (Battersby, McDonald, Williams & Wurziger, 1977).

The porphyrin molecule presents severe difficulties for a classical biosynthetic study by means of radioactive isotopes because the symmetry of the ring makes it impossible to design a degradation that will specifically remove selected atoms. The beauty of ^{13}C-labelling experiments is that this stable isotope of carbon can be located by means of magnetic resonance spectroscopy (Chapter 11 and Simpson, 1975). The ^{13}C nucleus has a spin of $\frac{1}{2}$ and the chemical shifts of almost every atom in a given molecule are different. Therefore, if you can assign the peaks of a ^{13}C n.m.r. spectrum to an individual carbon atom, you can tell whether a ^{13}C label is present at that position or not. Battersby tackled this problem by identifying the key carbon atoms to be followed as the porphobilinogen (PBG) CH$_2$ and the unsubstituted pyrrole carbon. He devised not only a method for the synthesis of labelled PBG but also syntheses of a series of porphyrins, each bearing a label at only one of the one-carbon bridges of the porphyrin ring. These compounds allowed him to assign the chemical shift of all the relevant carbon atoms. The labelled PBG was synthesised by means of an enzyme system, and from 90 % enriched starting material, PBG in which 81 % of the molecules contain two ^{13}C labels was obtained (Figure 7.9). These doubly labelled molecules are the ones to be followed. Battersby was able to use the information that the coupling constant of two ^{13}C nuclei in this environment is 4 Hz.

To ensure that the product porphyrins of the feeding experiment contained only two labelled atoms, the doubly labelled PBG was diluted with three parts of unlabelled compound. The mixture was then fed to a *Euglena gracilis* preparation. On isolation of the porphyrins, it was evident that the labelled atoms in rings A, B and C had not altered their position relative to each other, because they were still coupled by 4 Hz. However the labels associated with ring D showed a very large coupling of 72 Hz, which is characteristic of two adjacent labelled ^{13}C atoms in this environment. It is therefore clear that the CH$_2$ unit of the ring D transfers its allegiance from one carbon atom of its pyrrole ring to the other α carbon atom.

Figure 7.9. Any molecule of product contains only one pair of ^{13}C atoms, each pair is shown by a symbol (•, ■, ▲ and ★); only on ring D are there adjacent ^{13}C atoms.

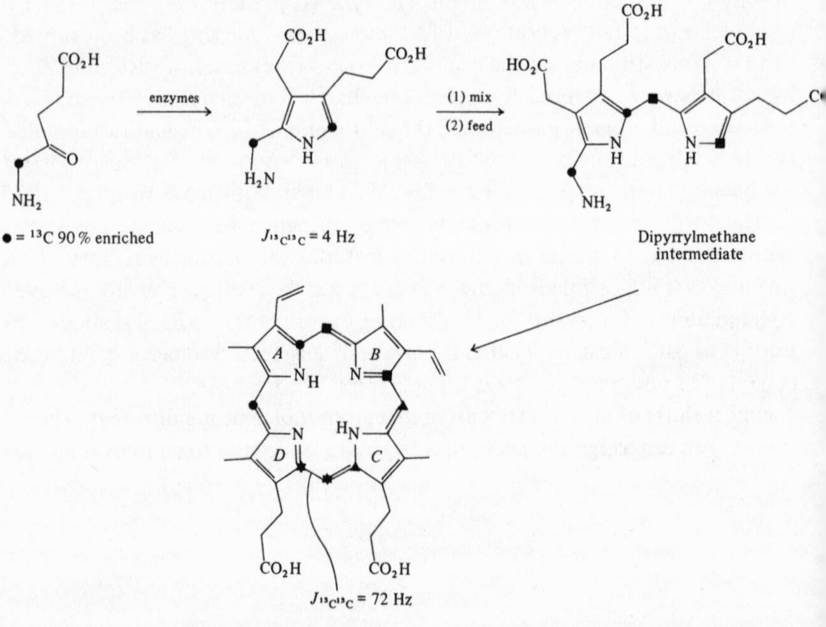

$\bullet$ = ^{13}C 90 % enriched $\qquad$ $J_{^{13}C^{13}C}$ = 4 Hz $\qquad$ Dipyrrylmethane intermediate

$J_{^{13}C^{13}C}$ = 72 Hz

The major remaining problem was to establish at what stage the enzymic 'switch' of the pyrrole ring D occurs. One strong possibility is that the acyclic tetrapyrrole is formed in a head to tail fashion and that the 'switch' of ring D takes place during cyclisation. There is only one direct test of this hypothesis namely to synthesise a labelled acyclic tetrapyrrole, known as a *bilane* and to investigate its incorporation into porphyrins during biosynthesis. It is no easy task to prepare the required bilane, and Battersby's synthesis, which brings together many classical procedures of porphyrin synthesis, is an excellent example of control of reactivity by protecting groups.

As we have already seen, pyrroles are highly reactive towards electrophiles and PBG contains an aminomethyl group that can readily give rise to an electrophile. The prime need, therefore is to protect the aminomethyl group of the required bilane (Figure 7.10). Fortunately, the amine can easily be masked by the carboxylate on the neighbouring acetic acid substituent as a cyclic amide or lactam; in fact the pyrrole starting material (7.10.1) is usually prepared as the lactam. It can then be reacted with a dipyrrylmethane (7.10.2) in an electrophilic substitution reaction in which protonation of the aldehyde affords the electrophile. The product (7.10.3) contains a dipyrrylmethene unit. Now is the time to introduce the ^{13}C label; as usual, an expensive label is introduced late in synthesis. Deprotection of the t-butyl ester of (7.10.3) with trifluoroacetic acid (cf. peptide synthesis, Chapter 3), and concomitant decarboxylation opens the way to introducing the label at the crucial γ-carbon atom through pyrrole (7.10.4) in a further electrophilic substitution. This gives rise to a biladiene (7.10.5), which on hydrogenation of the double bonds and hydrolysis of the methyl esters and amides affords the key precursor bilane (7.10.6).

As expected, the bilane underwent ready cyclisation in the absence of enzyme to give type I porphyrins (Figure 7.8) as shown by analytical and preparative liquid chromatography. In contrast, after incubation with the purified enzyme preparation from *Euglena gracilis*, the chromatographs showed a vast enhancement of the type III isomer, the natural functional porphyrin isomer. It was also possible to characterise the product by ^{13}C n.m.r. and to show that the label was in the expected position.

The conclusion is clear. Natural porphyrins are synthesised in a stepwise, linear manner and ring D undergoes rearrangement at the cyclisation stage. Chemists must now offer some rationalisation for the mechanism of rearrangement and then suggest experiments to test the mechanism. One currently favoured scheme is shown in Figure 7.10. If the enzyme causes the substrate bilane, (7.10.6) to fold such that electrophilic substitution occurs at the already substituted position, a spiro-intermediate (7.10.7) will be formed. Spiro centres are carbon atoms that are at a common apex of two rings and they behave as a pivot for the molecule's reactions. If bond b in 7.10.7 pivots round to the opposite unsubstituted position of the pyrrole ring D, the product will be a

Figure 7.10.

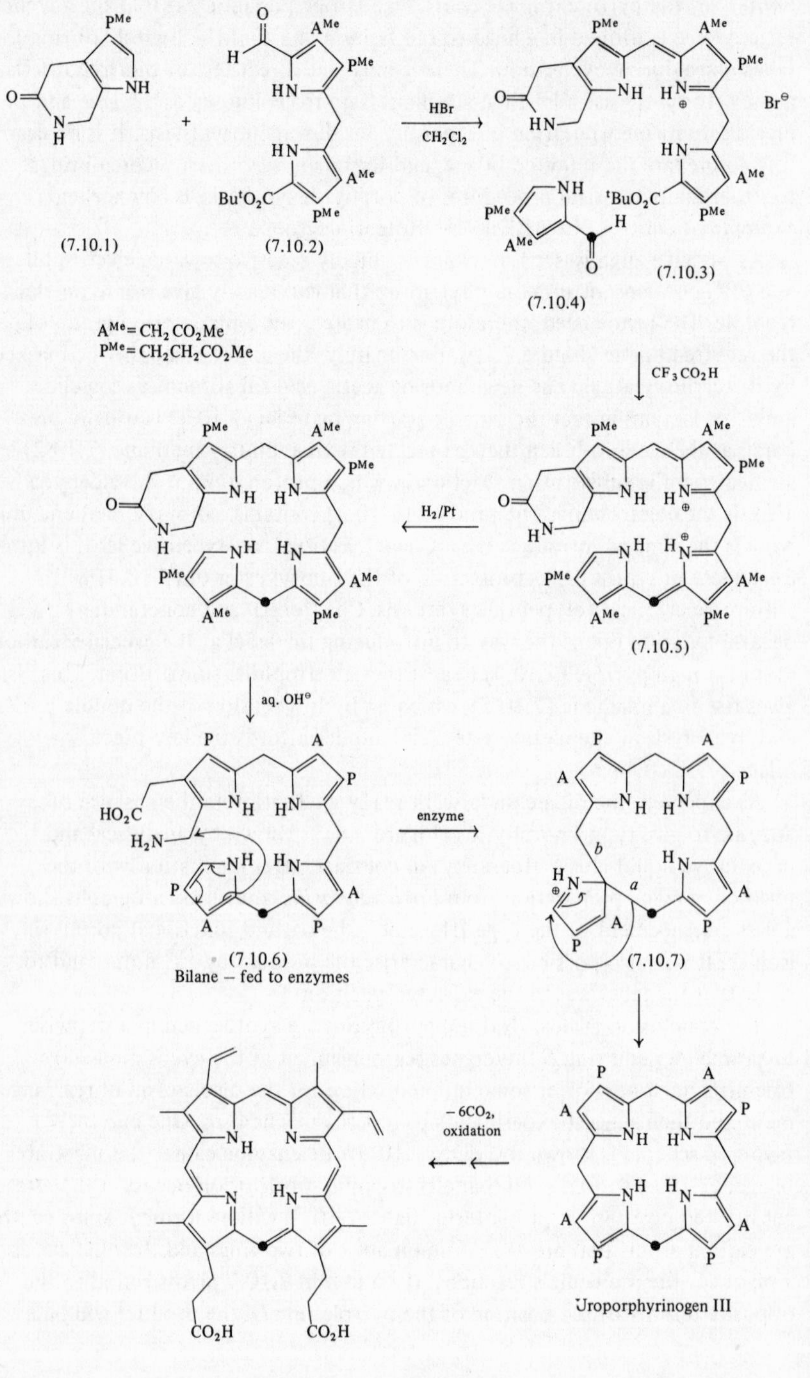
(7.10.1)
(7.10.2)
(7.10.3)
(7.10.4)
(7.10.5)
(7.10.6)
Bilane – fed to enzymes
(7.10.7)
Uroporphyrinogen III
A^Me = CH_2CO_2Me
P^Me = CH_2CH_2CO_2Me
HBr
CH_2Cl_2
CF_3CO_2H
H_2/Pt
aq. OH^⊖
enzyme
– 6CO_2, oxidation
Br^⊖
Bu^tO_2C
HO_2C
H_2N
CO_2H
CO_2H

type I porphyrinogen. However if bond *a* moves, the natural type III isomer will result. Similar rearrangements have been shown to occur in the biosynthesis of some alkaloids (Section 7.4).

7.2.3. *Some other examples of electrophilic substitution in biosynthesis*

In pyrroles, nature chose a substrate highly reactive towards electrophiles, and all the molecules that undergo biosynthetic transformations via electrophilic substitution reactions are much more reactive than benzene. Phenolic compounds are typical cases. The electrophile for substitution of phenols is very commonly an allylic pyrophosphate such as geranyl pyrophosphate or a homologue. These reactions represent the confluence of two biosynthetic streams, isoprenoids with either acetate or shikimate metabolites. The biosynthesis of tetrahydrocannabinol illustrates the former and the ubiquinones the latter (Figure 7.11).

Figure 7.11.

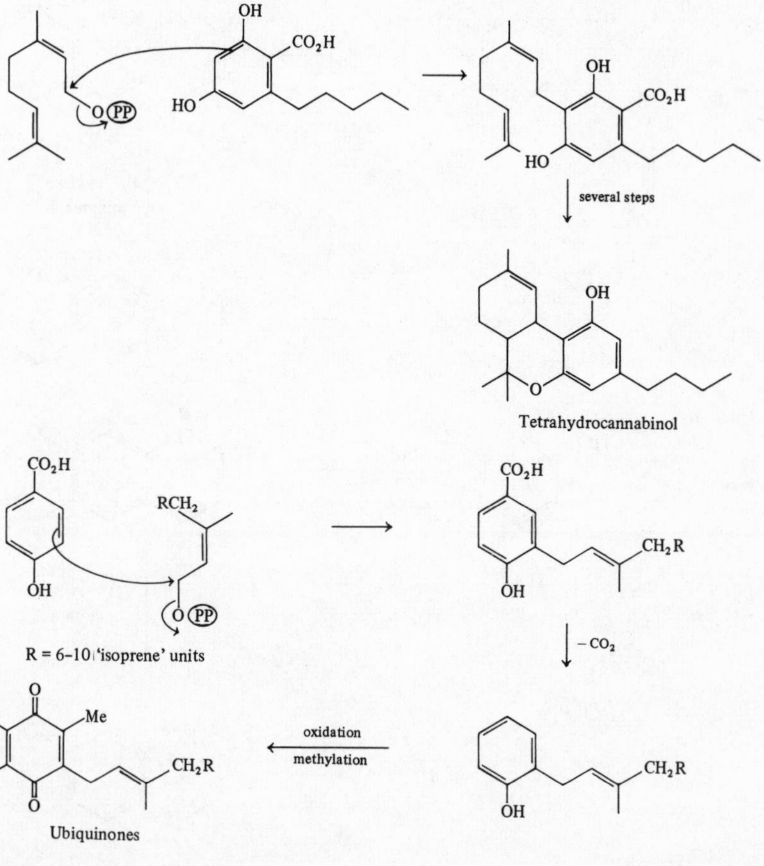

Tetrahydrocannabinol

R = 6–10 'isoprene' units

Ubiquinones

Problem 7.1.

The natural product *A* could conceivably be biosynthesised by the two routes shown in Figure 7.12.

(1) Comment upon the chemical feasibility of the two routes and decide which is the more probable.

(2) Suggest approaches that would permit an experimental differentiation to be made between the two routes.

Figure 7.12.

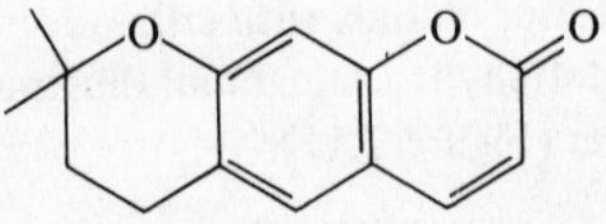

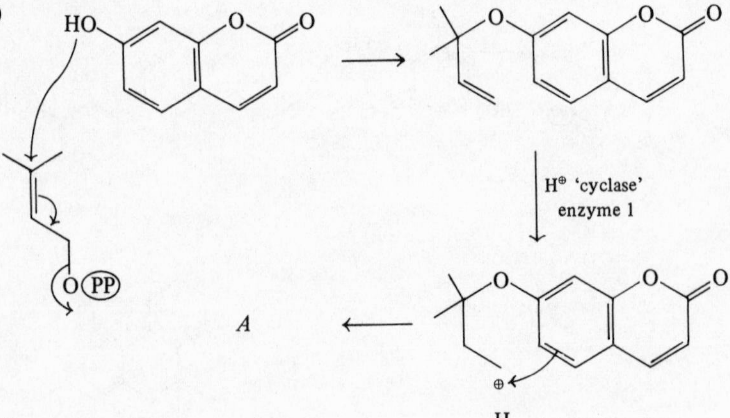

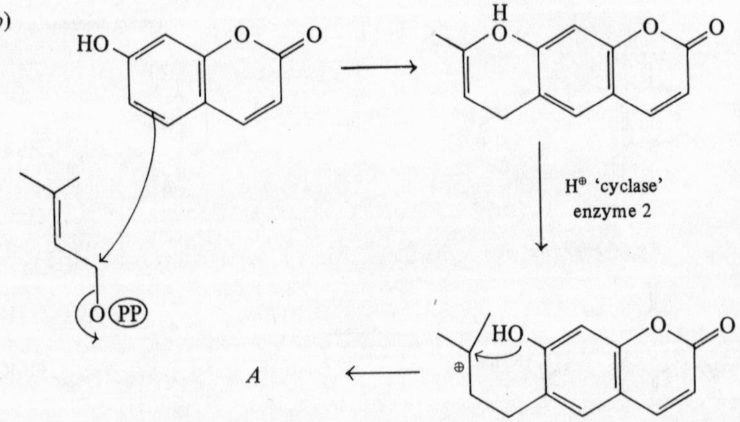

Problem 7.2.

The hallucinogenic alkaloid lysergic acid is thought to be derived from tryptophan and dimethylallyl pyrophosphate via the derivative *B* (Figure 7.13).

(1) Notice at what position tryptophan has undergone substitution. Is this to be expected?

(2) Substantiate your answer to (1) by considering the relative stability of the intermediate in substitution of indole at the 2, 3 and 4 positions.

(3) You should have shown that the most favoured position is 3. To understand how the alkyl substituent ends up at C-4, you will probably need to look up the Claisen rearrangement in your organic chemistry textbook.

Figure 7.13.

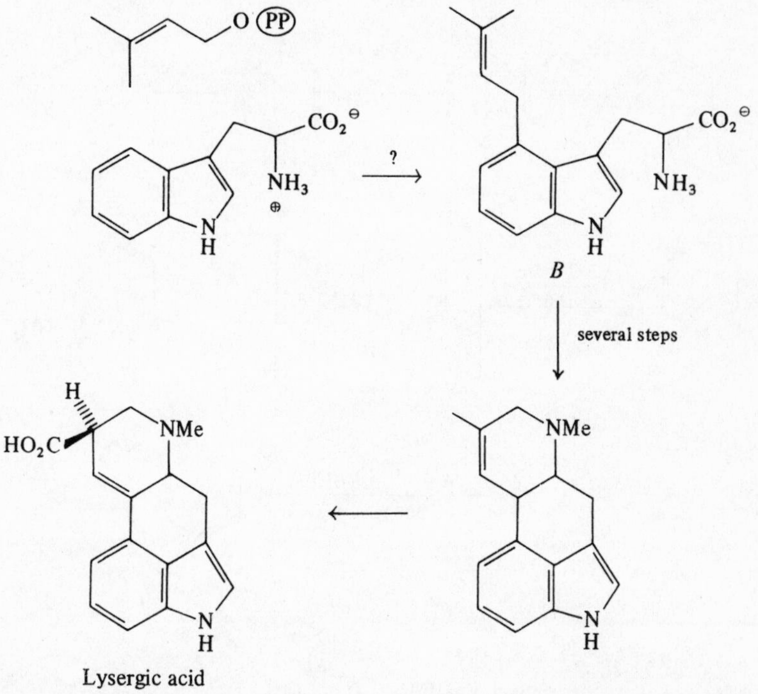

B

several steps

Lysergic acid

Another biosynthetic example of the Claisen rearrangement can be found in the shikimic acid pathway, the formation of prephenic acid from chorismic acid. Such rearrangements are examples of electrocyclic reactions (Tedder, Nechvatal, Murray & Carnduff, 1972, pp. 126, 389).

7.3.   Nucleophilic aromatic substitution

If we examine pyridine in the same way as we looked at pyrrole in the previous section, we see at once that the non-bonded electrons of the pyridine nitrogen atom are not required to make up the necessary six π-electrons for aromaticity. Therefore, pyridine will be basic, unlike pyrrole, and the effect of the nitrogen atom in the ring will be to withdraw electrons from the ring both inductively and through resonance (Figure 7.14).

Pyridine and all six-membered rings containing nitrogen are thus deficient of π-electrons relative to benzene. Rather than react with electrophiles, we would expect pyridine and its congeners to undergo substitution more readily with nucleophiles. The examples in Figure 7.14 show that this is indeed the

Figure 7.14. It should be noted that H⁻ is *not* a good leaving group: an acceptor is required and this may be provided by the work-up procedure, air or another reactant.

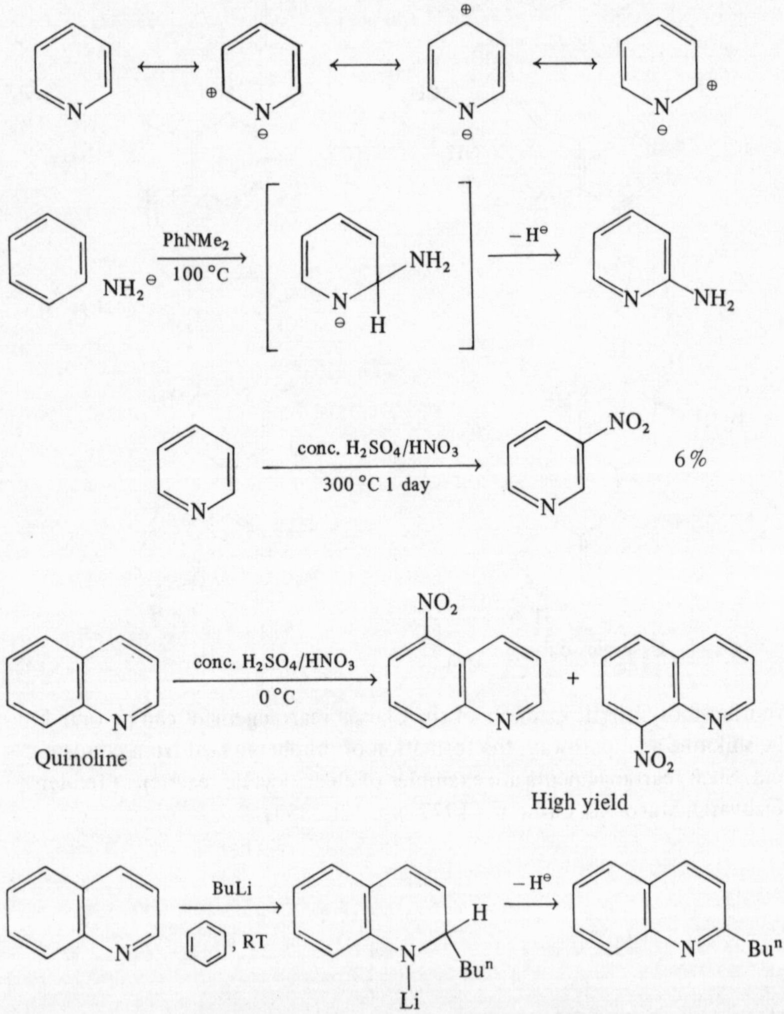

Quinoline

High yield

case, although pyridine is not especially reactive towards nucleophiles. Electrophilic substitution in quinoline takes place in the benzene ring, but nucleophilic substitution in the pyridine ring.

Problem 7.3.

Pyridine is a basic molecule. How will this property influence the reactivity of pyridine to electrophiles? (Hint: How does the smallest electrophile react with pyridine?) (Joule & Smith, 1978, p. 45.)

7.3.1. Applications of compounds reactive to nucleophilic aromatic substitution to biological chemistry

As we mentioned in the introduction to this chapter, the presence of a halogen as a leaving group greatly facilitates nucleophilic substitution and reactivity can be enhanced by adding further electron-withdrawing groups. Sanger took advantage of the combined effects of two nitro groups in 2,4-dinitrofluorobenzene in his famous method for determining the amino end group of peptides (Figure 7.15 and Chapter 11). The amino group of the peptide attacks the benzene ring in a nucleophilic substitution with displacement of fluoride ion. This affords a dinitrophenyl-substituted peptide which, on hydrolysis, liberates the yellow dinitrophenyl derivative of the N-terminal amino acid. This derivative is then easily identified by chomatography.

Trichlorotriazine contains three nitrogen atoms in a six-membered ring and is very reactive towards nucleophiles. All three chlorine atoms may be displaced

Figure 7.15.

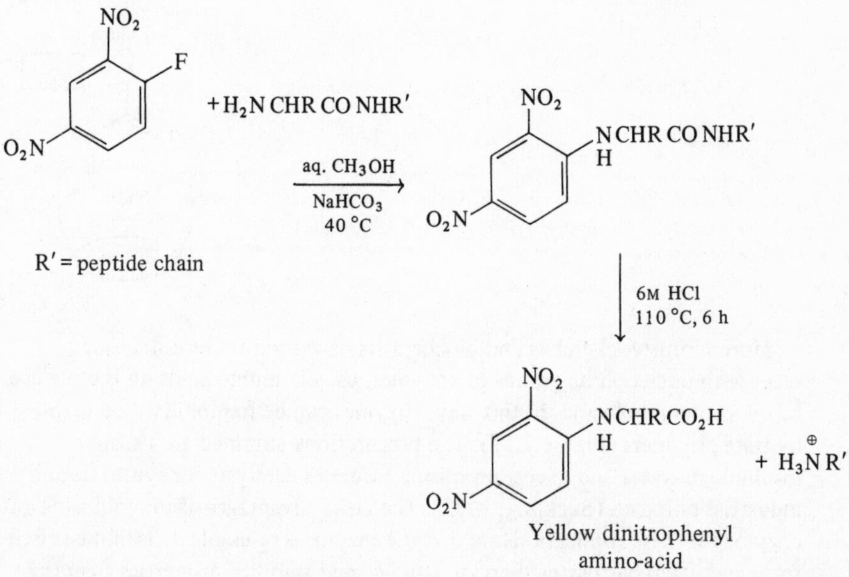

R′ = peptide chain

Yellow dinitrophenyl amino-acid

successively by nucleophilic ligands; the first application to take advantage of these chemical properties was the fibre-reactive dye (Figure 7.16). A dye bearing an amino substituent was treated with trichlorotriazine to produce the fibre-reactive dye itself. Many textile fibres contain hydroxyl groups which are part of polysaccharide chains, and under suitable conditions these hydroxyl groups can be made to react with the remaining two chlorine atoms of the triazine. The conditions are a little more vigorous than the substitution of trichlorotriazine because one electron-donating group, the amino group of the dye, has already been incorporated into the molecule. A wide range of other heterocyclic halogen derivatives have been applied to the same purpose. The dyed fibres are very stable to washing.

Figure 7.16.

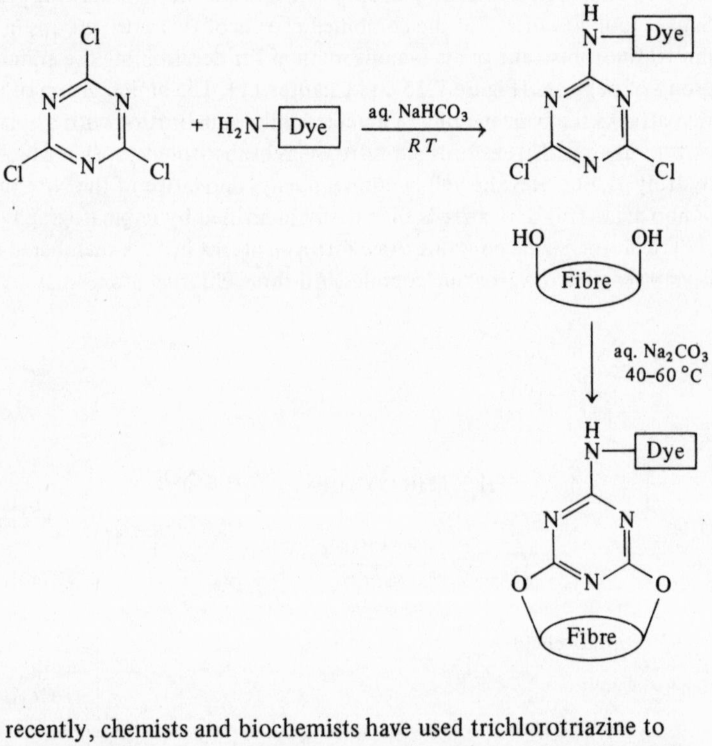

More recently, chemists and biochemists have used trichlorotriazine to react with nucleophilic groups of enzymes, usually amino acids on the surface of the protein molecule. In this way, enzymes can be immobilised on carbohydrate polymers (Figure 7.17). The preparations obtained are usually insoluble in water and have a promising future as catalysts for synthetic and industrial purposes (Suckling, 1977). The chief advantages of immobilising an enzyme for these applications are that the enzyme is re-usable in its immobilised form and often has better thermal, storage, and stability properties than the enzyme in solution. This field can be anticipated to become of increasing economic importance.

Figure 7.17.

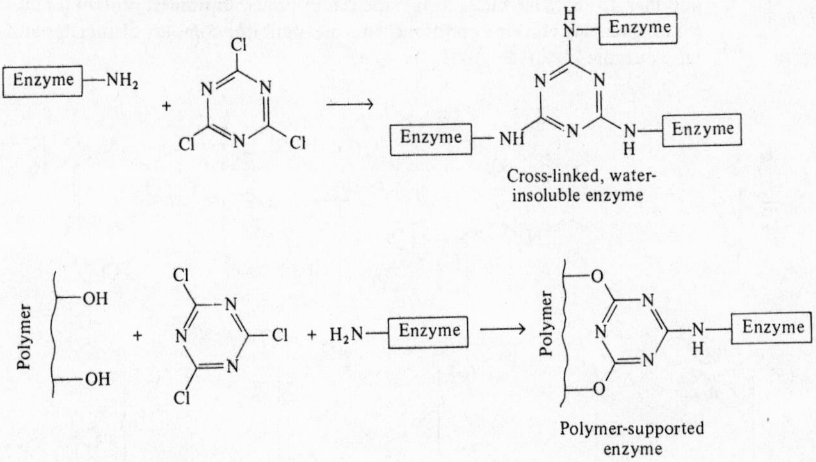

Cross-linked, water-
insoluble enzyme

Polymer-supported
enzyme

A third application of trichlorotriazine is in the purification of enzymes by affinity chromatography (Lang, Suckling & Wood, 1977). This technique is illustrated diagrammatically in Figure 7.18. It requires a ligand for the enzyme (often an inhibitor) to be covalently attached to a polymer. When a carbohydrate polymer is being used, a very stable link between polymer and ligand can be made with substituted triazines (Figure 7.18).

Problem 7.4.
Explain why 3-chloropyridine is inert to nucleophilic substitution by ethoxide anion at 40°C, whereas both 2- and 4-chloropyridine react rapidly to form ethyl ethers under these conditions. (Joule & Smith, 1978, p. 52.)

Problem 7.5.
Predict the ease of electrophilic and nucleophilic substitution in the nitrogen-containing heterocyclic compounds shown in Figure 7.19. (Joule & Smith, 1978, pp. 123, 299, 325.)

Figure 7.18. (*a*) Mixture of proteins applied to column of inhibitor-bearing polymer; (*b*) enzyme binds to specific inhibitor and unwanted protein is eluted; (*c*) on changing elution conditions, enzyme–inhibitor complex dissociates and pure enzyme is eluted.

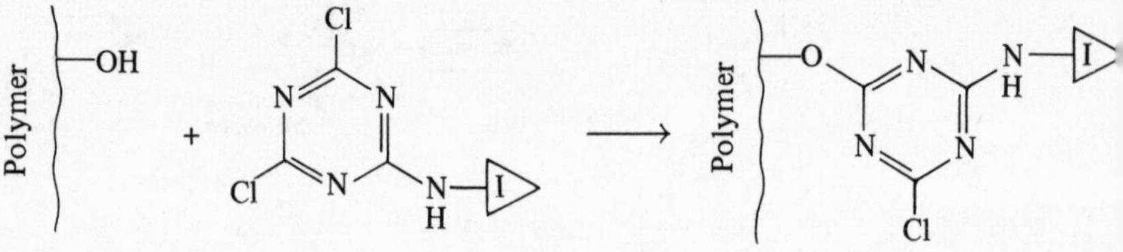

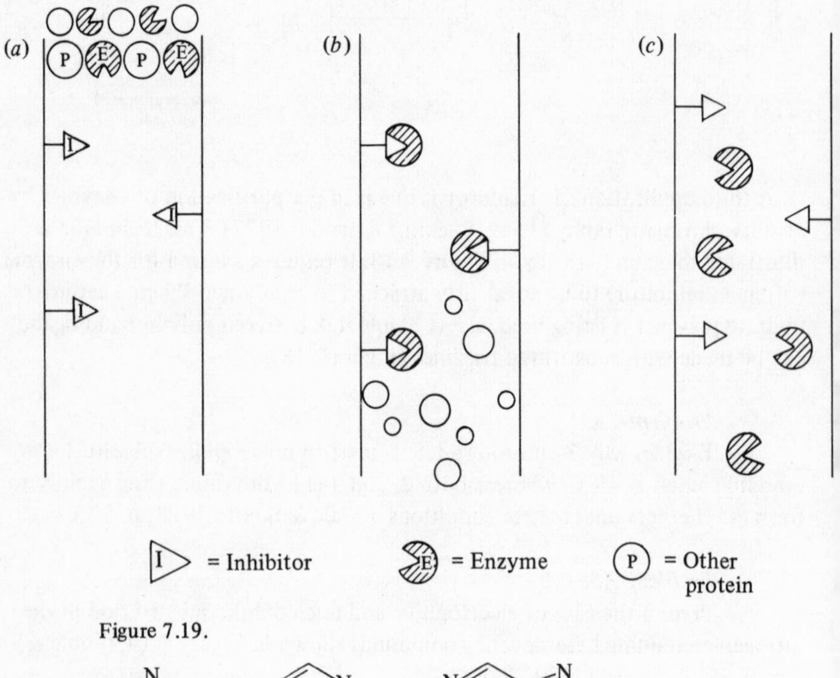

= Inhibitor   = Enzyme   (P) = Other protein

Figure 7.19.

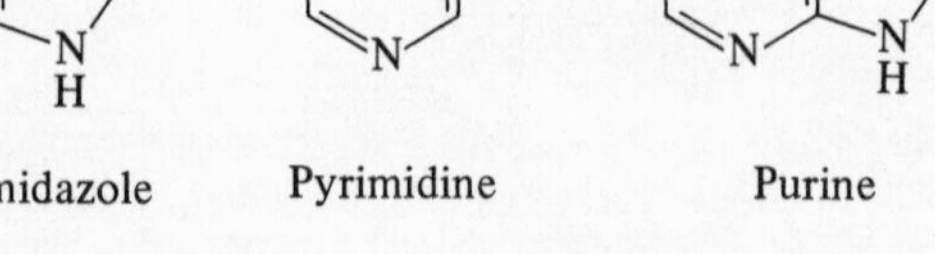

Imidazole   Pyrimidine   Purine

7.3.2. *Pyrimidine and purine interconversions*

With an understanding of nucleophilic aromatic substitution reactions we can now return to consider how the purine and pyrimidine nucleotides that were formed by condensation reactions (Chapter 4) are interconverted by the transformation of an oxo-substituted compound into an amino-substituted derivative. The cases shown in Figure 7.20 are typical: notice the involvement of the triphosphate cofactor.

Figure 7.20. GTP = guanosine triphosphate, the guanine analogue of ATP; GDP = guanosine diphosphate; Gln = glutamine (glutamic acid δ-amide); Glu = glutamic acid.

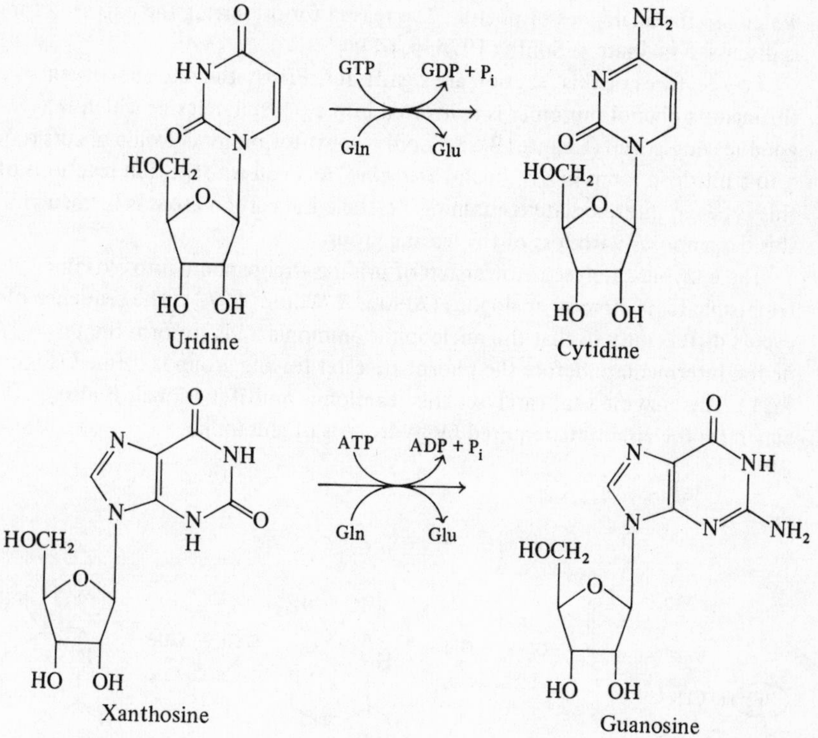

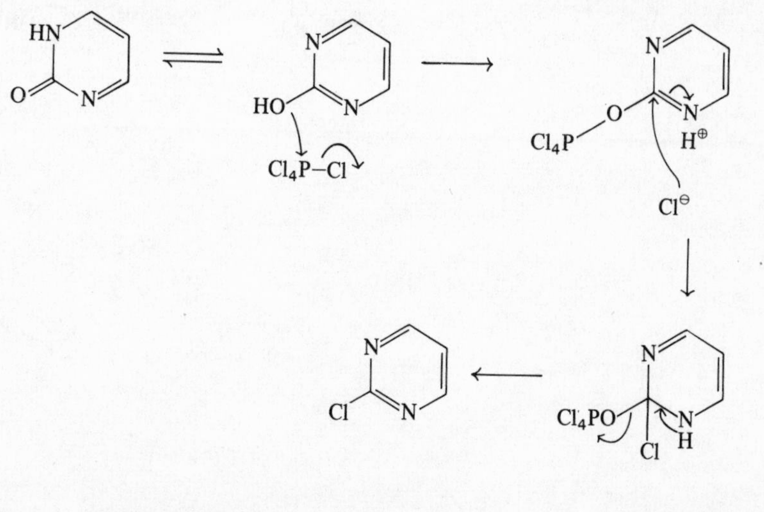

All of these reactions can be regarded formally as nucleophilic substitution reactions, and there are close chemical analogies. One of the most instructive from the biosynthetic point of view is the chlorination of oxo compounds by phosphorus pentachloride and phosphorus oxychloride. The mechanism is shown at the foot of Figure 7.20.

The oxo derivatives are tautomers of corresponding hydroxy compounds which are the analogues of phenol. The reason for preferring the oxo structure is discussed in Joule & Smith (1978, p. 61).

Two features of this reaction are significant. Firstly the oxo substituent through its phenol tautomer is converted into a phosphate ester which is a good leaving group (Chapter 3). Secondly, substitution by chloride occurs α or γ to a nitrogen atom in a six-membered ring. As in all substitution reactions of this type, an intermediate containing a tetrahedral carbon atom is formed and this decomposes with loss of the leaving group.

The enzyme-catalysed conversion of uridine triphosphate into cytidine triphosphate, is directly analogous (Akhtar & Wilton, 1973). The sequence of events differs only in that the nucleophile ammonia adds to form the tetrahedral intermediate before the phosphate ester leaving group is formed (Figure 7.21). The enzyme that catalyses this reaction is multifunctional. It also generates the ammonia required by hydrolysis of glutamine.

Figure 7.21.

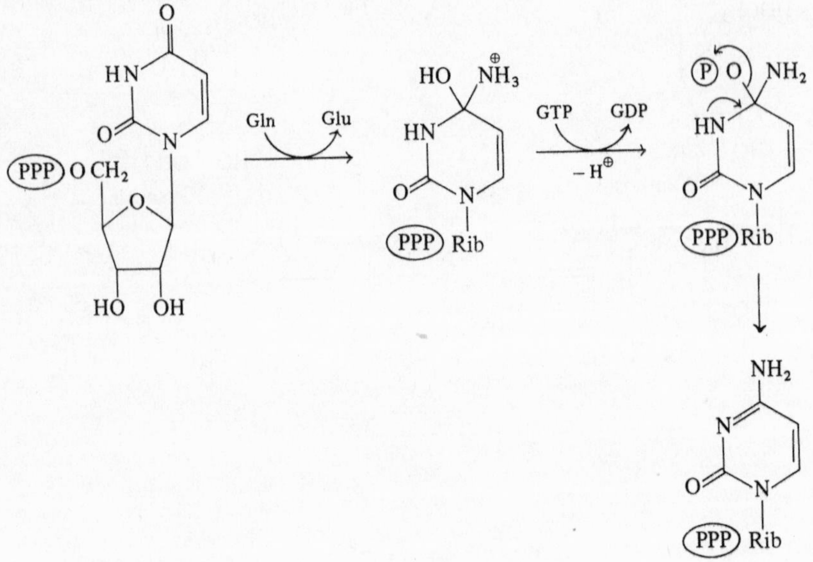

Similar sequences are thought to occur in the biosynthesis of adenylic acid from inosylic acid. The effectiveness of 6-mercaptopurine as an antitumour drug is probably due to blockage of amination of the thionucleotide analogue of inosylic acid; the thiophosphate may not be a sufficiently good leaving group for the enzyme-catalysed reaction (Figure 7.22).

Figure 7.22.

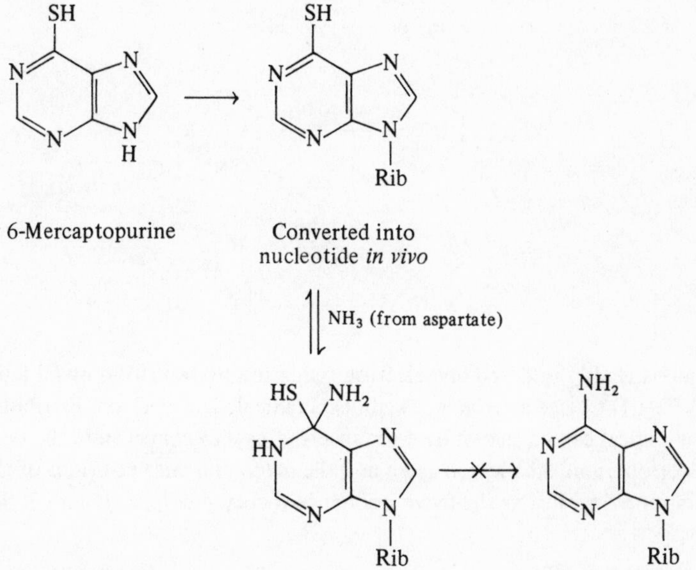

6-Mercaptopurine

Converted into
nucleotide *in vivo*

NH_3 (from aspartate)

7.4.   Aromatic substitution via radical intermediates

Whereas electrophilic substitution is characteristic of electron-rich aromatic compounds and nucleophilic substitution characteristic of electron deficiency, substitution via radical intermediates can occur in either class of compounds. The attacking group is a radical, a group containing an unpaired bonding electron. Radicals can be prepared by several methods in the laboratory, the most important of which are homolysis of thermally or photochemically labile bonds. In nature, a transition metal cation is frequently involved in radical generation. The reactions in Figure 7.23 are illustrative of laboratory radical substitution reactions in aromatic compounds and find no parallel in nature. There are however, a number of reactions known collectively as oxidative coupling reactions that are important in secondary metabolism in plants and are possible in the laboratory.

The chief substrates for oxidative coupling reactions are phenols. Phenolate

Figure 7.23.

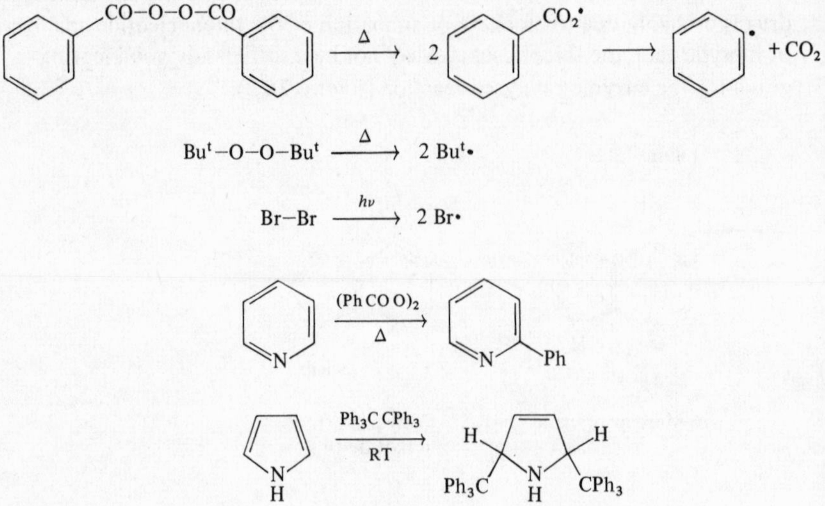

anions readily undergo one-electron oxidation by transition metal ions such as Fe(III) to give a resonance-stabilised radical. The electron distribution in the radical can be shown by e.s.r. spectroscopy to concentrate the unpaired electron upon the oxygen atom and the *ortho* and *para* positions of the benzene ring, just as the naive resonance theory predicts. (Figure 7.24, *top*).

Figure 7.24.

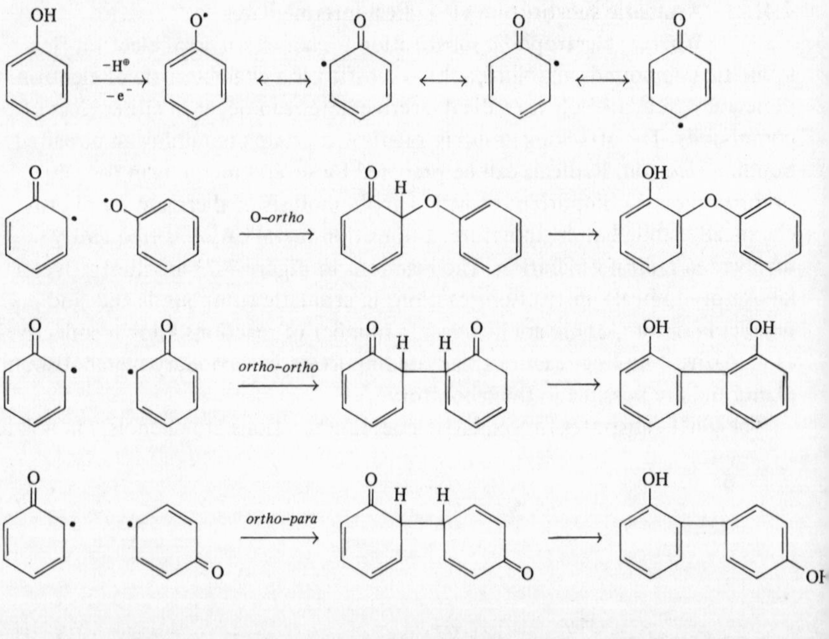

This means that the radical can react through any of these positions. When phenol or similar compounds are oxidised to radicals and the radicals pair, an oxidative coupling reaction has taken place. Three of the possible modes are illustrated in Figure 7.24. If these ideas and reactions are applied to the biosynthesis of many phenolic natural products, it becomes easy to see how the complex ring structures are built up (Figure 7.25).

Figure 7.25. Biosynthesis of some polyphenols via oxidative coupling.

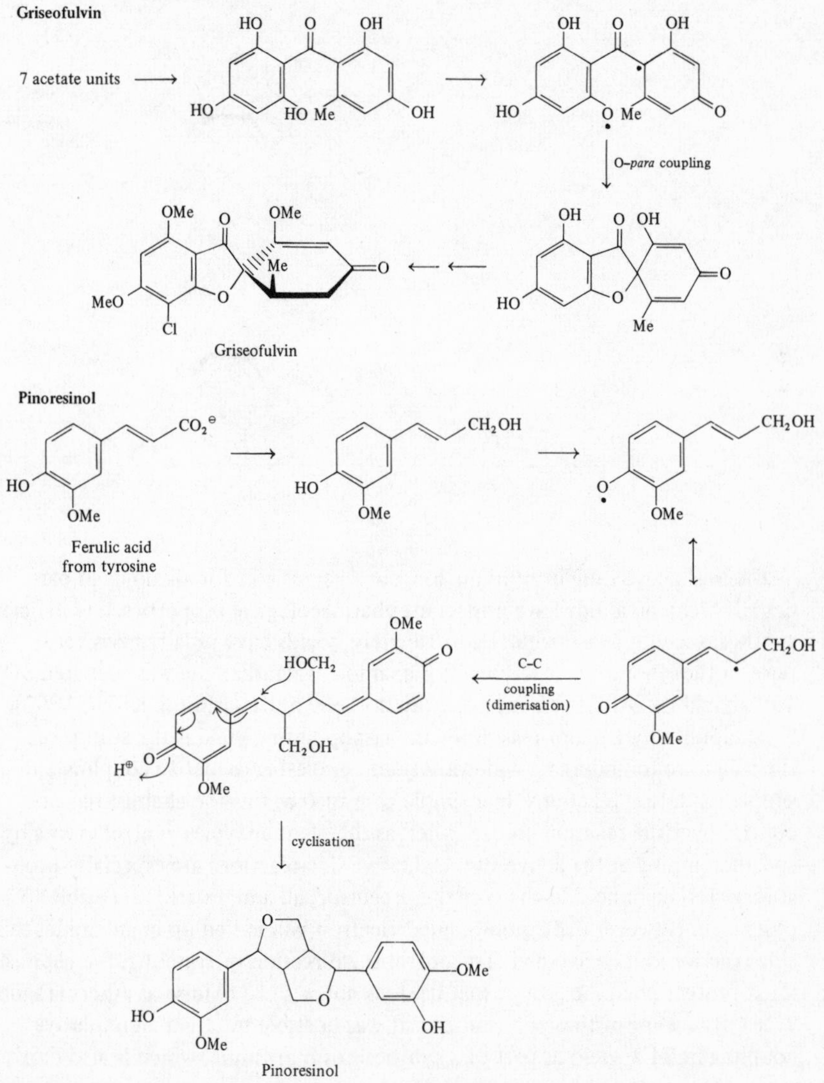

Griseofulvin

7 acetate units →

O–*para* coupling

Griseofulvin

Pinoresinol

Ferulic acid
from tyrosine

C–C
coupling
(dimerisation)

cyclisation

Pinoresinol

Figure 7.25. (continued). Biosynthetic derivatives of reticuline.

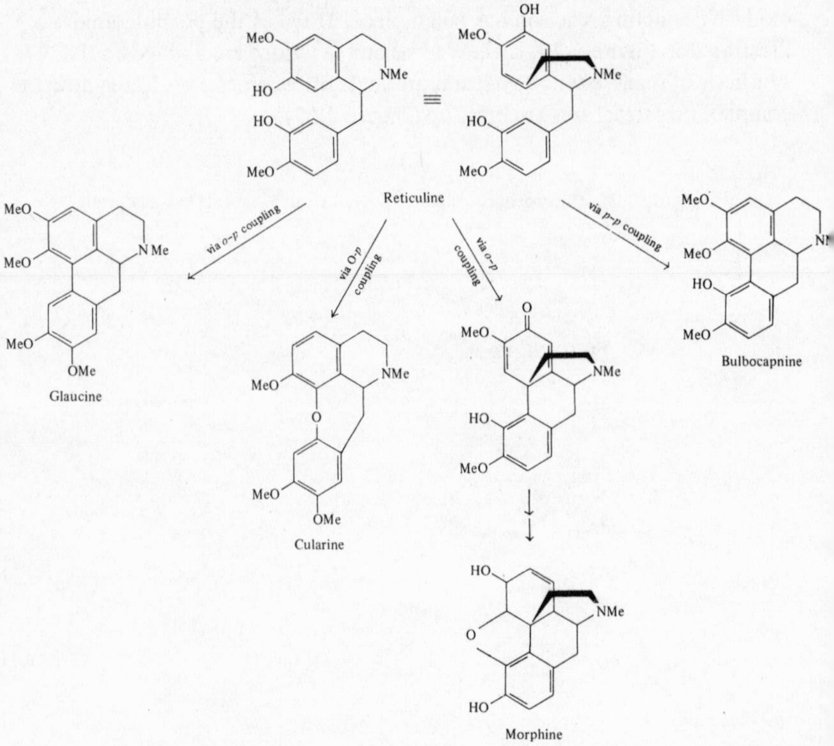

Chemists have sought to mimic biogenetic syntheses for alkaloids in particular. Many alkaloids have important pharmacological properties, and efficient syntheses would be valuable. Unfortunately, yields have usually been very poor in the laboratory reactions. For example, galanthamine was prepared in 1.4 % yield by oxidative coupling from norbelladine (Barton & Kirby, 1962).

Undoubtedly a major reason for the disappointing yield is the ability of the substrate to undergo a wide variety of competing oxidative coupling and simple oxidation reactions. In a simple case such as this the chemist has no control over the reaction course, whereas the plant enzymes control events by specific binding at the active site. Oxidative side-reactions are especially probable with the amine side-chain of the precursor; all amines are susceptible to oxidation. However if the non-bonded electron pair is tied up in an amide, this side-reaction can be avoided. To control the direction of coupling, the chemist must protect phenolic groups that he does not wish to oxidise as ethers (Figure 7.26). By means of these precautions, it was possible to effect an oxidative coupling in 24 % yield as part of a synthesis of maritidine, which is also derived from norbelladine (Schwartz & Holton, 1970). Despite improvements such as these, the best source for most alkaloids of importance is isolation from an appropriate plant.

Figure 7.26.

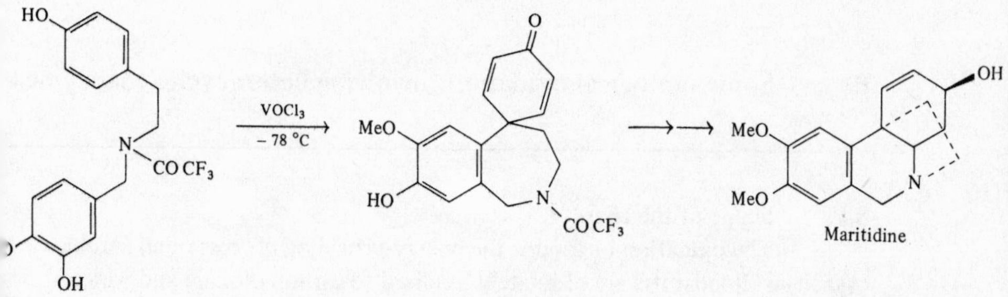

Norbelladine methyl
ether trifluoroacetate

8.1.   Scope of the chapter

The oxidation of food is the energy-providing process of all aerobic organisms. Foodstuffs are ultimately oxidised to carbon dioxide and water, the fate of all organic compounds. Between the ingested food and the final oxidation products lie many steps of oxidation, few of which involve molecular oxygen itself. In aerobic organisms, molecular oxygen is the terminal oxidant, but the assistance of many enzymes and coenzymes is required before the high electron-affinity of oxygen finally is satisfied and the oxidation is complete. Oxidation is in general best defined as occurring when a substrate loses one or more electrons to an oxidising agent. This definition is easy to see in action in redox reactions of metal cations such as iron (II), a reaction important in many oxidising enzymes (see Chapters 6, 9 and 15; Figure 8.1). At first sight, it is less obvious in oxidations of organic substrates, for example the formation of carbonyl groups from alcohols. This reaction is best considered as a dehydrogenation, a removal of hydrogen from the substrate. Figure 8.1 illustrates how dehydrogenation fits into the general definition of oxidation.

Figure 8.1.

One-electron oxidation

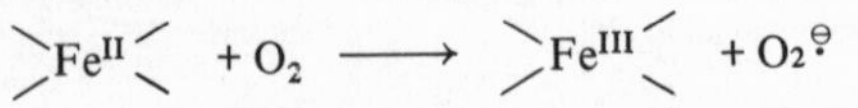

$$\text{>Fe}^{II}\text{<} \ + O_2 \ \longrightarrow \ \text{>Fe}^{III}\text{<} \ + O_2^{\ominus}$$

Two-electron oxidation

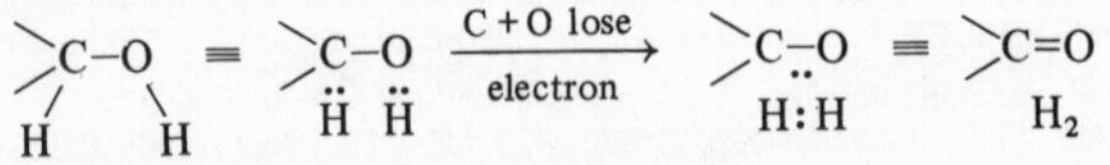

$$\text{>C–O}_{/H \quad \backslash H} \ = \ \text{>C–O}_{\overset{..}{H} \quad \overset{..}{H}} \ \xrightarrow[\text{electron}]{\text{C + O lose}} \ \text{>C–O}_{H:H} \ = \ \text{>C=O}_{H_2}$$

In this chapter, two areas of biological oxidations have been selected for discussion, both involving the mediation of coenzymes. Nicotinamide and

Figure 8.2. Some oxidative conversions involving coenzymes.

Nicotinamide-dependent

$$RCH_2OH \underset{\text{dehydrogenase}}{\overset{\text{alcohol}}{\rightleftharpoons}} RCHO$$

$$CH_3CH(OH)CO_2H \underset{\text{dehydrogenase}}{\overset{\text{lactate}}{\rightleftharpoons}} CH_3COCO_2H$$

Flavin-dependent

$$RCH_2NH_2 \underset{\text{oxidase}}{\overset{\text{amine}}{\longrightarrow}} RCHO$$

$$RCHO \underset{\text{oxidase}}{\overset{\text{aldehyde}}{\longrightarrow}} RCO_2H$$

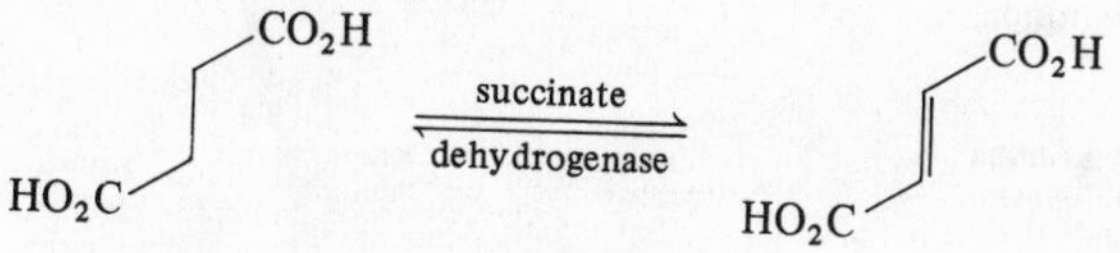

$$\underset{\text{dehydrogenase}}{\overset{\text{succinate}}{\longrightarrow}}$$

Flavin and nicotinamide-dependent

$$RS{-}SR \underset{\text{reductase}}{\overset{\text{glutathione}}{\rightleftharpoons}} 2RSH$$

flavin coenzymes (Figures 8.4 and 8.13) are involved in a very wide range of biological oxidations, with alcohols, amines, aldehydes, alkenes and thiols as substrates, amongst others (Figure 8.2).

Most of these reactions do not require molecular oxygen as a reactant: the coenzymes themselves are the oxidants or reductants. This statement implies that mechanisms exist for oxidising and reducing the coenzymes as required by the organism. Usually there exists a sequence of redox enzymes that will oxidise a reduced coenzyme, and such a system is known as an electron-transport chain (Figure 8.3a). In many organisms, the synthesis of ATP is coupled to the electron-transport chain, and it is the main way that energy from oxidation of foods is converted for metabolic use (Chapter 2).

There are some enzymes, known as oxidases, that will oxidise a substrate directly with molecular oxygen. A large class of these enzymes is found in mammalian liver and they are responsible for much of the metabolism of drugs (see Chapters 9 and 15); these enzymes are known as cytochromes P 450. They contain a porphyrin–iron complex as prosthetic group and are made up of a group of proteins that constitute a miniature electron-transport chain (Figure 8.3). A number of other oxidases require both flavins and metal cations for activity. Aldehyde oxidase and xanthine oxidase are examples of enzymes employing both molybdenum and iron (Figure 8.3c).

Finally, there are three important enzymes that protect cells of aerobic organisms against indiscriminate attack by potent oxidising species that may easily be generated adventitiously in an aerobic environment. Catalase and peroxidase reduce hydrogen peroxide, and superoxide dismutase catalyses the disposal of the superoxide radical anion, the product of a one-electron transfer to molecular oxygen (Figure 8.3d).

It is impossible to do justice to all of these reactions in the space of one chapter, and we have chosen to concentrate attention upon the reactions mediated by nicotinamide and flavin coenzymes. Biological oxidations figure also in Chapters 6, 7, 9 and 15 in this book.

Figure 8.3. (a) A mitochondrial electron-transport chain; (b) cytochrome P 450; (c) xanthine oxidase, an enzyme catalysing oxidation by molecular oxygen; (d) three enzymes that catalyse the disposal of potent oxidising species (D is an electron donor).

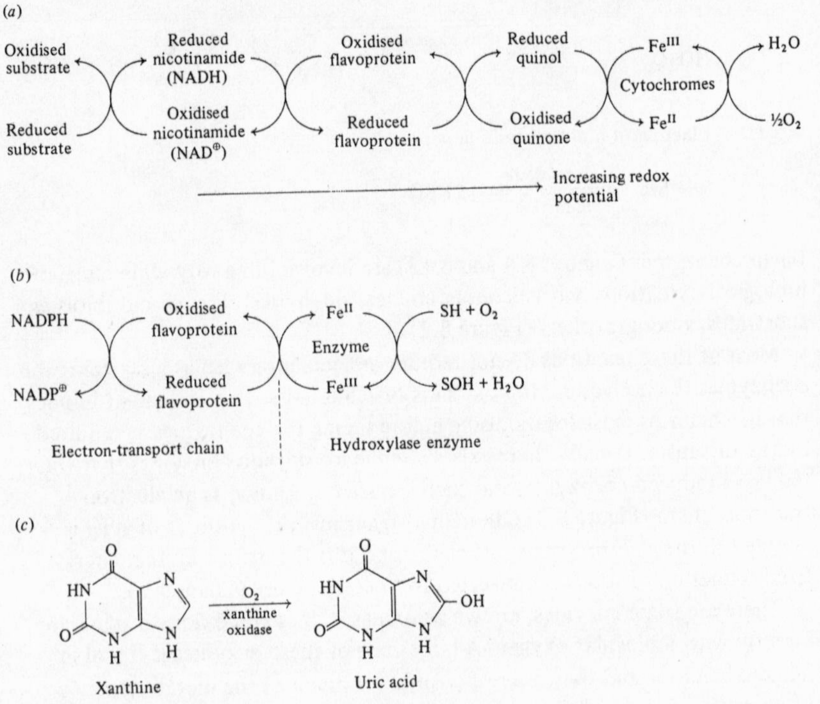

(a)

(b)

(c)

(d)

$$2H_2O_2 \xrightarrow{\text{catalase}} 2H_2O + O_2$$

$$H_2O_2 + H_2D \xrightarrow{\text{peroxidase}} 2H_2O + D$$

$$2O_2^{\ominus} + 2H^{\oplus} \xrightarrow[\text{dismutase}]{\text{superoxide}} O_2 + H_2O_2$$

8.2. Nicotinamide coenzymes

There are two coenzymes that contain the carboxamido-substituted pyridine ring nicotinamide, and they differ only in that one is phosphorylated on the 3'-hydroxyl group of one ribose fragment. This extra phosphate makes no difference to the essential chemical reactivity of the coenzyme, but it has a profound effect upon the ability of enzymes to bind the coenzymes at their active site. Indeed most enzymes that require the phosphorylated coenzyme will not accept its unphosphorylated relative at all.

The oxidised forms of the coenzyme are abbreviated by the letters NAD^+ and $NADP^+$, which stand for nicotinamide adenine dinucleotide (phosphate). In the older American literature, the abbreviations DPN (diphosphopyridine nucleotide) which is NAD^+, and TPN (triphosphopyridine nucleotide) which is $NADP^+$ (Figure 8.4) are used.

The NAD^+ molecule is quite large, and to understand its chemistry we must consider segments of the molecule in turn. The adenine nucleotide section and the additional phosphate are primarily concerned with binding to an enzyme's active site. In most enzymes, the coenzyme must be bound to the active site before the substrate itself can bind. Much effort has been expended in determining the sequence of binding to enzymes requiring coenzymes. Kinetics is the chief tool by which these binding sequences can be established (Chapter 12). Biochemists often call the binding sequence the mechanism of the enzyme, and indeed it is a very important aspect of the mechanism of action of the enzyme as a whole. However, a feature that has great chemical interest is the transformation that the substrate and coenzyme undergo when they are bound together at the active site, and it is this mechanistic step that we shall concentrate upon in this section. The fragment of the coenzyme that is involved in this step is the N-alkyl pyridine ring, the nicotinamide. On reduction, NAD^+ is converted into NADH, a 1,4-dihydropyridine. Hydrogen is transferred from

Figure 8.4. Nicotinamide adenine dinucleotide (phosphate), NAD^+ ($NADP^+$).

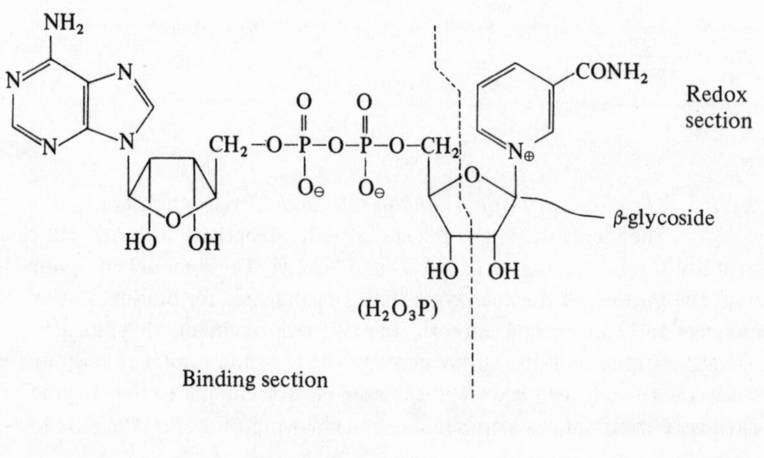

the substrate to the 4-position of the nicotinamidium ring; this has been amply demonstrated by deuterium-labelling experiments, as summarised in Figure 8.5.

Figure 8.5. (*a*) At equilibrium, D will be distributed between NAD$^+$ and the alcohol; (*b*) mixture isolated and deuterium content determined; no loss of deuterium on degradation. The conclusion: the enzyme does not transfer hydrogen to C-2 or C-6 of the coenzyme.

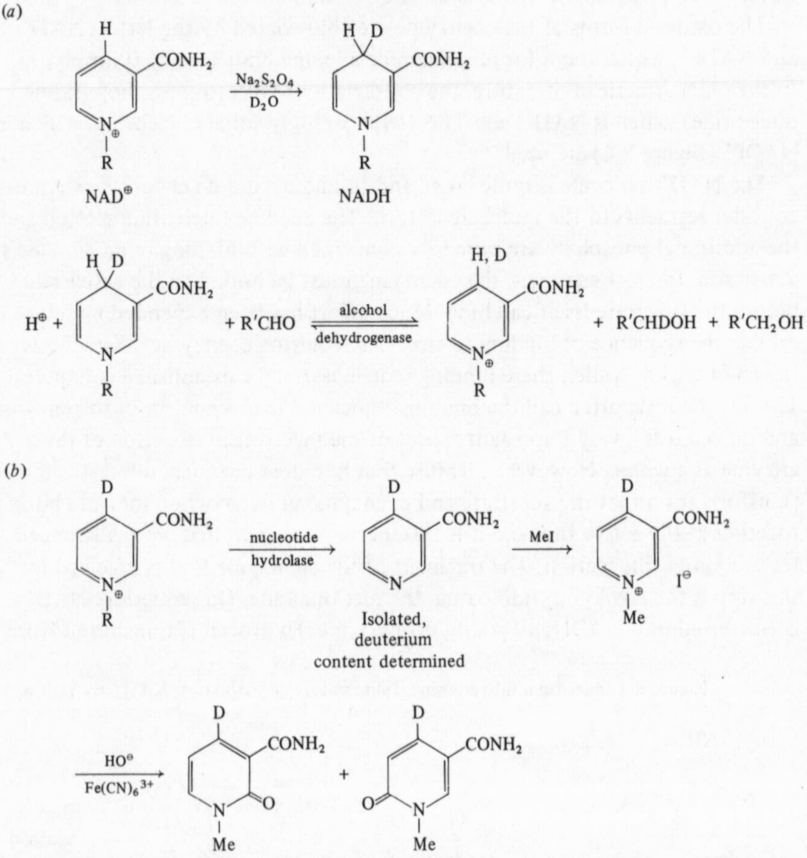

(*a*)

(*b*)

8.2.1.   The chemistry of pyridinium salts and dihydropyridines

The identification of the site of redox reactivity in NAD$^+$ led chemists to study simpler analogues of NAD$^+$ and NADH. These model compounds omit the portion of the coenzyme that is responsible for binding to the enzyme, and concentrate upon the redox-active fragment: they are also *N*-alkyl pyridinium salts. Unfortunately, many of the results of early model studies in the late 1950s were not related closely enough to the enzyme's chemistry itself, and as a consequence of this, and of more detailed knowledge of the structures of the enzymes, there has recently been a renaissance of interest in model systems for NAD-mediated reactions. We now have a much more balanced view of the chemistry of these compounds but, as we shall see, there are important points of mechanistic detail that still remain obscure.

We saw in Chapter 7 that pyridine derivatives readily undergo nucleophilic substitution reactions. Since a pyridinium salt bears a positive charge, we would expect that the addition of a nucleophile to a pyridinium ring would occur extremely easily, and this is observed experimentally. Delocalisation of the positive charge shows that addition could occur at either the 2-, 4- or 6-positions (Figure 8.6). Depending upon the nucleophile chosen and the reaction conditions, any or all of these reactions may occur. In the absence of a hydride ion acceptor, these additions are often reversible.

An adduct of a nucleophile and a pyridinium salt contains a carbon atom that has changed from trigonal to tetrahedral; the aromatic conjugation of the ring has been lost and the ring reduced. Such adducts are known as substituted dihydropyridines. A dihydropyridine itself is formed when the nucleophile that adds is hydride ion from a reagent like sodium borohydride or lithium

Figure 8.6. Reactions of pyridinium cations (see Joule & Smith, 1978, p. 68).

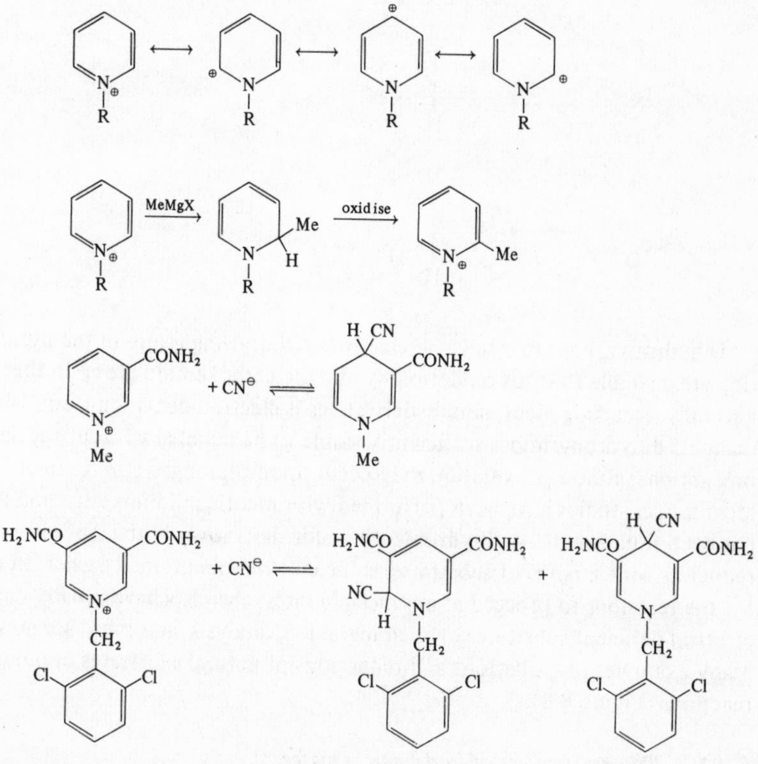

aluminium hydride. With sodium borohydride a 1,2- and 1,6-dihydropyridine are usually formed, whereas lithium aluminium hydride, a stronger reducing agent, normally generates a mixture of all three dihydro-isomers. An analogy can be drawn between this addition reaction to C=N and the reduction of aldehydes or ketones by metal hydrides (Figure 8.7).

Figure 8.7.

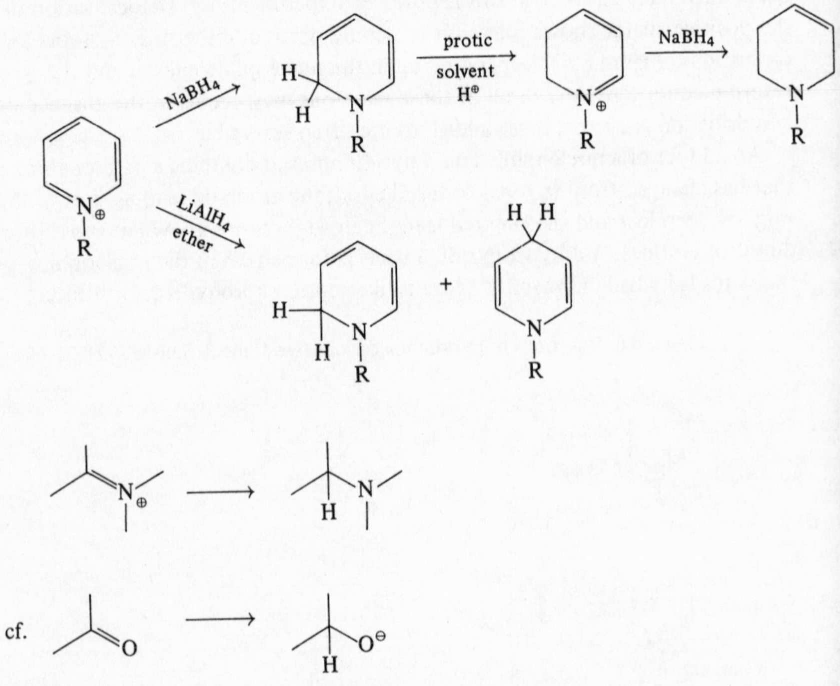

Dihydropyridines that have no electron-withdrawing group in the pyridine ring are unstable towards oxidation by air: the carboxamido group in the naturally occurring nicotinamide derivatives is electron-demanding enough to make its dihydropyridines sufficiently stable to be handled without special precautions, although oxidation may occur upon prolonged storage in air. Most model studies have been performed with nicotinamidium salts, and it has been found that 1,4-dihydronicotinamide derivatives in the laboratory will reduce as wide a range of substrates as the enzymes mentioned earlier. In order for the reactions to proceed at appreciable rates, chemists have usually chosen electron-deficient substrates or used metal ion catalysis, and there are now many examples of oxidations and reductions of natural substrates in model reactions (Figure 8.8).

8.2.2. The mechanism of hydrogen transfer

All of the above model reactions and the enzyme-catalysed reactions

Figure 8.8. Some model reactions of dihydronicotinamides. Notice the similarity in substrates between these model reactions and the enzyme-catalysed reactions described in the text.

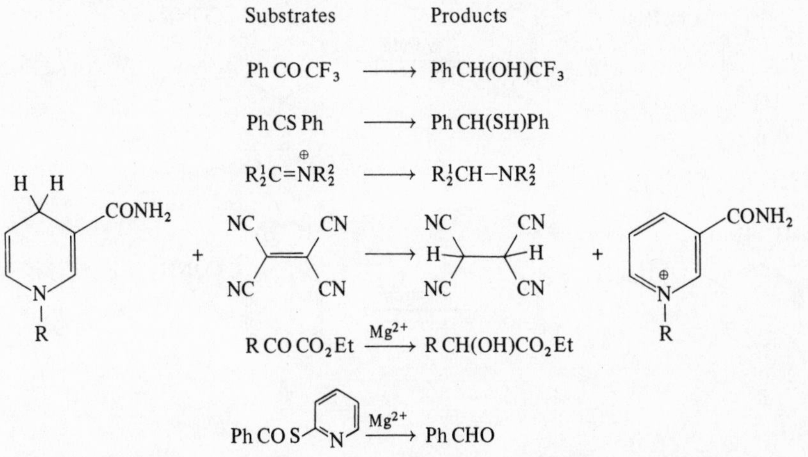

Substrates   Products

$Ph\,CO\,CF_3 \longrightarrow Ph\,CH(OH)CF_3$

$Ph\,CS\,Ph \longrightarrow Ph\,CH(SH)Ph$

$R_2^1 C{=}\overset{\oplus}{N}R_2^2 \longrightarrow R_2^1 CH{-}NR_2^2$

$R\,CO\,CO_2Et \xrightarrow{\;Mg^{2+}\;} R\,CH(OH)CO_2Et$

$Ph\,CO\,S{-} \xrightarrow{\;Mg^{2+}\;} Ph\,CHO$

involve a transfer of hydrogen from substrate to coenzyme and vice versa. It is an important chemical problem to find out what the hydrogen-transfer mechanism is. Two possibilities are currently being considered. Firstly, hydrogen could be transferred as H^-, hydride ion, crudely analogous to metal hydride reactions (Figure 8.7). Alternatively, a hydrogen atom, $H\cdot$, plus an electron could be involved (Figure 8.9). There is much model evidence to show that electron transfer from dihydropyridines to suitable substrates such as nitro compounds and transition metal ions is a facile process. Radical intermediates can also be detected by e.s.r. spectroscopy. However, no studies of the enzyme-catalysed reactions have yet provided evidence for radicals, and, unfortunately, kinetic techniques including isotope effects and substituent effects have been unable to distinguish between H^- and $H\cdot$ transfer. It is still an open question, therefore, what the mechanism of hydrogen transfer by nicotinamide coenzymes is (Hamilton, 1971, Dunn, 1975; Staunton, 1978).

8.2.3. Stereochemical aspects of nicotinamide coenzyme dehydrogenases

The detailed tertiary structures of many dehydrogenases are known from X-ray diffraction studies, and this means that we can discuss not only the stereochemistry of the reaction from the point of view of the substrate, but also from the point of view of the enzyme-coenzyme complex. We have already considered several aspects of substrate stereospecificity in the early chapters: in Section 5.3.1. we met cases of stereospecific reduction of aldehydes by liver alcohol dehydrogenase. Dehydrogenases also show stereospecificity with regard to the *face* of the nicotinamide ring to which the hydrogen atom is transferred. Lactate dehydrogenase, for example, transfers

Figure 8.9.

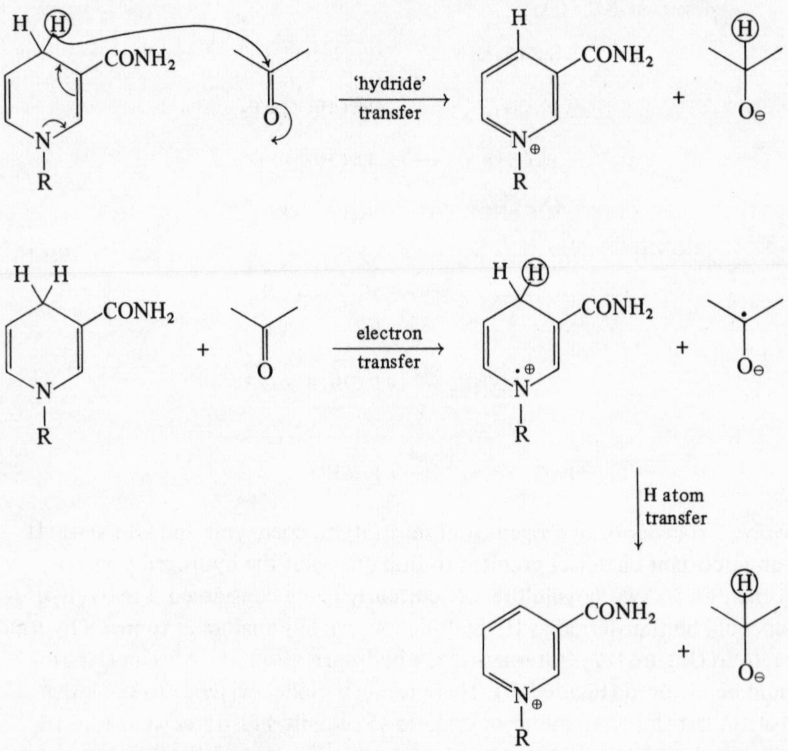

hydrogen to what is known as the A-face of the coenzyme (Figure 8.10). In contrast, glyceraldehyde 3-phosphate dehydrogenase reacts with the coenzyme receiving hydrogen on the opposite face, the B-face. The X-ray structures have shown that the conformation in which the enzyme binds the coenzyme is responsible for this stereospecificity. The binding in the case of lactate dehydrogenase is illustrated diagrammatically in Figure 8.11. Notice the forces involved in binding the coenzyme to the enzyme. There is a detailed complementarity between the various portions of the coenzyme molecule and the amino acids that are found in its immediate vicinity. Thus arginine 101, which is cationic, interacts by ionic bonds with the anionic portions of the bridging diphosphate link between the two nucleotides. (Asparagine, an amide, hydrogen bonds with the amide of the nicotinamide in glyceraldehyde 3-phosphate dehydrogenase.) Finally, the relatively non-polar adenine ring is surrounded chiefly by non-polar amino acids, such as phenylalanine. These forces are typical of those that hold interacting biological systems together (Chapters 12 and 16). Indeed, the X-ray structures have shown that there is a high degree of homology in the regions of several dehydrogenases that bind the adenine

Figure 8.10. The enantiotopic hydrogens of NAD(P)H are conventionally labelled H_A and H_B, as indicated.

Lactate dehydrogenase

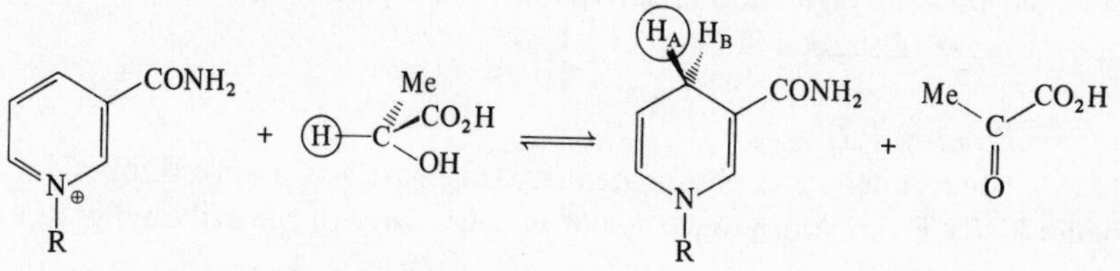

Glyceraldehyde 3-phosphate dehydrogenase

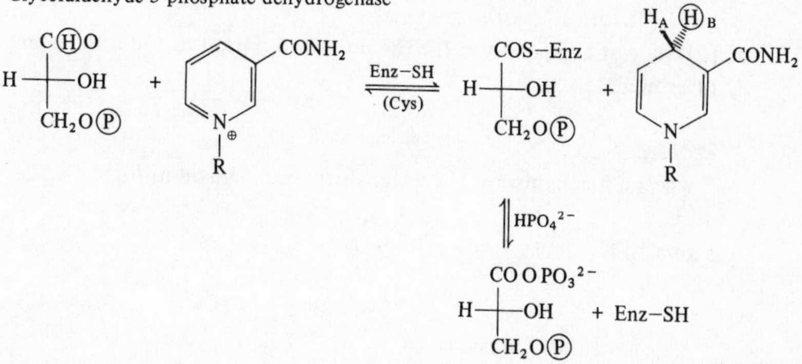

Figure 8.11. Binding at the active site of lactate dehydrogenase.

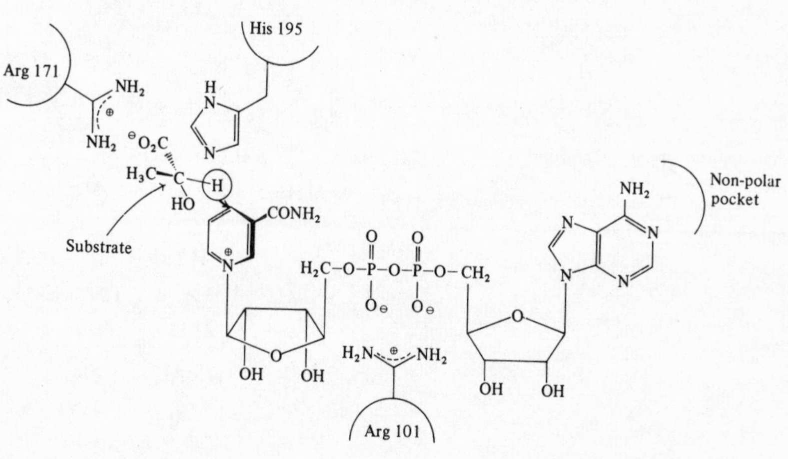

portion of NAD. Not only are these detailed structural features important in understanding the stereochemical course of the enzyme-catalysed reactions, they are also a signficant beginning in studying the conformational changes associated with the regulation of activity of the enzyme. Some of these control mechanisms are discussed in Chapter 13.

Problem 8.1.

Glyceraldehyde 3-phosphate dehydrogenase acts upon a thiol ester (Figure 8.10). From your knowledge of the chemistry of the carbonyl group and its derivatives:

> (1) From what species does hydrogen transfer to NAD$^+$ occur and how is it formed by the enzyme?
> (2) Suggest a mechanism for the phosphorylation of the thiol ester intermediate.

Problem 8.2.

Suggest a mechanism for the transformation shown in Figure 8.12.

Figure 8.12.

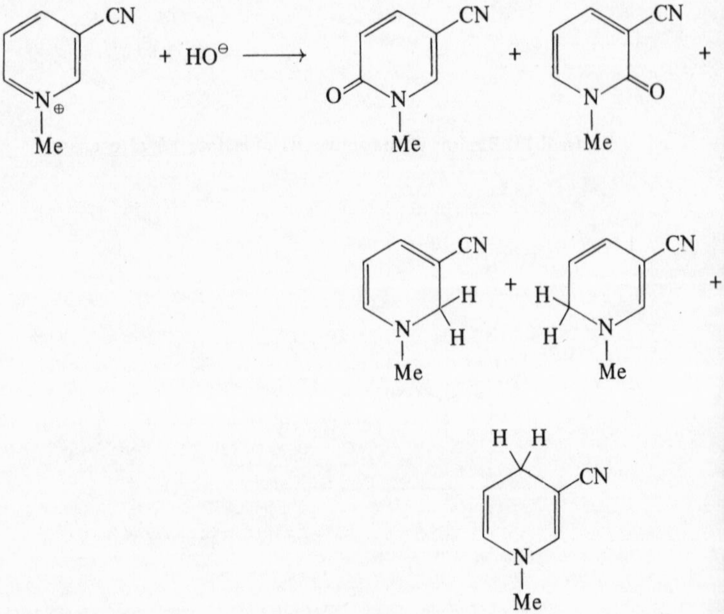

Problem 8.3.

It was noted in Figure 8.8 that many model reactions of dihydropyridines and carbonyl compounds are catalysed by metal ions. In contrast, iminium salts require no catalysis. By comparison with metal ion catalysed ester hydrolysis (Chapters 3 and 9), suggest a function for the metal ion and why metal ions are not necessary in iminium salt reduction (Staunton, 1978.)

8.3. Flavin coenzymes

With the nicotinamide coenzymes, it was possible to discuss the chemistry of hydrogen transfer in terms of relatively simple reactions of model pyridinium salts and dihydropyridines. Although ambiguities remain concerning the details of the enzyme-catalysed reactions, it is true to say that the background chemistry of pyridinium salts is fairly well understood. However, if you look at the structure of the heterocyclic system that is the redox reactive centre of flavin coenzymes, a much more formidable chemical problem is posed (Figure 8.13). Such a large ring system is too complex for the conceptual methods of organic chemistry to choose, *a priori*, what reaction paths will be followed, but it is relatively easy to decide which sites of the molecule are

Figure 8.13. Flavin coenzymes.

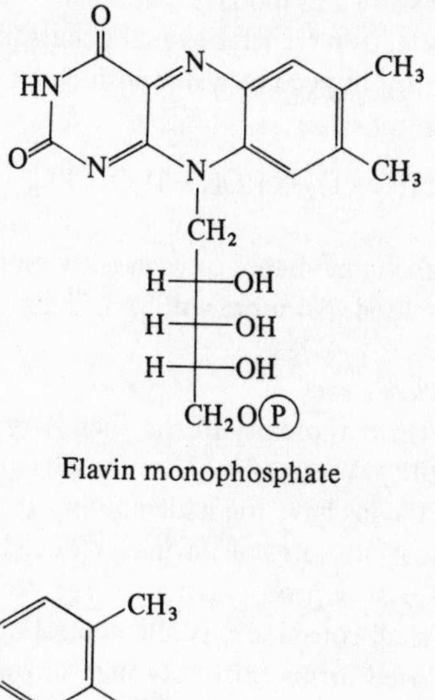

Flavin monophosphate

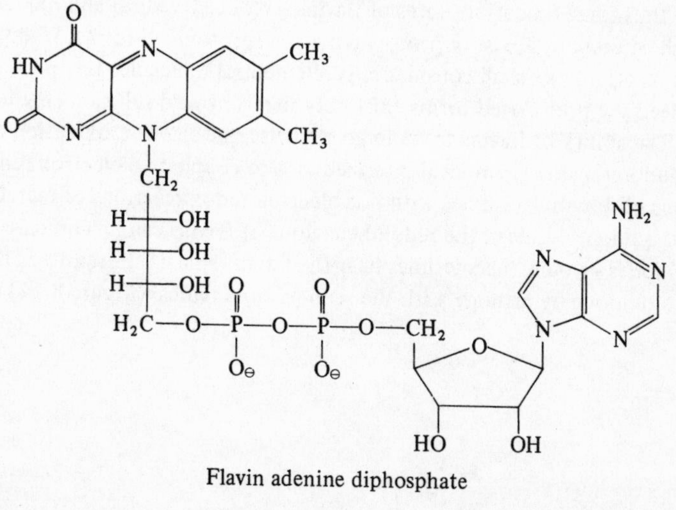

Flavin adenine diphosphate

likely to react with a reagent and to set up experiments that will test the suggestion. In this section, we shall look at some examples of the exploratory chemistry that has been carried out on the flavin system and see how much light the results shed upon the mechanism of action of flavin dependent dehydrogenases (Hamilton, 1971; Bruice, 1976, Hemmerich, 1976). There are three classes of flavoprotein-catalysed reactions, and some aspects of the first two will be discussed in the sequel.

(1) One-electron transfer reactions with quinones or metal ions as partners. Flavoproteins do not transfer one electron to oxygen.
(2) Two-electron transfer processes in which either oxygen or an organic molecule can be the electron donor, e.g.

$$RCH_2NH_2 + Fl_{ox} + O_2 + H_2O \rightarrow RCHO + Fl_{red}H_2 + H_2O_2$$

$$2RSH + Fl_{ox} \rightarrow RSSR + Fl_{red}H_2$$

These are known as **oxidase** reactions.
(3) Four-electron transfer overall occurs in the flavin **oxygenases**, which reduce oxygen to water with concomitant hydroxylation of an organic substrate:

$$RH + Fl_{red}H_2 + O_2 \rightarrow ROH + H_2O + Fl_{ox}$$

The chemistry of flavin-mediated oxygen activation is extremely complex and not yet well understood. No more will be said about it here.

8.3.1. Flavin redox states

Much of the complexity of the chemistry of flavins arises from their exceptional versatility. Whereas pyridinium salts are virtually condemned to add nucleophiles, flavins have the added ability to form stable radicals. There are thus three oxidation states of flavins – reduced, radical and oxidised – and each of these can exist in protonated and deprotonated forms. When discussing the reactions, we shall consider only the neutral molecules for simplicity: indeed the protonated forms exist only in strong acid solution (Figure 8.14).

The ability of flavins to undergo stepwise one-electron oxidation and reduction makes them ideal intermediates to couple two-electron redox reactions of dihydropyridines with one-electron redox reactions of metal cations and quinones. Indeed the redox behaviour of flavins can be compared with quinones – you will sometimes hear the flavin radical referred to as the flavin semiquinone by analogy with the semiquinone radical (Figure 8.14).

Figure 8.14.

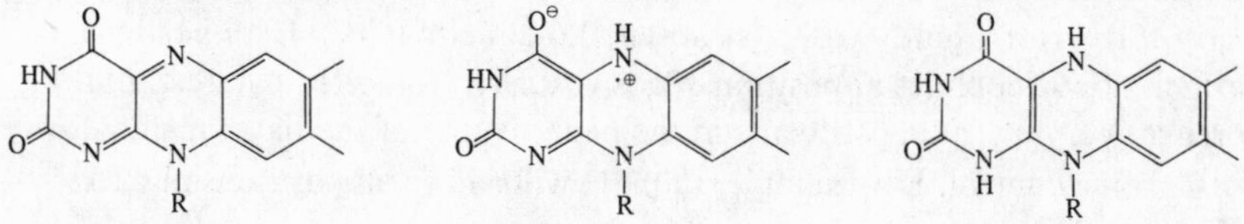

Fl_{ox}	$FlH\cdot$	$Fl_{red}H_2$
Oxidised (quinone)	Radical (semiquinone)	Reduced

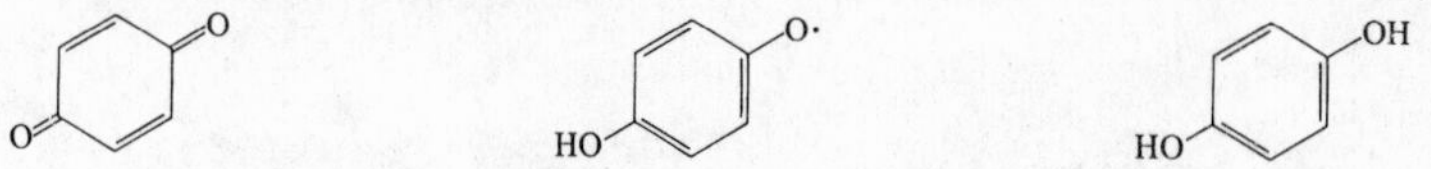

8.3.2. Reactions of the oxidised flavin coenzymes

The oxidised flavin ring system is a planar tricyclic heterocycle that has many sites open to nucleophilic attack. Like all nitrogen heterocycles with six-membered rings containing nitrogen, the C=N fragments can be compared with an isolated C=O. Thus in the flavin system, we must consider addition of a nucleophile to be possible at positions 4a, 5, 10a, and also the vinylogous positions in the benzene ring, 6 and 8 (Figure 8.15).

The only way to sort out a question of multiple reactivity such as this is by carefully controlled model studies (Bruice, 1976; Hemmerich, 1976). To make the chemistry tractable, the natural coenzyme is transformed into a simpler analogue. Sugars always present difficulties in handling and therefore

Figure 8.15.

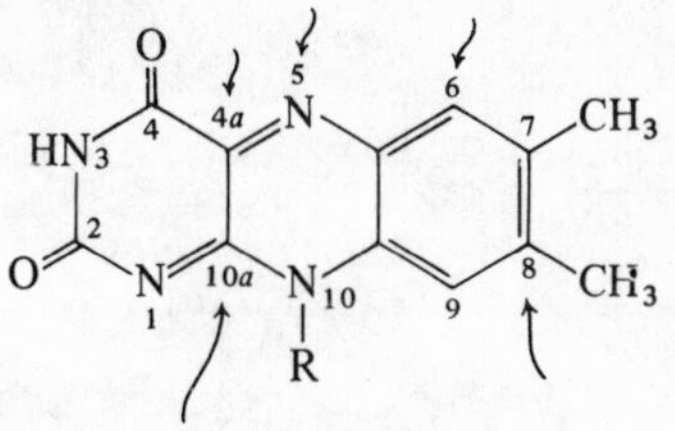

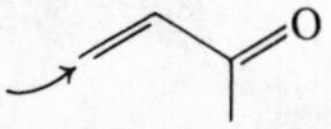

is the 'vinylogue' of

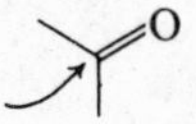

the group at N-10, ribityl in nature, is often changed to a phenyl derivative or a simple alkyl side-chain. The choice of this group can also simplify the chemistry further. If a bulky group is chosen, 2,6-dimethylphenyl, for example, attack of nucleophiles at position 10a is virtually prohibited by steric hindrance. Varying the substituents in the benzene ring of the flavin might also offer some control. For example, sulphite will add to all positions but 10*a* (Figure 8.16).

Most strong nucleophiles, like hydroxide, sulphide and carbanions, are found to add to either positions 4*a* or 5, and it is thought that addition to one

Figure 8.16.

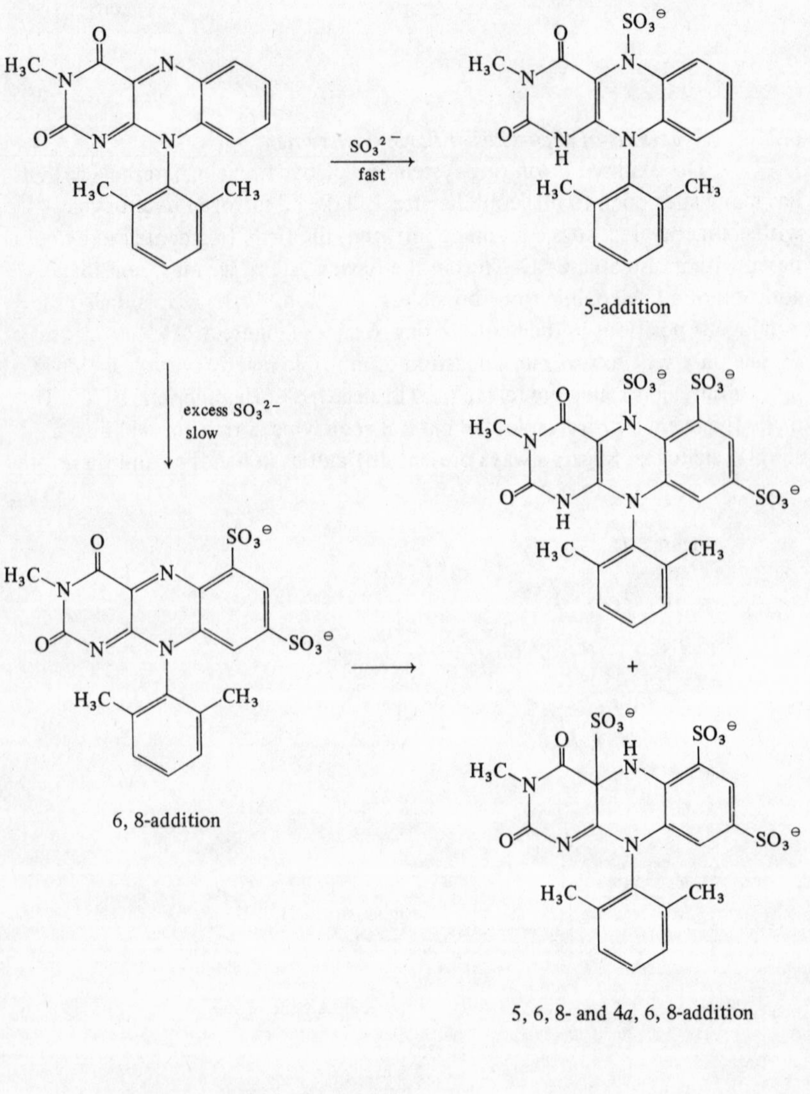

5-addition

6, 8-addition

5, 6, 8- and 4*a*, 6, 8-addition

Figure 8.17.

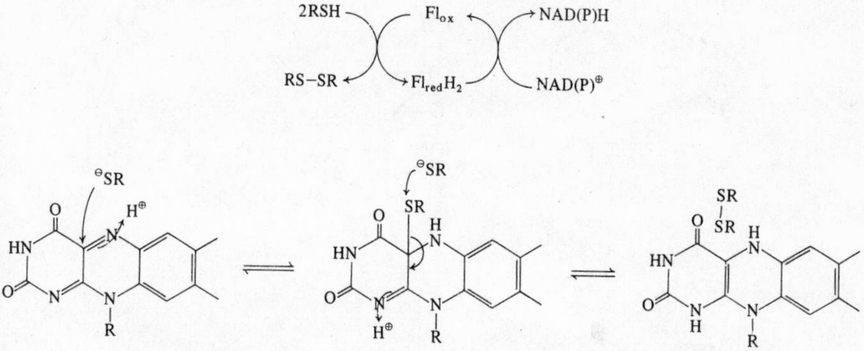

4*a*-addition

of these positions is most probable in the enzyme-catalysed dehydrogenations. For example, glutathione reductase and lipoamide oxidase (Chapter 13) are enzymes that catalyse the oxidation of thiols to disulphides through the intermediacy of a flavin coenzyme. Nicotinamide coenzymes are also required to return the flavin to its original oxidation state (Figure 8.17). With a model system it was found that the reaction rate showed a dependence upon pH that could be interpreted as due to general acid catalysis (Chapter 3). No evidence in favour of a radical mechanism was found. It is known that flavoquinone is initially protonated at N-1, but in an acid-catalysed step, and the nucleophilic thiol anion was added to C-4*a*. The attack of a second thiol anion upon the first sulphur atom causes the flavin adduct to fragment leading to the disulphide and the reduced flavin. Unfortunately, flavin-dependent enzymes are much larger than nicotinamide nucleotide dependent enzymes and consequently little is known about the detailed enzyme/coenzyme/substrate interactions that will control the site of addition of nucleophiles. It is known, however, that the flavin coenzymes are often covalently attached to their enzymes, unlike the nicotinamide coenzymes. A bond is formed between the carbon atom substituent at C-8 and an amino acid side-chain such as a histidine or a cysteine (Figure 8.18).

Figure 8.18.

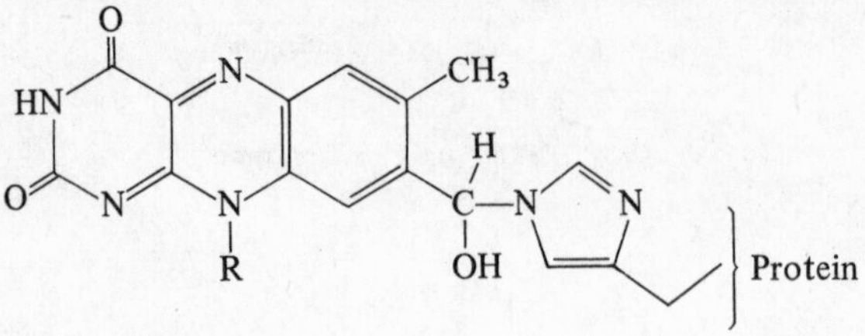

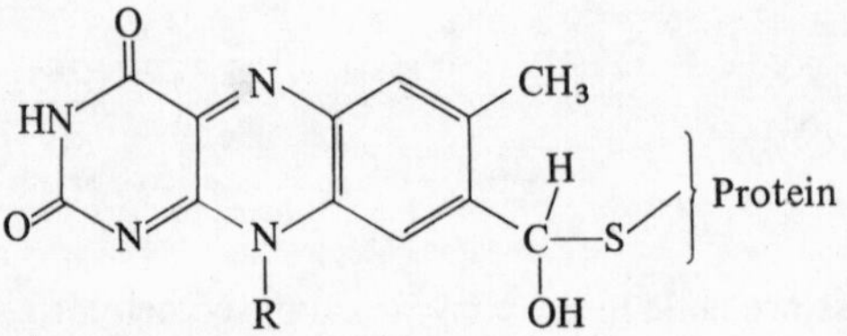

Problem 8.4.

Succinate dehydrogenase catalyses the reaction in Figure 8.1. It requires flavin coenzymes, and the substrate is thought to be oxidised directly by the flavin. Model reactions have studied analogous methyl esters.

> (1) Bearing in mind the acid/base properties of esters, and the reactivity of the flavin that we have just discussed, what intermediate form of the ester is probable in the dehydrogenation reaction?
> (2) How might this intermediate interact with the coenzyme?
> (3) Suggest a mechanism for the model reaction.

8.3.3. Reactions of flavin radicals

Two isomeric radicals can be obtained by one-electron oxidation of flavohydroquinone. Structurally they differ in whether N-1 or N-5 is protonated and they show distinct spectroscopic properties. The 1-H isomer is red but the 5-H isomer is blue (Figure 8.19a). Analogous properties are found for N-1 or N-5 alkyl derivatives. The most striking difference between the isomers, however, is that the N-1 substituted radicals are unstable and short lived whereas the N-5 isomer is much more stable.

It has been found that flavoproteins fall into two groups characterised by the type of radical that they produce. One-electron transfer proteins characteristically generate blue radicals that are essential to the enzymic reaction. On the other hand, oxidases and other two-electron enzymes give rise to red radicals that have no part in the biological reaction. There must clearly be some structural device by which these proteins make the flavin react in the way that they require. A plausible suggestion (Hemmerich, 1976) is that protonation of the appropriate nitrogen atom (N-1 or N-5) by an acidic group at the enzyme's active site could control which radical will form. An oxidase would protonate on N-1 to avoid the formation of a stable radical. On the other hand, a one-electron transferring protein would protonate N-5 and at the same time force the flavin towards planarity (Figure 8.19).

Figure 8.19.

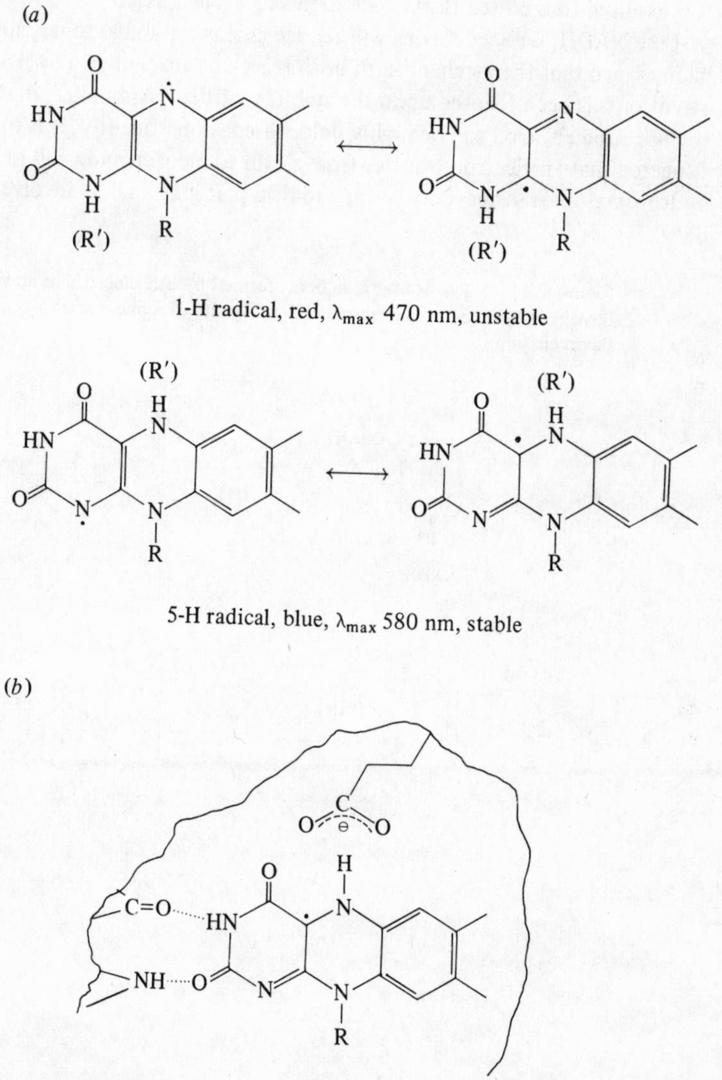

(a)

1-H radical, red, λ_{max} 470 nm, unstable

5-H radical, blue, λ_{max} 580 nm, stable

(b)

8.3.4. *Reactions of reduced flavin coenzymes*

Flavins are, of course, redox reagents, and so the discussion of their chemistry in terms of oxidised and reduced states is somewhat arbitrary and assists only to concentrate upon the mechanism of a reaction in a given direction. Many flavin reactions are irreversible and a substrate is oxidised leaving the reduced flavin along with fragments of the substrate. Examples include amino acid oxidase, lactate oxidase and glucose oxidase. The mechanisms of action of these enzymes are still highly controversial but there is more general agreement about the mechanism by which the reduced flavin is reoxidised to oxidised flavin. The oxidising agent is NAD^+.

Like NADH, reduced flavins will reduce aldehydes and ketones, and it has been argued that the mechanism of both reactions are similar. In particular, attention has been focused upon the stability of the flavin radical in which the unpaired electron can be highly delocalised. Consequently, it is to be expected that an electron transfer from flavin to nicotinamide will precede hydrogen atom transfer: once again a radical pair may well be involved (Figure 8.20).

Figure 8.20 (*a*) Two isomeric radicals formed by one-electron oxidation of flavohydroquinone; (*b*) hypothetical active site of a one-electron-transferring flavin enzyme.

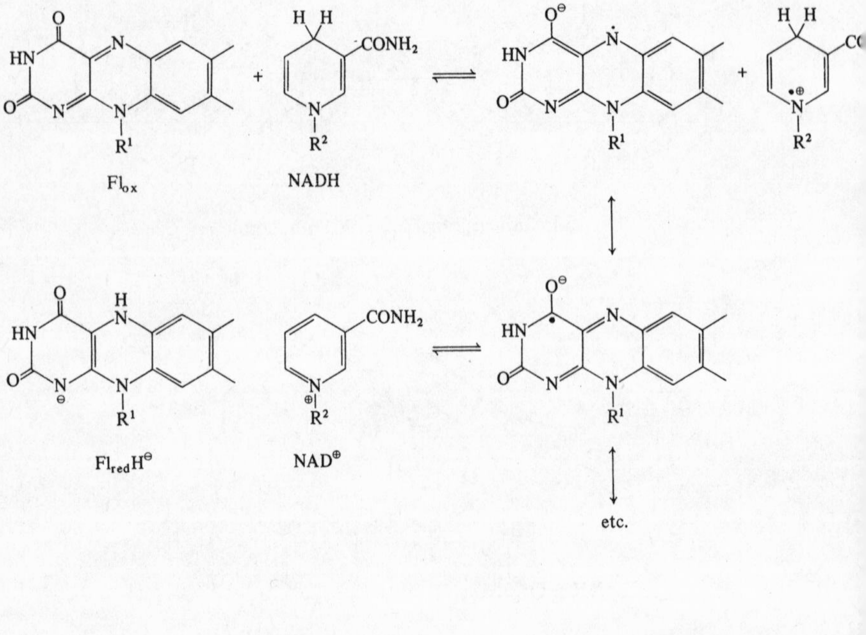

9   Metal ions in biochemistry

9.1.   The importance of metal ions in biological systems

Although the reactions that we have discussed so far concerned organic molecules as substrates, we have come across several in which metal ions behave either as reagents or as catalysts. For example, ATP interacts strongly with Mg^{2+}, and Zn^{2+} catalyses the phosphorylation of pyridine-2-aldoxime (Chapter 3). Zn^{2+} also acts as an acid catalyst in alkaline phosphatase, and we have noted in the last chapter the importance of iron complexes in electron transport. In addition to these ions, Na^+, K^+, Ca^{2+}, Co^{2+}, Cu^{2+}, and Mo^{5+} all have important biological functions. The purpose of this chapter is to show how the chemical properties of some of these metal ions suit them to their biological function in general terms, and to examine in more detail the role of zinc in carboxypeptidase and iron in haemoproteins.

It is only in the last ten years that the importance of metal ions in biochemistry has become fully apparent. Without metal ions, no cell could retain its integrity or carry out metabolism. Shortages of metals in the diet can lead to serious metabolic disorders such as some anaemias (due to iron shortage) or impairment of nerve function and bone formation, especially in infants (calcium deficiency).

9.1.1.   Classification of metal ions

Metal ions are classified according to their electronic structure. **Main group** elements have either s- or p-orbitals as their highest occupied atomic orbital, and their most stable ion is formed by the loss of these outermost electrons leaving a complete electronic shell. Thus lithium ($1s^22s^1$) readily loses one electron to afford the stable cation Li^+ ($1s^2$). The removal of a further electron from a complete shell requires more energy than is available in ordinary chemical reactions, although in a mass spectrometer and in some other instruments sufficient energy can be provided. The stability of these closed-shell configurations means that redox reactions between different ions of main group elements are rare.

In contrast to main group elements, **transition metals** have d-orbitals as their highest occupied level, and several electrons must be lost in order to attain closed-shell electronic configurations. In many cases, very highly charged ions are unstable. A balance is therefore struck between charge and the formation of closed shells, and several oxidation states become available (Cotton & Wilkinson, 1972; Figure 9.1). We shall discuss later some of the factors that control the stability of the oxidation states available to transition metal ions. The important feature to note now is that there is not much difference in energy between the highest occupied orbital of a transition metal ion (a 3d-orbital) and the lowest unoccupied orbital (a 3d- or 4s-orbital). There are thus low-energy acceptor orbitals that can readily form coordinate bonds with suitable ligands: in other words, transition metal ions readily form complexes. On the other hand, the lowest unoccupied orbitals of main group metal ions are of much higher energy than the highest occupied levels. This means that main group ions prefer to interact via ion–ion and ion–dipole mechanisms to form coordinated species.

This contrast of basic inorganic chemistry is reflected in the biological functions of the two groups of ions. Whereas transition metal ions are chiefly concerned with catalysing reactions either through ligand bonding to metal or via redox reactions between the available oxidation states, main group ions function as charge carriers, as structure forming ions or as carriers of messages.

Figure 9.1. Relative energies of 3d- and 4s-electrons in the first transition series.

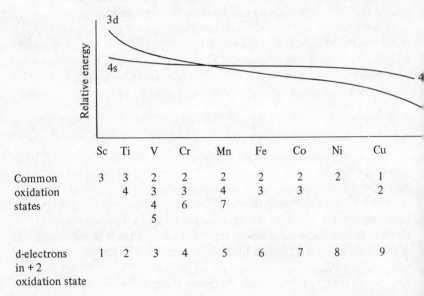

	Sc	Ti	V	Cr	Mn	Fe	Co	Ni	Cu
Common oxidation states	3	3	2	2	2	2	2	2	1
		4	3	3	4	3	3		2
			4	6	7				
			5						
d-electrons in +2 oxidation state	1	2	3	4	5	6	7	8	9

Table 9.1 summarises the biological occurrence of some important metal ions, and some of their functions.

Table 9.1. *Metal ions of biological importance*

Ion	Class and configuration	Occurrence and function
Na^+	Main group I ...$2p^6$	Low concentration inside cells; high concentration in blood; Charge carrier.
K^+	Main group I ...$3p^6$	High concentration inside cells; low concentration in blood; Charge carrier; controls activity of some enzymes.
Mg^{2+}	Main group II ...$2p^6$	High concentration inside cells. Controls geometry of ionic ligands, e.g. ATP; chlorophyll and photosynthesis.
Ca^{2+}	Main group II ...$3p^6$	Very low concentration inside cells. Trigger of nerve and muscle action; component of bone and other hard tissue.
Fe^{2+}	Transition series ...$3p^6 3d^6$	O_2 binding in haemoglobin; electron transport via cytochromes and iron–sulphur proteins; oxygenases, e.g. cytochrome P 450.
Co^{2+}	Transition series ...$3p^6 3d^7$	Coenzyme B_{12} mediated reactions.
Cu^{2+}	Transition series ...$3p^6 3d^9$	Oxidases and oxygenases.
Zn^{2+}	Transition series ...$3p^6 3d^{10}$	Acid catalyst in alcohol dehydrogenase, alkaline phosphatase and carboxypeptidase.
Mo^{5+}	Transition series ...$4p^6 4d^3$	Xanthine oxidase, nitrogenase.

9.1.2. *An approach to bioinorganic chemistry*

It is possible to extend the simple discussion of the preceding paragraphs to begin to understand the mechanism of action of metal ions in biological systems. Just as our discussion of organic reactions in the earlier chapters relied upon an understanding of the basic classes of functional groups and their reactions, so the biochemistry of metal ions can best be approached with a knowledge of their properties and compounds.

Unfortunately, as we shall see, it is not easy to find good models for many metal ions in natural systems by studying simple chemical compounds, but other techniques have been developed. Chief among these is the method known as **isomorphous replacement**. It is often possible to remove a metal ion from its biological ligands by dialysis: the metal-free preparation can then be supplied with a different metal ion of similar charge and size (isomorphous) to the naturally occurring ion. A study of the properties of the system with the replacement ion then allows something to be deduced about the mechanism of action of the metal in its host environment. Substitute ions can be chosen either to modify the chemical properties of the biological system or to act as a spectroscopic probe, and we shall meet examples of both situations in this chapter. In either case, a thorough background knowledge of the chemistry of both natural and replacement ions is essential if the results are to be interpretable.

9.2. Main group ions

When the ion content of living systems is analysed, some significant differences between the concentrations of chemically similar ions are observed (Table 9.2).

Table 9.2. *Ionic composition of cells (Williams, 1970)*

Tissue	Ion concentration (mmol kg^{-1})			
	Na$^+$	K$^+$	Mg^{2+}	Ca^{2+}
Human red blood cells (wet)	11.0	92.0	2.5	0.1
Yeast cells (dry)	10.0	110.0	13.0	1.0
Euglena cells (wet)	5.0	103.0	4.8	0.3
Skeletal muscle	27	92	22	3

All of the cells have high concentrations of potassium and magnesium, but low concentrations of sodium and calcium in comparison with the fluid that bathes the cells. Since the cells do not exist in isolation but live in an aqueous medium, separated from it by a membrane, there will tend to be a diffusion of ions across the membrane so that the concentrations of each ion inside and outside the cell become equal. In order to maintain the required osmotic imbalance for it to function, the cell takes active measures to expel sodium and calcium and to accumulate magnesium and potassium. This requires work.

The correct ion balances (Table 9.2) are maintained by ion pumps which use ATP. Indeed, the bulk of the work that is done by a cell is directed towards maintaining the correct ion balances. Potassium left inside the cell associates with negatively charged functional groups; magnesium stabilises polyanionic structures by chelation and acts as a weak acid. The extrusion of calcium leads to its deposition in shells and bones and the small remaining concentration ($c.$ 10^{-7} M) acts as an initiator of structural changes in proteins. The large difference between sodium and potassium concentrations within and outwith the cell causes a substantial electric potential across the cell membrane. If the membrane is reasonably permeable, as most non-bacterial membranes are, then it is possible for the organism to make use of the potentials for nerves, muscles and brains. The question that must now be answered is: what is the chemistry behind the biological selectivity and function?

9.2.1. Group IA ions

Group IA cations are characterised by high lability of complexes. In other words, a water molecule hydrating a sodium cation in solution is very rapidly replaced by another water molecule. The only ligands that can form stable complexes with these ions have many donor atoms in the same molecule – they are multidentate chelating ligands. The geometry of the ligand determines the relative position of the donor atoms and hence defines the space into which the complexed cation must fit. Selectivity arises because sodium forms a smaller monocation than potassium; the ionic radii are 0.95 and 1.33 Å respectively. These radii can be related to the number of ligands that each ion prefers to coordinate around it, the coordination number. For Na^+, six donor atoms are the best fit; the larger K^+ prefers eight. It is easy to demonstrate the effect of ionic radius with a series of synthetic ligands known as macrocyclic polyethers.

Macrocyclic polyethers, consisting of several oxygen atoms joined together in a ring through ether linkages, form relatively stable complexes with group IA metal cations (Figure 9.2). When complexed with a suitable ether, a sodium or potassium salt can become soluble in an organic solvent like toluene. It then becomes possible to use ionic reagents in non-polar solvents, a valuable aid to synthesis. Ether 9.2.1 has the smaller cavity and binds Na^+ ten times more strongly than K^+; 9.2.2, in contrast, has space for a larger ion – it chelates with K^+ but Na^+ slips through. Such observations emphasise the important generalisation that the stability of chelate complexes is highly sensitive to the size of the cation. It is probable that the sodium and potassium pumping systems that control ion balances in cells make use of these size differences to distinguish between the ions. As we shall see shortly, a similar argument applies to magnesium and calcium.

Figure 9.2.

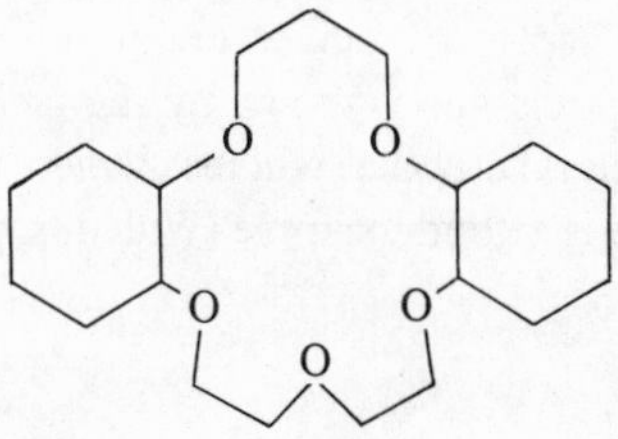

9.2.1 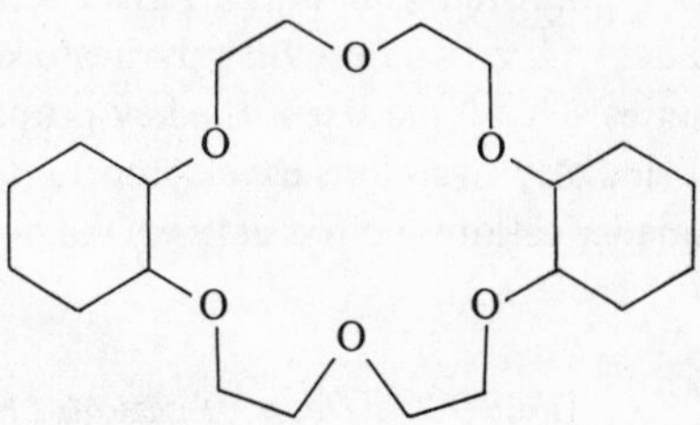   9.2.2

Interestingly, some antibiotics are macrocyclic chelates. The donor groups are ether oxygens or carbonyl groups, and the external surface of the ring is hydrophobic. Naturally occurring ligands that bind main group ions are known as ionophores. Thus valinomycin (9.2.3) binds potassium cations in preference to sodium; its antibiotic activity may well be due to interference with the ion balances of bacterial cells. Indeed, it has been found that valinomycin will transport potassium ions efficiently through bacterial cell wall membranes (Fenton, 1977).

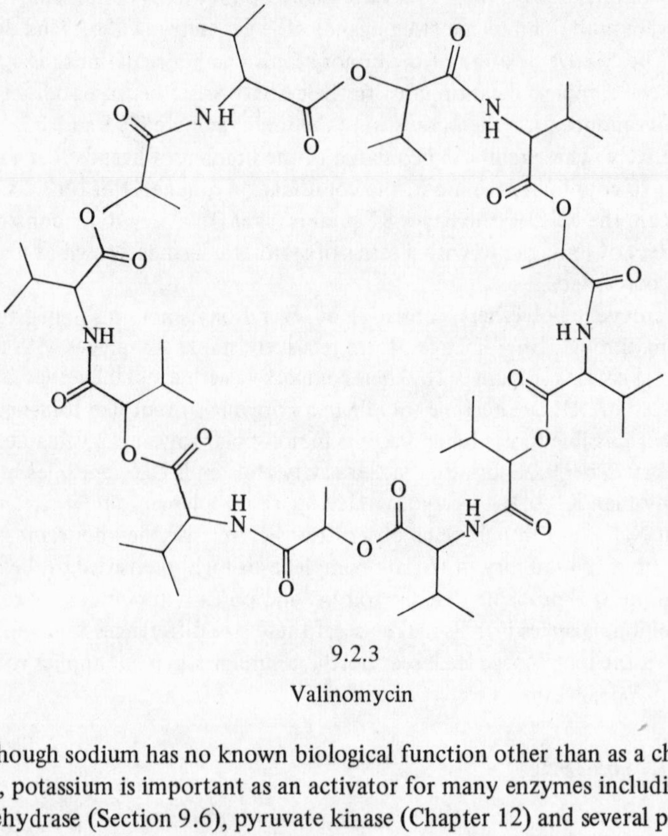

9.2.3

Valinomycin

Although sodium has no known biological function other than as a charge carrier, potassium is important as an activator for many enzymes including diol dehydrase (Section 9.6), pyruvate kinase (Chapter 12) and several phosphatases. That ionic size is the key property of potassium is shown by the fact that similarly sized ions have comparable biological activities, whereas larger or smaller cations are less active. Diol dehydrase is a good example (Williams, 1970; Table 9.3).

Table 9.3. *Effects of cations on activity of diol dehydrase*

Cation	Ionic radius (Å)	Relative activity
Li^+	0.6	0
Na^+	0.95	6
K^+	1.33	10
Tl^+	1.40	10
NH_4^+	1.45	10
Rb^+	1.48	10
Cs^+	1.69	6.5

If we want to study how potassium activates an enzyme, we run into a problem. Potassium has no readily observable spectroscopic properties, neither u.v. nor e.s.r. However, an obvious acceptable choice is to substitute one of the 'isomorphous' ions that give activity for potassium. For spectroscopic studies thallium is a good choice, firstly because it has an electronic excitation ($7s \to 7p$) that is measurable in the u.v. (between 215 and 250 nm, depending upon the complex) and secondly because the ^{205}Tl nucleus has a nuclear spin quantum number of $\frac{1}{2}$, which makes it suitable for study by nuclear magnetic resonance spectroscopy. The use of such probe ions is invaluable in bioinorganic studies provided that enough model chemistry is known. Some of the work using thallium in the study of pyruvate kinase is discussed in Chapter 12.

9.2.2. Group IIA ions

The divalent cations of group IIA form much more stable complexes than the monovalent group I elements; this is well shown by the stability of their EDTA complexes:

Ion	Na^+	K^+	Mg^{2+}	Ca^{2+}
log stability constant	1.7	0	8.9	10.7

The biological function of the group IIA cations takes advantage of the increase in stability of complexes: the structures of many biological chelating agents can be modified or controlled by the action of magnesium or calcium ions. Once again, the size of the ions increases as we descend the group. Mg^{2+} has an ionic radius of 0.65 Å compared with 0.99 Å for calcium, and this large difference provides the basis for biological selectivity between magnesium and calcium. It is difficult to pack large anions like carboxylate, phosphate or sulphate around a small magnesium cation. Magnesium complexes are accordingly less stable in competition with water ligands than the corresponding calcium compounds. For this reason, calcium salts are generally less soluble in aqueous solution and it is thought that this property, combined with their stability, has led to the selection of calcium carbonates and phosphates for shell and bone materials through evolution.

The study of the mechanisms of action of the group IIA cations in biological systems has again been advanced by means of the isomorphous replacement technique. Mn^{2+} (ionic radius 0.8 Å) has been widely used as a probe for magnesium in phosphate-transferring enzymes (Chapter 12), although the fit is not perfect. However, the Mn^{2+} ion (d^5) has the advantage of excellent spectroscopic properties because of its paramagnetism. Although Mn^{2+} might be judged a good probe for Ca^{2+} on the basis of its size, it has been found that the stability of manganese complexes more closely resembles those of magnesium than calcium, notably in the preference for neutral, basic donor ligands (amines, imidazole) over acid anions (sulphate, phosphate, carboxylate). Consequently, the rare earth cations, especially the lanthanides, which have ionic radii between 1.15 Å (La^{3+}) and 0.93 Å (Lu^{3+}), are used as substitutes for calcium.

Their steric requirements and coordination number (maximum 8) are the same as calcium, and these factors more than outweigh the effect of the larger charge. Biologically, these similarities are demonstrated by the ability of lanthanum(III) to block mitochondrial calcium transport, to inhibit bacterial nucleases and to compete for calcium sites of bone proteins.

One of the most subtle tasks which calcium ions undertake is the triggering of nerve impulses and muscle contraction. The ability of calcium to form reasonably stable complexes with phosphate and carboxylate ligands is the chemical basis for these functions. The transmission of an impulse from one brain cell to another and from nerve to muscle is a membrane phenomenon. Figure 9.3 is a schematic drawing of a **synapse**, that is the small space between two nerve cells. On the arrival of an impulse from the left, a transmitter substance is released from vesicles near to the presynaptic membrane. Acetyl choline is a common transmitter in the peripheral nervous system and the brain itself uses several amino acids and amino phenols (catecholamines) to communicate between cells. The sequence of events is summarised in the legend to Figure 9.3. Calcium ions trigger the events which cause the synaptic vesicles to rupture and expel the transmitter substance.

To anticipate the discussion of Chapter 14, the main components of membranes found in the nervous system are fatty acids and phosphate esters of fatty acid glycerates (phospholipids). The mechanism by which the protein receptors alter the conformation of such membranes is a mystery, as is the way in which calcium causes rupture of synaptic vesicles. The following analogy may point the way to the solution of these problems. Sodium salts of fatty acids are, of course, soaps, and the effect of calcium salts in water (hard water) upon soaps is well known: the calcium salts of the fatty acids precipitate as a scum. It may be that by binding to the charged ends of lipids in the membrane, calcium ions cause a temporary 'solidification' or rupture of the vesicle membrane, thereby allowing the contents to be expelled.

Many other biological events are triggered by calcium ions. Hormone release from an endocrine cell into the bloodstream can be considered in a similar way to the release of a neurotransmitter. The mechanism by which calcium ions trigger muscle contraction is somewhat better understood (Cohen, 1975). Muscle cells receive their cue to contract when acetylcholine is released by the nerve that stimulates the muscle. Acetylcholine binds to a receptor on the muscle cell and this causes calcium ions to be released within that cell. Ca^{2+} then binds to a small protein known as troponin that is attached to the long muscle fibre, the protein actin. When calcium binds, the conformation of troponin is changed. This in turn causes a further protein, tropomyosin, to prompt the muscle fibre to contract. When the nerve impulse ceases, pumping mechanisms sequester the calcium ions from troponin, tropomyosin returns to its resting position, and the muscle relaxes. It has also been established that proteins involved in blood coagulation (fibrinogen and factor-x) undergo a conformational change in the presence of Ca^{2+}.

9.3. Transition metal ions and complexes

The chief chemical characteristics that distinguish transition metal cations from main group cations are variable oxidation state, the ability to form very stable complexes and the existence of readily observable spectroscopic

Figure 9.3. (*a*) At rest both cells actively expel Na^+ and Ca^{2+}; K^+ is allowed in. (*b*) The membrane of the excited transmitting cell now permits Na^+ to diffuse in and depolarise it; Ca^{2+} also moves in and interacts with the membrane of the synaptic vesicles, causing the transmitter substance (T) to be released. (*c*) Transmitting cell returns to rest by pumping out Na^+ and Ca^{2+}; receiving cell is excited by the binding of the transmitter to the receptor.

(*a*) **Transmitting cell**   **Receiving cell**

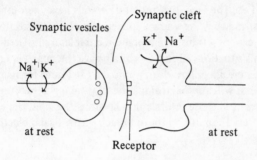

(*b*)

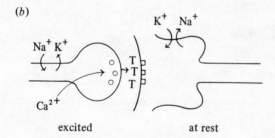

(*c*)

transitions. All of these properties are consequences of the behaviour of electrons in d-orbitals, which are the highest-energy occupied orbitals in transition metal cations. d-Orbitals (and also p-orbitals) are not spherically symmetrical like s-orbitals or a filled shell of electrons as found in main group cations, and consequently it is important to know which of the d-orbitals are of lowest energy if we are to understand bonding in transition metal compounds. The relative energies of d-orbitals are strongly dependent upon the stereochemistry of the complex. The stereochemical aspect of transition metal complexes is very important in metalloenzymes.

Figure 9.4 shows the shape of d-orbitals; two distinct classes can be defined. Firstly there are the two orbitals (d_{z^2} and $d_{x^2-y^2}$) whose lobes lie along the axes of the cartesian coordinates chosen: in group theory notation, these orbitals are called e_g. The second group of d-orbitals have their lobes lying between the axes, and they are known collectively as t_{2g}. Let us now consider the hypothetical case of a transition metal ion completely isolated and uncomplexed. In this situation, all five d-orbitals have the same energy; they are degenerate. If we now allow ligands to approach the ion to form an octahedral complex, the ligands will approach along the x-, y- and z-axes. Consequently, any electrons present in those orbitals that lie along the axes, the e_g orbitals, will experience a repulsion due to the approach of the ligand electrons. The

Figure 9.4.

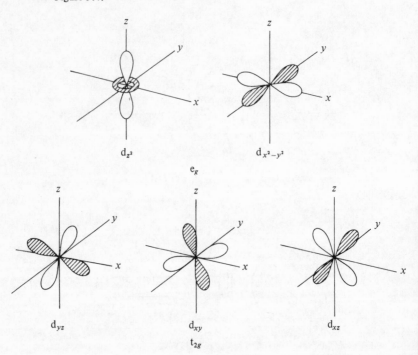

energy of the e_g orbitals will thus be raised with respect to the t_{2g} orbitals which point between the approaching ligands. We say that the d-orbitals are split by the ligand field (Figure 9.5). The magnitude of the splitting, Δ, depends upon the chemical character of the ligand: the better the donor, the greater the splitting is a crude but generally useful approximation. The ligands that produce the largest splittings, the so-called strong-field ligands, are those that are capable of σ-bonding between their p-orbitals and the metal d-orbitals. A list of ligands important in biology in approximate decreasing order of splitting ability follows: CO, CN$^-$, porphyrin, phenanthroline, ethylene diamine, ammonia, RS$^-$, H$_2$O, F$^-$, RCO$_2^-$, OH$^-$, Cl$^-$, Br$^-$, I$^-$. The first four of this series are strong-field ligands, ethylene diamine to fluoride are intermediate in strength and the remainder are weak-field ligands. This order is more or less independent of the metal ion.

Returning to the octahedral complex, if we remove the ligands that lie along the z-axis (often called the axial ligands) we obtain a square planar complex. The e_g pair of orbitals are no longer equivalent and are further split so that the $d_{x^2-y^2}$ becomes the highest-energy d-orbital. By a similar argument, it is easy to see that in a complex of tetrahedral symmetry, the t_{2g} orbitals point towards the ligands and become the higher-energy group of d-orbitals; this time, the lower-energy orbitals are e_g. The ligand field splitting for a tetrahedral complex is usually smaller than for an octahedral complex.

The characteristic electronic spectroscopic properties of transition metal complexes in the visible and near-i.r. regions of the spectrum chiefly reflect the energy of the lowest occupied orbitals of the complex; these orbitals can

Figure 9.5

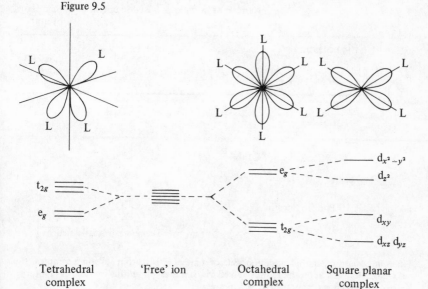

Tetrahedral complex   'Free' ion   Octahedral complex   Square planar complex

be identified with the aid of ligand field theory. Electronic and magnetic properties of transition metal ions provide useful information about the stereochemistry of the complexes. We shall use this in discussing biological examples shortly.

The above discussion is a very simplified picture of the important concepts of ligand-field and molecular-orbital theory of transition metal complexes and we now need to see how it helps us understand the chemistry that is observed. Iron complexes have been chosen as examples because they provide us with a useful background for consideration of a wide range of important biological complexes, the haemoproteins.

9.3.1. *Redox chemistry of some iron complexes*

Iron has two common oxidation states, +2 and +3, which have six and five d-electrons respectively.* Higher oxidation states are also known and are probably involved in biological oxidation reactions involving haemoproteins and molecular oxygen. Lower oxidation state compounds can also be prepared provided that π-bonding ligands are present. Normally, however, either iron(II) or iron(III) is found. The relative stability of the two oxidation states can be measured by the redox potential for a given complex; some pertinent data are given in Table 9.4.

Table 9.4. *Effects of ligands on redox potentials of iron complexes*

$Fe^{2+} \rightarrow Fe^{3+} + e^-$

Ligand	Number of ligands per iron	Redox potential (V)	Spin state of iron
1,10-phenanthroline	3	−1.10	Low
H_2O	6	−0.77	Low
CN^-	6	−0.36	Low
$EDTA^{4-}$	1	0.12	High

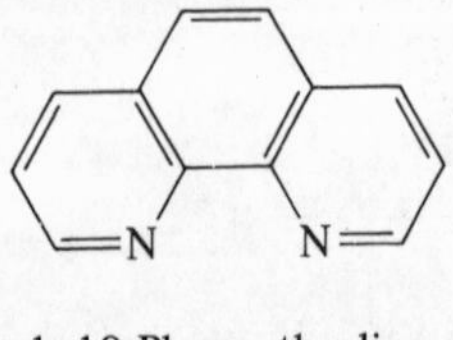

1, 10 Phenanthroline

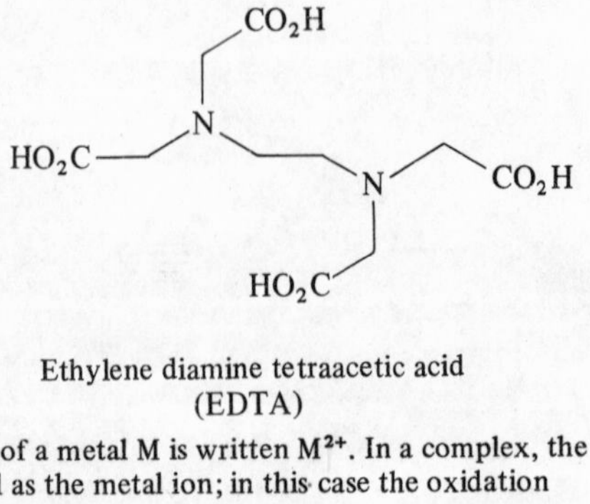

Ethylene diamine tetraacetic acid
(EDTA)

*In ionic equations, the +2 oxidation state of a metal M is written M^{2+}. In a complex, the net charge depends upon the ligands as well as the metal ion; in this case the oxidation state of the metal is represented as M^{II} or metal(II).

It is obvious from these data that the ligands have a very large influence upon the redox behaviour of the system, and their effects can be rationalised by considering the splitting of the d-orbitals and how they are occupied by the electrons. All the complexes in Table 9.4 are octahedral. Therefore the appropriate arrangement of d-orbitals is the e_g orbitals of higher energy than the t_{2g} orbitals (Figure 9.5). For each oxidation state, we can arrange the electrons in one of two ways – a low-spin state in which as many electrons as possible have paired spins, and a high-spin state in which the converse is true (Figure 9.6).

Figure 9.6.

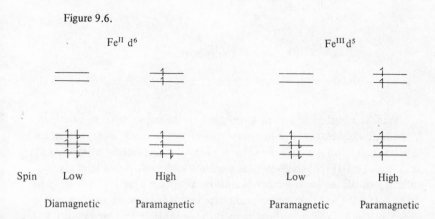

	Fe^{II} d^6		Fe^{III} d^5	
Spin	Low	High	Low	High
	Diamagnetic	Paramagnetic	Paramagnetic	Paramagnetic

At the foot of Table 9.4 is shown the weak-field ligand EDTA. The energy difference between the e_g and t_{2g} orbitals in this complex is small and it costs little energy to place electrons in the higher-energy e_g orbitals. In this and similar cases, the expenditure of energy in using e_g orbitals is less than the energy required to pair the spin of the electrons in the t_{2g} orbitals; the pairing of electrons in the same orbital always requires energy because of their mutual repulsion (Hund's rule). We thus obtain the complex with the greatest number of unpaired electrons possible, the so-called high-spin state (Cotton & Wilkinson, 1972).

EDTA is an anionic ligand and the redox potential tells us that under the standard conditions of the measurement, the iron(III) oxidation state is the more stable of the two. Since iron(III) is more highly charged than iron(II) it is not surprising to find that anionic ligands prefer to complex with the more positively charged cation. Both iron(II) and iron(III) EDTA complexes are high spin.

In contrast to EDTA, at the top of the table is the neutral ligand, 1,10-phenanthroline. Being uncharged, this ligand will not favour iron(III). But, because of its ability to π-bond, it stabilises the +2 oxidation state (Figure 9.7). π-Bonding occurs here by donation of the p-electrons of the ligand into the e_g metal orbitals which have, of course, compatible symmetry. Obviously this donation will be easier in the oxidation state that has empty e_g orbitals. In the case of iron, this criterion is best met by the iron(II) low-spin oxidation state. Maximum delocalisation of electrons takes place and we can generalise that iron(II) low spin is favoured by strong π-bonding ligands. Also, since the splitting of the d-orbitals is now relatively large, it is energetically preferable to pair spins of electrons in the t_{2g} orbitals than to place them in the higher-energy e_g level. Low-spin complexes are preferred for this reason also.

Figure 9.7.

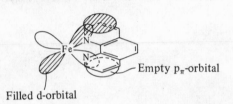

Filled d-orbital

Water is a neutral non-π-bonding ligand; cyanide is an anionic π-bonding ligand. Consequently the redox potentials of their complexes are determined by a combination of the factors that we have just discussed. Fe(II) high-spin and both Fe(III) spin states possess unpaired electrons. Accordingly, complexes with these electronic configurations are paramagnetic. Their magnetic moments and their electron paramagnetic resonance spectra are measurable (see Chapter 11). Both of these properties are valuable for the insight that they give into the reactivity and stereochemistry of metal complexes.

Similar arguments to the above can be used to rationalise the redox behaviour of many transition metal complexes. Try the following examples for yourself.

Problem 9.1.
The reaction

$$Cr^{2+}(H_2O)_6 \rightarrow Cr^{3+}(H_2O)_6 + e^-$$

has a redox potential of +0.41 V.

(1) Suggest a reason for the greater stability of chromium(III).
(2) How would the trisbipyridyl complexes be expected to behave in a similar redox reaction? Hint: first classify the ligand. (Cotton & Wilkinson, 1972, pp. 802, 834.)

2, 2'-Bipyridyl

Problem 9.2.

In contrast to iron, π-bonding stabilises the cobalt(III) oxidation state as shown by the data in Table 9.5.

Table 9.5. *Effects of ligands on redox potentials of cobalt complexes*

$Co^{2+} \rightarrow Co^{3+} + e^-$			
Ligand	Redox potential (V)	Spin states	
		II	III
$(H_2O)_6$	-1.84	High	Low
$(NH_3)_6$	-0.10	High	Low
$(CN)_6$	0.80	Low	Low

Explain this phenomenon, and suggest an explanation for the spin states that are observed. Begin by looking at the electron configurations of the four possible spin states. Which of these states are paramagnetic? (Cotton & Wilkinson, 1972, pp. 802, 875.)

9.4. Biologically important iron complexes

Two major functions of iron complexes in biology take advantage of the ability of iron to transfer electrons and thus to undergo redox reactions. Such complexes are made up of iron and protein ligands or porphyrin ligands which are bound in some way to a protein. The protein–porphyrin iron complexes, the haemoproteins, have a wide range of important biological functions. The redox proteins of the electron transport chain, cytochromes a, b, and c, shuttle between the iron(II) and iron(III) oxidation states (Figure 9.8). Cytochrome P 450 is an oxidase and a very important enzyme in drug metabolism (Chapter 15); it converts molecular oxygen into a highly reactive species that can hydroxylate organic molecules. To do this, cytochrome P 450 makes use of several oxidation states of iron. Other haemoproteins are oxygen-transporting compounds. The two major compounds, haemoglobin and myoglobin, are active in the iron(II) oxidation state only, and their structures have evolved to inhibit oxidation to the iron(III) state. In addition to haemoproteins, there

Figure 9.8.

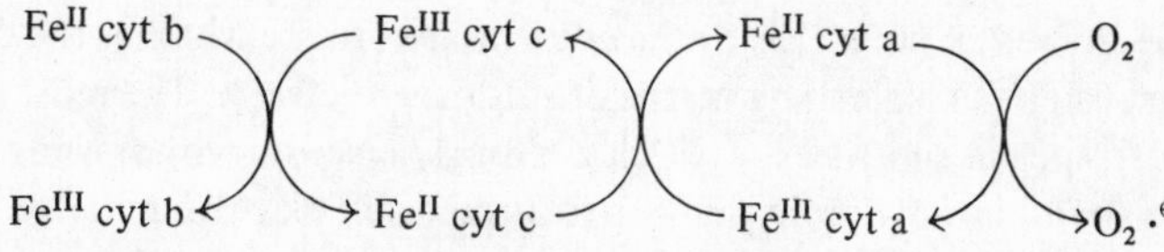

Fe^{II} cyt b — Fe^{III} cyt c — Fe^{II} cyt a — O_2

Fe^{III} cyt b — Fe^{II} cyt c — Fe^{III} cyt a — $O_2 \cdot^{\ominus}$

exists another group of iron-containing redox proteins, the ferredoxins which are especially important in oxidase enzyme complexes.

Obviously it is important to understand why these different iron complexes show their distinctive redox behaviour. This question takes on an added force when it is realised that haemoglobin, myoglobin and all the cytochromes have iron complexed by porphyrins. The most reasonable way to approach this problem is to simplify the system by considering initially how the structure of the porphyrin influences the redox properties of iron and secondly what effect the protein has.

Figure 9.9.

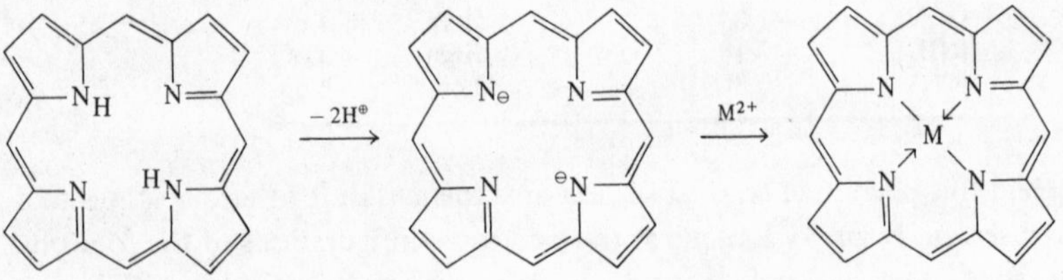

9.4.1. *Redox properties of metalloporphyrins*

The porphyrin ring is a square planar tetradentate ligand (Furhop, 1974; Williams & Moore, 1977; Figure 9.9). It forms complexes through its dianion and, because of its unsaturated structure, a porphyrin ligand can also undergo $p_\pi d_\pi$-bonding with a metal ion. We might reasonably expect that the stability of the various oxidation states of metalloporphyrins will be sensitive to the ability of a substituted porphyrin to donate electrons through its anion (σ-bonding) or to delocalise them through π-bonding. It is easy to study these properties using the readily available *meso*-tetraphenyl porphyrins as models for the naturally occurring compounds. Some pertinent data are shown in Figure 9.10 (Kadish *et al.*, 1976). From these figures we can see that the more electron withdrawing the ligand, the more stable the iron(II) oxidation state is. The magnitude of the effect here is smaller than in the purely inorganic complexes that we looked at earlier (Section 9.3) but this is not surprising when the substituent in the benzene ring is remote from the site of the redox reaction.

The porphyrins found in **cytochromes a, b, and c** are not the same – they differ in the substitution pattern of the ring (Figure 9.11). We must now examine the redox potentials of the cytochromes to see whether the differences observed can be explained by means of substituent effects. The most significant difference in substituents is that a strongly electron-withdrawing formyl group is found in the porphyrin of cytochrome a in place of a methyl group in the other two. Much smaller effects would be expected to be caused by changing a vinyl (cytochrome b) for a thio- or hydroxyethyl group (cytochromes c and a). It is probable that the formyl group in haem a makes a contribution to stabilising the iron(II) state in cytochrome a, just as the cyano

Figure 9.10.

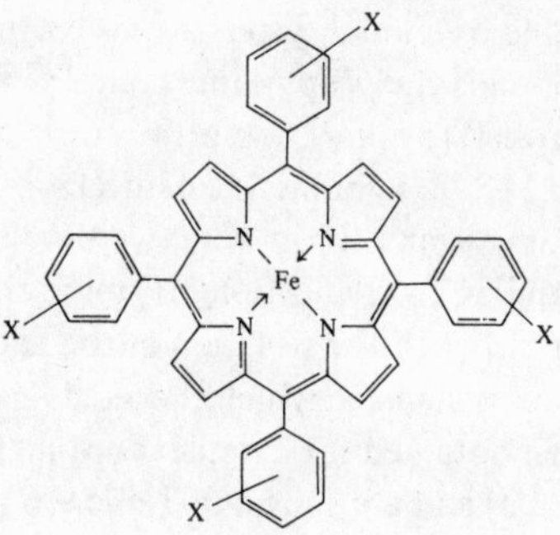

$$[Fe^{II}] \rightleftharpoons [Fe^{III}]^{\oplus} + e^{\ominus}$$

X m CH$_3$ H p Cl m Cl p CN

⟶ Increasing electron-withdrawing ability

$E^{\frac{1}{2}}$ 0.31 V   0.30 V   0.25 V   0.20 V   0.17 V

⟶ Increasing stability of Fe^{II} state

group did in the tetraphenylporphyrin model compounds. In contrast, there must be another factor responsible for the difference between cytochromes b and c, since substituent effects are small. It is the protein that brings about the change, both through further ligation of the iron in the axial positions and through the gross chemical environment that it provides for the haem.

Figure 9.11.

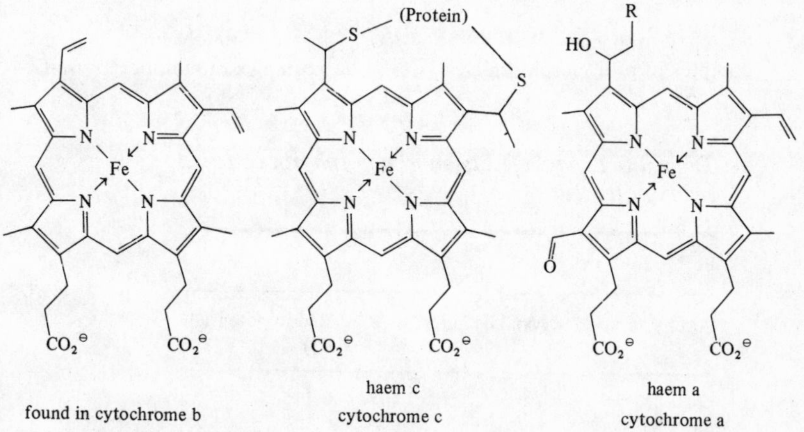

found in cytochrome b

haem c
cytochrome c

haem a
cytochrome a
R=prenyl side chain

E' -0.02 V -0.25 V -0.29 V

Typical redox potentials (Fe^{II} → Fe^{III} + e$^-$)

9.4.2. The effect of the protein on the properties of the metalloporphyrin

Both the effect of axial ligands and of haem environment on the redox potential are easily demonstrated by model systems. For example in the case of protoporphyrin IX compounds the *bis*-pyridine complex has $E_0' = 0.015$ V for the oxidation of iron(II) at pH 9.6, but the corresponding figure for the *bis*-cyano complex is 0.183 V. Pyridine is a neutral π-bonding ligand and cyanide an anionic π-bonding ligand. From what we have already seen, we would expect the cyano complex to favour iron(III) more than the pyridine complex. We can get a little closer to the situation in the naturally occurring cytochromes by using model systems in which the axial ligands are those found in nature. Data have been obtained for complexes of mesohaem and protohaem derivatives (Figure 9.12) and are shown in Table 9.6 (Kassner, 1972).

Figure 9.12.

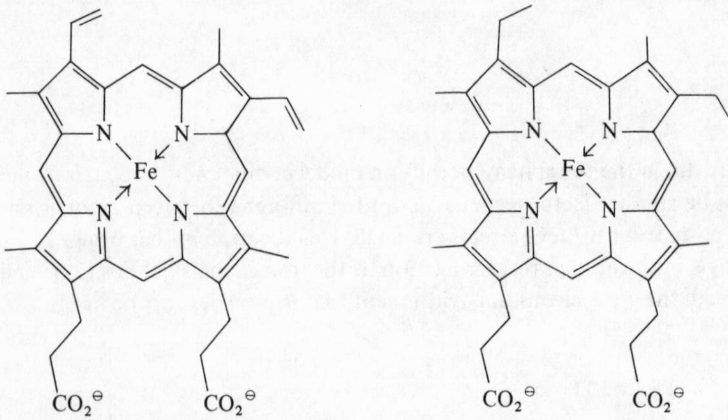

Protohaem
Fe^{II} complex of protoporphyrin-IX

Mesohaem
Fe^{II} complex of mesoporphyrin-IX

Table 9.6. *Ligand effects on iron-porphyrin complex redox potentials*

$Fe^{2+} \rightarrow Fe^{3+} + e^-$

Porphyrin	Axial ligands	Redox potential (V)
proto IX	–	0.115
meso III	–	0.158
meso III	His$_2$	0.22
meso III	His Met	0.11
meso III	Met$_2$	−0.02

The major effect that can be seen from these figures is due to the different character of the two axial ligands. The imidazole ring of histidine is a good σ-donor of electrons, and would thus be expected to favour the higher oxidation state. The sulphur of methionine (a thioether) is also an effective donor but in addition possesses the ability to accept electrons through π-bonding with its empty 3d-orbitals. Therefore this ligand should favour the lower oxidation state. It is indeed found that cytochromes with a methionine ligand have more negative redox potentials (favouring the iron(II) state) than those in which both the axial ligands are histidines. Cytochrome c is an example of the former class and cytochrome b_5 an example of the latter.

The remaining influence on the redox potential that we must consider is the environment that the protein provides for the haem. In the cytochromes it has been suggested that small differences in bond lengths of axial ligands can account for the various redox potentials observed in cytochromes with the same ligands (Williams & Moore, 1977). Of greater significance, however, is the fact that the oxygen-carrying haemoproteins **haemoglobin and myoglobin** do not undergo ready oxidation to the iron(III) state in which they are biologically inactive, despite the presence of oxygen itself in the complex. Unoxygenated haemoglobin has the unusual square pyramidal stereochemistry (Figure 9.13) and is a high-spin, paramagnetic complex. This stereochemistry is unknown in complexes in solution because high-spin tetracoordinated or hexacoordinated complexes are thermodynamically more stable. When oxygen binds to a haemoprotein, a hexacoordinate, low-spin, diamagnetic iron(II) complex is formed (Figure 9.13; Pratt, 1975).

Figure 9.13.

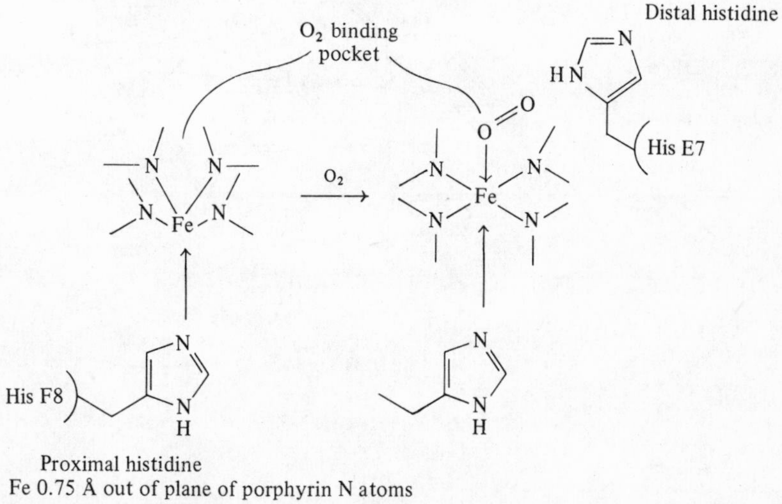
Distal histidine
O_2 binding pocket
His E7
O_2
His F8
Proximal histidine
Fe 0.75 Å out of plane of porphyrin N atoms

There are thus two major unusual features of the coordination chemistry of haemoglobin: firstly, the pentacoordinate iron(II) complex, and secondly the stability of the oxygen adduct to oxidation. The protein in which the porphyrin is bound makes both possible. The precise structure of the protein favours the pentacoordinate geometry and makes available a vacant site for oxygen (and other small ligands such as CO, N_3^-, and CN^-) to complex with iron. The other axial ligand is the imidazole ring of the so-called proximal histidine. Apart from a second (distal) histidine, too far away from the iron to coordinate with it, the remainder of the immediate environment of the iron and oxygen in oxyhaemoglobin is composed of non-polar amino acids (Figure 9.14), as X-ray crystallography has shown (Perutz, 1978). This non-polar environment inhibits the oxidation of oxyhaemoglobin through several mechanisms, two of which we shall consider.

Figure 9.14. V, $-CH=CH_2$; P, $-CH_2.CH_2.CO_2H$; a, ligand binding site; b, proximal histidine; c, distal histidine; $---$, approximate boundary of protein chains. The named amino acids all fall within Van der Waals' contact distance with the haem, and are all in their approximate binding positions. All, except for threonine (Thr), are non-polar.

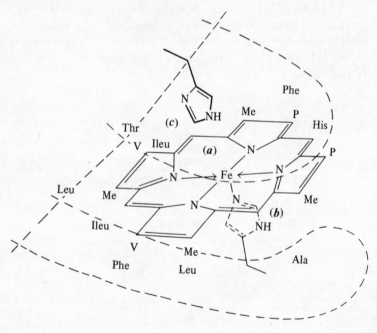

Metal–oxygen complexes readily decompose in the presence of acid catalysts (see cytochrome P 450 below) or by dimerisation (Figure 9.15). The protein excludes acid catalysts from the vicinity of the bound oxygen by closing the approach route with the distal histidine, which hydrogen bonds to the oxygen (Figure 9.13). In doing so, dimerisation is prevented also. The second effect of the non-polar environment concerns the redox potential of the iron. It has been found that the polarity of solvent greatly affects the redox potential of porphyrin iron complexes (Kassner, 1972). The *bis*-pyridine mesoporphyrin iron complex has a redox potential of −0.399 V in benzene (non-polar solution), which compares with a value of −0.096 V in aqueous solution. This shows that the lower oxidation state is substantially stabilised by the apolar environment. Presumably a factor in this behaviour is the unfavourability of generating a positively charged iron(III) complex in a non-polar solvent.

Figure 9.15.

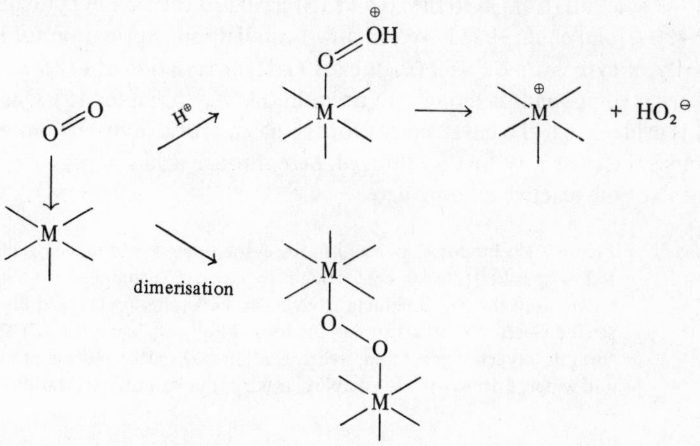

In marked contrast to haemoglobin in which oxidation reactions are inhibited, **cytochrome P 450** is a haemoprotein that has evolved to generate a highly reactive hydroxylating species from molecular oxygen.The porphyrin is the same in each case and it is therefore the protein that is responsible for the antithetic behaviour of the two. Figure 9.16 shows the operative components of the adrenal steroid hydroxylating cytochrome P 450 system. The hydroxylating species is generated from the iron(II) state of the cytochrome, which is formed by reduction of the iron(III) oxidation state. Reducing power is supplied by a series of proteins and cofactors that make up an electron-transport

Figure 9.16. The cytochrome P 450 system of adrenal steroidogenic tissue.

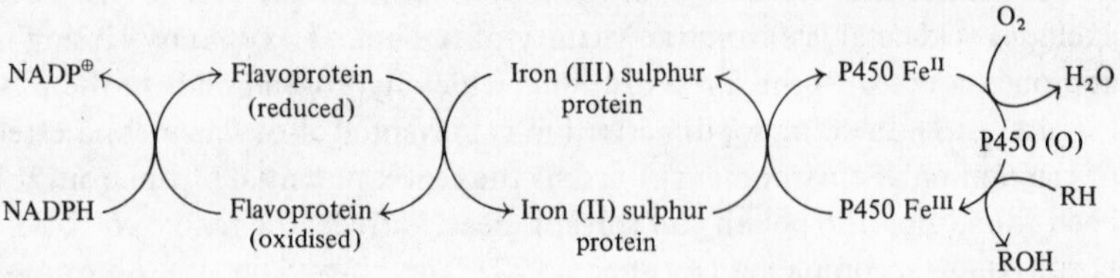

NADP$^\oplus$ $\rightarrow$ Flavoprotein (reduced) $\rightarrow$ Iron (III) sulphur protein $\rightarrow$ P450 Fe^{II} $\rightarrow$ O$_2$ $\rightarrow$ H$_2$O

NADPH $\rightarrow$ Flavoprotein (oxidised) $\rightarrow$ Iron (II) sulphur protein $\rightarrow$ P450 Fe^{III} $\rightarrow$ P450 (O) $\rightarrow$ RH, ROH

chain (see Chapter 8). We shall have more to say about the iron–sulphur protein, ferredoxin, shortly. Most cytochromes P 450 are membrane-bound enzymes that are difficult to isolate in pure form. Consequently, few crystalline derivatives are available for X-ray analysis. It is therefore necessary to investigate the chemistry of oxygen activation by careful comparison of the electronic and electron paramagnetic spectra of the natural cytochrome with carefully chosen model compounds. The results of a great deal of work suggest that a thiolate anion (from cysteine) is an axial ligand of the iron in cytochrome P 450 (Holm *et al.*, 1975). We can now formulate an explanation for the reactivity of cytochrome P 450 (Figure 9.17). Hydroxylation of drugs and other foreign compounds is thought to occur in this way (Chapter 15). Once again, it is evident that the environment for the haem provided by the protein determines the chemistry that is observed, here through acid catalysis and stabilisation of the reactive intermediate.

Figure 9.17. Proposed mechanism for cytochrome P 450 (*a*) Substance S binds to low-spin Fe^{III} P 450, converting it to a high-spin complex. (*b*) One electron enters from the P 450 reductase, giving an Fe^{II} state. (*c*) Oxygen binds and a second electron enters, forming the formal Fe^{II}–O$_2^-$ complex. (*d*) The active form of oxygen is generated, hydroxylation takes place and the products, SOH and water, diffuse off the complex, leaving it in its initial oxidation state.

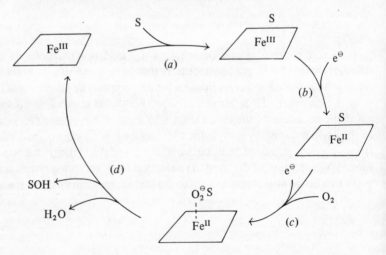

Problem 9.3.

Some mutant haemoglobins have tyrosine substituted for the proximal histidine. Thus the phenolic oxygen becomes the fifth ligand for the porphyrin iron complex. If tyrosine binds through its anion, what effect on reversible oxygen binding would you expect? Hint: compare the chemical character of imidazole and phenolate as ligands, then deduce the effect of the change upon the redox properties of the iron. (Pratt, 1975, p. 128).

9.4.3. *Iron–sulphur proteins*

To complete our survey of biologically important iron complexes, we should mention iron–sulphur proteins. These are low molecular weight electron-transfer proteins that are associated with oxidation/reduction systems in photosynthesis, cytochrome P 450 enzymes, nitrogen fixation and others. Iron-transferring proteins ferritin and transferrin are also iron–sulphur proteins. Plant and bacterial electron-transferring iron–sulphur proteins are known as ferredoxins, and similar proteins from mammalian adrenal glands are called adrenodoxins. The amino acid composition of these proteins indicates that they evolved very early in the development of life. The ferredoxins themselves have low molecular weights (*c.* 11 000) and contain iron bonded to sulphur ligands. Both sulphide ions and cysteine residues from the protein donate to iron, and there is an equivalent amount of iron and non-protein sulphur. The small size of these proteins has permitted the determination of some of the most accurate X-ray structures for macromolecules: resolution of 1.5 Å has been obtained (Coleman, 1974). The results show that iron is part of a cluster in which it is approximately tetrahedrally coordinated by four sulphur atoms (Figure 9.18). A range of redox potentials has been found for the ferredoxins (−0.3 to −0.5 V) but it is not yet clear what the cause of the variable redox potential is. Neither is the electron-transfer mechanism properly understood, although there are a number of synthetic clusters of similar properties to the ferredoxins (Coleman, 1974).

Figure 9.18.

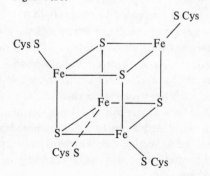

9.5.   **Transition metal ions as acid catalysts in enzyme-catalysed reactions – carboxypeptidase**

The notion that any electron-accepting species can be acidic is familiar to chemists in Lewis's definition of acids and bases. Synthetic chemists frequently use boron, aluminium, tin and zinc compounds, usually as their anhydrous halides, in reactions that require the generation of an electrophile (Chapter 7). Where a Lewis acid is required, nature chooses zinc. The example of alkaline phosphatase has already been cited (Chapter 3), and zinc cations are also important as acid catalysts in the enzymes carboxypeptidase and carbonic anhydrase. In this section we want to look at some of the ways in which zinc-containing enzymes have been studied, and especially to bring out the importance of coordination stereochemistry in transition metal complexes and metalloenzymes.

9.5.1.   The zinc cation and its substitute, cobalt

Zinc has an outer shell of $3d^{10}$- and $4s^2$-electrons and behaves as a very electropositive metal, losing its two 4s-electrons readily to form a dication. The dication has filled d-orbitals. Consequently only the zinc(II) oxidation state is important, and redox reactions do not occur readily. Nevertheless, the 4s- and 4p-levels are available to accept electrons, thereby causing many anhydrous zinc compounds to be Lewis acids.

The zinc dication is small (radius 0.74 Å), similar to lithium (0.60 Å) and magnesium (0.65 Å). Like these elements, four coordinate complexes are usual, the ligands taking up an approximately tetrahedral geometry, for example $Zn(CN)_4^{2-}$. In contrast, the larger d^{10}-cations cadmium and mercury can show hexacoordination, although linear and tetrahedral compounds of each are known. In none of these ions with full d-shells do ligand field effects influence the coordination geometry.

Unfortunately for the study of zinc proteins, zinc(II) has no useful spectroscopic properties because there are no unpaired electrons and there are no d–d transitions possible. It is therefore necessary to find a suitable substitute probe transition metal ion and cobalt(II) has proved ideal (Lindskog, 1970). There has been intensive study of many zinc proteins in which zinc has been replaced by cobalt. The ionic radius of the cobalt dication is 0.74 Å, the same as zinc. A very wide range of complexes of cobalt are known and the spectroscopic properties are characteristic of the stereochemistry of the complex (Figure 9.19). Although cobalt(II) prefers octahedral coordination with all but the weakest field ligands, tetrahedral compounds similar to zinc derivatives have been prepared. If the ligand itself has strong stereochemical demands, cobalt's preferences can be set aside and square pyramidal and square planar complexes, can be formed. Such cobalt derivatives have importance also in vitamin B_{12} chemistry, as we shall see in Section 9.6. With background information such as this it is possible to study the coordination chemistry of zinc enzymes using cobalt(II) ions as probes.

9.5.2. Carboxypeptidase A

Carboxypeptidase A is formed from its precursor protein by trypsin-catalysed hydrolysis in the stomach (Chapter 13). It has a molecular weight of 34 000 and contains one zinc cation per molecule of enzyme. The natural substrates of carboxypeptidase are proteins and peptides, both of which are hydrolysed sequentially, one amino acid residue at a time, from the C-terminus. A great deal is known about its mechanism of action from X-ray crystallographic studies (Lipscomb & Quiocho, 1971) and also from chemical studies (Dunn, 1975). It is interesting to compare and correlate the results of both approaches.

The coordination of zinc

If a large enough series of different metal complexes is studied, trends in the measured stability constants allow suggestions to be made by extrapolation about the ligands that bind zinc in carboxypeptidase. Not only cobalt, but also nickel, manganese, cadmium, lead, mercury and copper derivatives of

Figure 9.19. Some cobalt complexes and their spectroscopic properties.

	λ_{max} (nm)	ϵ	
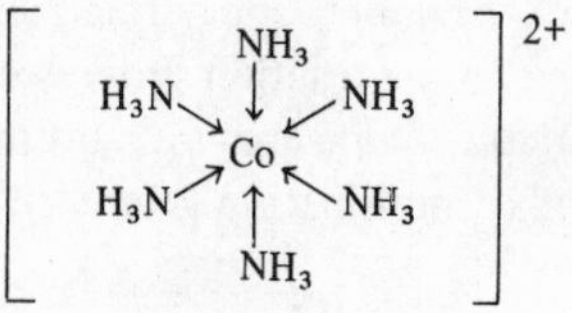 $[Co(NH_3)_6]^{2+}$	490	9	in H_2O
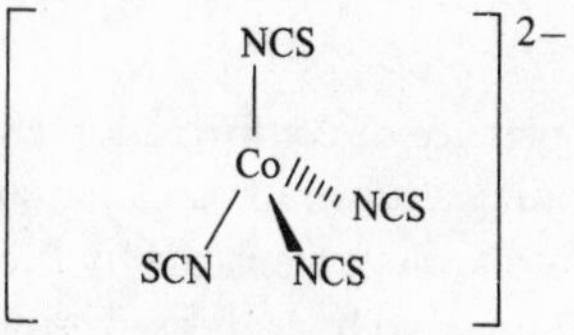 $[Co(NCS)_4]^{2-}$	620	1800	in acetone
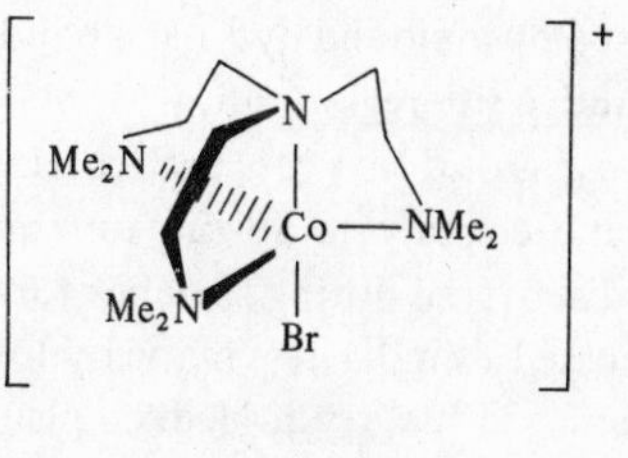 complex$^+$	830	35	
	650	120	in CH_2Cl_2
	500	110	
Co carboxypeptidase	572	160	
	555	160	in H_2O
	500	shoulder	

carboxypeptidase are known, and the stability constants of this series were found to parallel the known stability of the corresponding 2-mercapto-ethylamine complexes. Furthermore, reagents that readily react with thiol groups inhibit carboxypeptidase. It was therefore suggested that nitrogen and sulphur ligands bind zinc in carboxypeptidase. However, when the electronic spectrum of the cobalt enzyme was compared with typical cobalt complexes' spectra (Figure 9.19), no obvious correlation was found. This result implies that the zinc in the enzyme is bound in a very low-symmetry environment, and it came as no surprise when the X-ray studies showed the stereochemistry of the coordinated zinc to be a highly distorted tetrahedron with bond angles about the zinc from 143° to 86°. In contrast to this result, it was surprising to find that the ligands were two histidines, glutamic acid and a water molecule. No sulphur ligands were in range of zinc. It is probable that the discrepancy between chemical and X-ray results can be resolved if carboxypeptidase adopts a different conformation when complexed with the heavy metal cations from the normal in zinc and cobalt enzymes. Heavy metal ions (lead and mercury) are classified as soft cations and form their strongest complexes with soft ligands like sulphur. Much less stable complexes are formed with harder ligands like water and carboxylate. The risks in using chemical data to interpret enzyme mechanisms and structure are obvious. There is no way of telling when the protein is using an alternative dimension of behaviour to the normal, no way of detecting an abnormal conformation change during reaction. Interestingly, irregular coordination seems to be a characteristic of zinc enzymes; both carbonic anhydrase and thermolysin have similar coordination geometries to carboxypeptidase.

The mechanism of peptide bond hydrolysis

X-ray structure studies have revealed the importance of conformation changes in the chemistry of carboxypeptidase. On binding a substrate, a major conformational adjustment takes place – the water molecule is displaced by the carbonyl group of the peptide bond that is to undergo hydrolysis. The enzyme-substrate complex so formed is greatly stabilised by ionic and hydrogen bonding with other neighbouring amino acids in the active site region, but it is the interaction of the zinc cation with the carbonyl group that gets the reaction going. There are many chemical reactions that are catalysed by zinc ions acting as Lewis acids, and most of these have been discovered during searches for model systems for carboxypeptidase. We discussed a similar case in the hydrolysis of phosphates in Chapter 3. With this as an example, try to deduce plausible mechanistic schemes consistent with the following data.

Problem 9.4

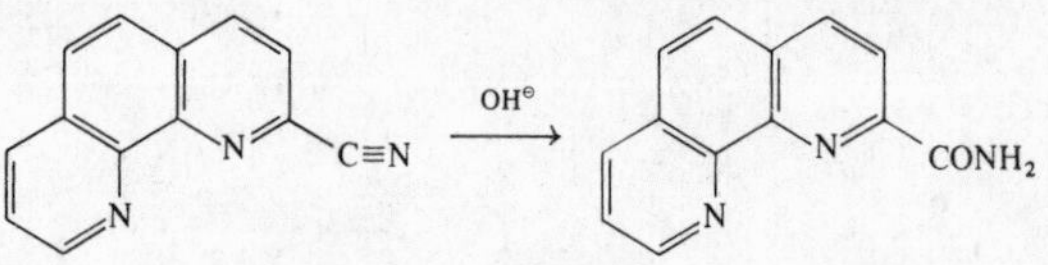

Rate $\propto [HO^-]$; rate greatly accelerated by metal dications (Zn^{2+}, Cu^{2+}, Ni^{2+}); when the concentration of metal cation equals the substrate concentration the reaction is first order in both. Hint: What ligands are available for the metal in the ground state of the reaction?

(1) Can all of these ligands coordinate with the metal at once?

(2) Suggest a geometry close to the transition state for the hydrolysis reaction and ask the same questions about the transition state. Remember that catalysis can be considered to occur by stabilising the transition state. (Breslow & Schmir, 1971.)

9.6. Coenzyme B_{12} – bioorganometallic chemistry

The B_{12} coenzymes are derivatives of cobalamin, which, in the form of its cyano derivative, is known as vitamin B_{12} (Figure 9.20). The term cobalamin applies to the whole of the complex ligand, the planar corrin ring and the benzimidazole nucleotide. Corrins are biosynthetically derived from porphobilinogen and are thus relatives of porphyrins; benzimidazole is related biosynthetically to riboflavin, one of the vitamins we met in the last chapter. The biologically reactive centre of B_{12} coenzymes is the complexed cobalt(II) ion.

Figure 9.20. Coenzyme B_{12}.

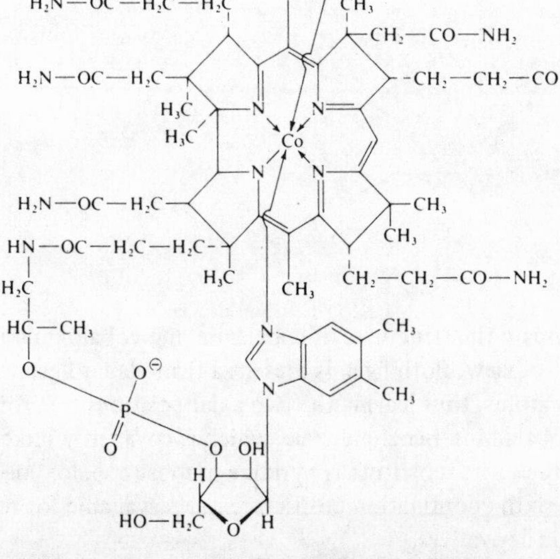

Coenzyme B_{12}

Figure 9.20. cont.

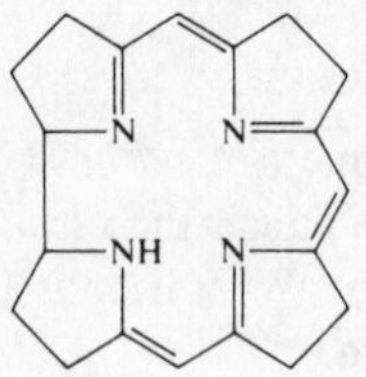

The corrin

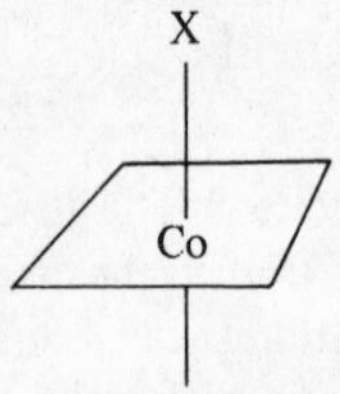

X

Abbreviated structure
for coenzyme B_{12}

Several types of enzymic reactions are mediated by B_{12} coenzymes through their different forms. Firstly, methyl cobalamins are involved in methane, methionine and acetate synthesis, in other words in one-carbon metabolism. A second class of reactions involves 5'-deoxyadenosine as the sixth ligand. 5'-Deoxyadenosyl cobalamin carries out a number of double 1,2-rearrangements, typical of which is the conversion of propane-1,2-diol into propanal by the enzyme diol dehydrase.

In both of the above reaction classes, intermediates with cobalt–carbon bonds are formed, hence the description of coenzyme B_{12} chemistry as bio-organometallic chemistry. To appreciate the behaviour of the coenzymes it is obviously important to understand the chemistry of simpler cobalt organometallic compounds. When the complex structure of vitamin B_{12} became clear from X-ray crystallography, it was apparent at once that mechanistic studies with this molecule would be both expensive and difficult. A suitable model compound was sought. Cobalt *bis*-dimethylglyoxime derivatives, known as cobaloximes, are easy to prepare, and much of the current understanding of B_{12} chemistry derives from model studies with this system (Schrauzer, 1977; Figure 9.21).

Figure 9.21.

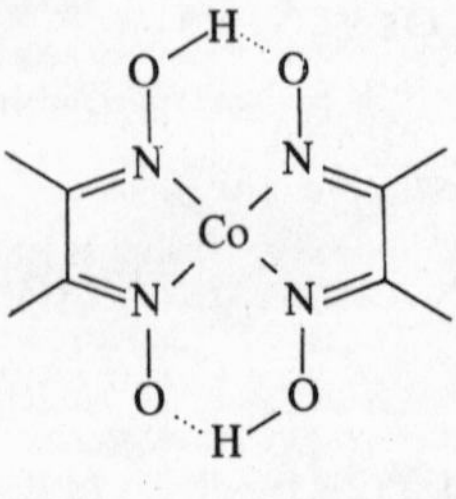

Let us compare the structures of cobalamin and cobaloxime from the cobalt's point of view. Both ligands present a firm planar ligand field through four nitrogen atoms, thus leaving the two axial positions free for further ligation. In cobalamins, benzimidazole, which is covalently linked to the corrin, takes one of these: its substitute is pyridine in most cobaloxime-based model systems. The sixth coordination position remains available for reaction via organometallic derivatives.

What possible modes of reaction are open to a cobalt–carbon bond in this environment? There are three possibilities, all of which can be demonstrated using cobaloximes, and three oxidation states of cobalt are involved (Figure 9.22). If cobalamins or cobaloximes are reduced (for example with sodium borohydride under nitrogen) a cobalt(I) anion is generated (Figure 9.22, *top*). It is one of the most powerful nucleophiles known and will react with many positively charged or polarised species from protons to Michael acceptors (Figure 9.23). You will notice a formal similarity to Grignard reagents.

Figure 9.22.

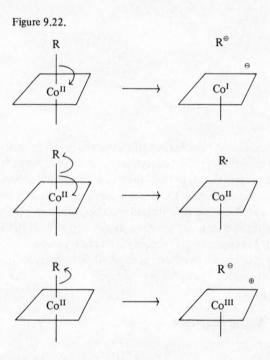

If an alkyl cobalamin or cobaloxime is photolysed, the second mode of reaction occurs. A paramagnetic d^7 cobalt(II) species and an organic radical are the products. The cobalt(II) compound can interact with suitable substrates to initiate rearrangement reactions, as we shall see shortly (Figure 9.22, *centre*). The third oxidation state of cobalt is attained by mild oxidation of the cobalt(II) compounds, thereby generating cobalt(III) derivatives which are cationic and reactive towards nucleophiles (Figure 9.22, *bottom*). Despite the known tendency for cobalt(III) compounds to be slow to undergo ligand exchange, cobaloximes and cobalamins in this oxidation state will react with Grignard reagents or other carbanions to form alkyl derivatives. Even with a knowledge of this background chemistry it has not been easy to find satisfactory mechanistic explanations for the enzyme-catalysed reactions.

Figure 9.23.

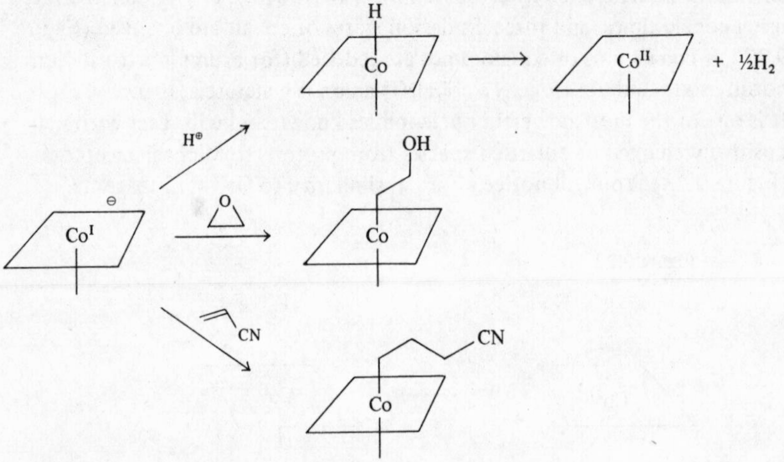

Let us firstly investigate the mechanism of C_1 metabolism. The other common C_1 donors in biology are S-adenosylmethionine and N-5-methyl tetrahydrofolic acid (Figure 9.24). Both of these contain methyl groups that are labile towards nucleophiles, and it is possible that they alkylate a cobalt(I) cobalamin through a nucleophilic substitution reaction. This now generates a cobalt(II) species. Both chemical and enzymic studies suggest that the methylation of thiols (such as homocysteine) occurs via a radical process – a reducing agent, as yet unknown in nature, is required. It is therefore inappropriate with the current state of knowledge to compare methyl cobalamin with methyl magnesium iodide.

Figure 9.24. Role of coenzyme B_{12} in C_1 transfers.

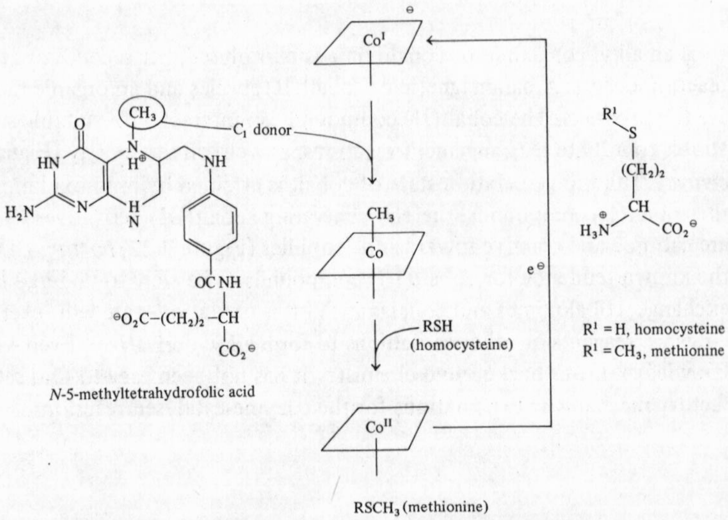

N-5-methyltetrahydrofolic acid

R^1 = H, homocysteine
R^1 = CH_3, methionine

$RSCH_3$ (methionine)

The second class of B_{12} mediated reactions, the rearrangements, are mechanistically amongst the most complex biological reactions known.

Diol dehydrase and its many relatives have been very thoroughly studied by isotopic labelling of the substrate. In this way it has been shown that the hydrogens that migrate do not exchange with the solvent, although they do interchange with the hydrogen atoms of the methylene group of 5'-deoxyadenosine (Figure 9.25.). ^{18}O studies have shown that oxygen transfer is also intramolecular with respect to the substrate. Any mechanism proposed must

Figure 9.25. A possible mechanism for diol dehydrase.

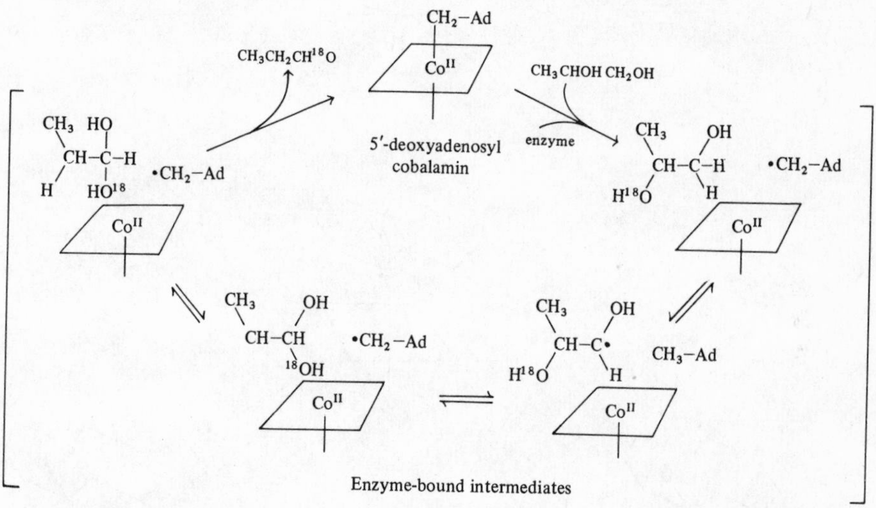

Enzyme-bound intermediates

therefore add no groups to the substrate or to the coenzyme, it may only rearrange the positions of the atoms already present.

Spectroscopic studies have shown that on binding to several of these isomerases in the presence of substrate, 5'-deoxyadenosyl cobalamin undergoes a dissociation to give an alkyl radical and the d^7 cobalt(II) coenzyme. The latter is easily characterised from its e.p.r. and electronic spectra. Together these results suggest a radical mechanism initiated by hydrogen abstraction from the substrate by the deoxyadenosyl radical. Rearrangement occurs in the immediate vicinity of the cobalt atom, which could well have the job of lowering the activation energy for radical rearrangement (Figure 9.25). Recently, a model reaction has been discovered that demonstrates such hydrogen migrations clearly (Breslow, 1976; Figure 9.26.). By alkylating a cobaloxime with cyclodecane, a group known to undergo facile intramolecular hydrogen shifts, it was possible to observe the rearrangement of hydrogen and cobalt by marking the starting position with deuterium. N.m.r. spectroscopy showed that the rate of hydrogen migration was the same as the rate of cobalt migration.

Figure 9.26. Hydrogen transfer in a coenzyme B_{12} model study.

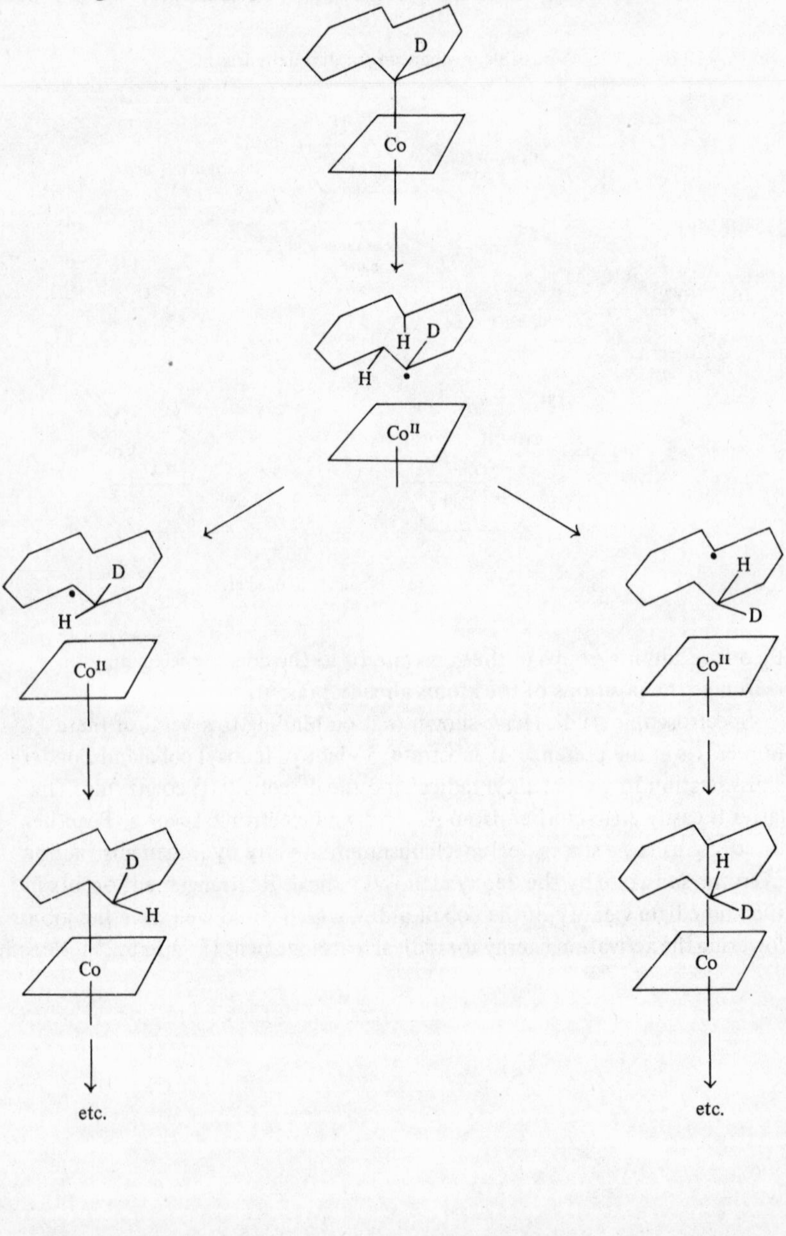

We have shown in the previous chapters that many parallels exist between enzyme-catalysed reactions and laboratory analogues. However, in general, the relationship between the two classes was formal; the similarity was limited to the gross outline of the reaction course and to general considerations of mechanism. Great differences were noted in the conditions required to carry out a reaction and, at a more subtle level, in details of reaction mechanism. In the laboratory, the simplest expedient to make a reaction progress at a reasonable rate is to heat it. In the constant-temperature environment of animals it is enzymes that keep metabolism moving. We have alluded to several mechanisms that enzymes employ in catalysis in discussing specific examples earlier, and it is the purpose of this chapter to weave together these threads into a continuous fabric.

A convenient approach to interpreting the chemical mechanisms of enzymic catalysis is by the equation for the rate constant of a reaction derived from transition state theory:

$$k = (k_B T/h) \exp(-\Delta G^{\ddagger}/RT)$$

where k_B is Bolzmann's constant and h is Planck's constant
Since $\Delta G^{\ddagger} = \Delta H^{\ddagger} - T\Delta S^{\ddagger}$,

$$k = (k_B T/h) \exp(-\Delta H^{\ddagger}/RT + \Delta S^{\ddagger}/R)$$

Thus the rate constant can be controlled by two terms, one an enthalpy term ($\Delta H^{\ddagger}$) reflecting the activation energy for the reaction, and the other an entropy term ($\Delta S^{\ddagger}$) which relates to the ease with which the reactants can come together. The equation shows us that a decrease in $\Delta H^{\ddagger}$ and an increase in $\Delta S^{\ddagger}$ will speed up the reaction. Both changes are used by enzymes. Wherever an enzyme stabilises a reactive intermediate, it is reducing the activation energy for reaction (the $\Delta H^{\ddagger}$ term). The very act of bringing the reactants together at the active site makes the entropy change for reaching the transition state more favourable: when they are close together, it is more probable that compounds will react. Detailed discussions of these ideas can be found in Jencks

(1969), but a more approachable exposition is provided by Bender & Brubacher (1973).

10.1. Catalysis by lowering activation energy

If a reaction involves a high-energy intermediate such as a carbanion, a carbenium ion or a radical, an effect that reduces the energy of the intermediate will almost always make the transition state for the formation of the intermediate more easily attained. This is the argument used to deduce which of a number of possible isomers will be preferred in an aromatic substitution reaction, for example. An analogous argument holds for catalysis by enzymes: if the enzyme can stabilise the intermediate in some way, the activation energy will be lowered and the reaction rate increased. There are several guises in which such effects are found and examples from the preceding chapters can be selected as illustrations. Many are cases of **covalent catalysis** in which the enzyme or coenzyme bonds directly with the substrate.

10.1.1. *Carbanion stabilisation*

Stabilising an anion requires electron-accepting groups. Enzyme active sites are not rich in such groups because many of the side-chains are either anions themselves or electron-rich aromatic rings. Cationic groups usually interact with anions such as carboxylate and phosphate anions, in other words, with stable anions. The problem for the enzyme, therefore, is to stabilise transient carbanions, yet it has no suitable functional groups to do the job. Coenzymes provide the solution. Coenzyme A, pyridoxal phosphate and thiamine pyrophosphate all make carbanions easier to form (Figure 4.21 and 4.24). In the case of coenzyme A, the thiol site is intrinsically more reactive than the oxygen analogue. The other two coenzymes stabilise anions by charge delocalisation in a similar way to carbonyl groups.

10.1.2. *Carbenium ion stabilisation*

Enzymes have at their disposal several groups capable of stabilising a positive charge. Potentially any nucleophile (O, S or N) will serve, but usually thiolate from cysteine is thought to be involved. For example, the biosynthesis of thymidilic acid from uridylic acid is thought to involve a stabilised Wheland intermediate (Chapter 7, Figure 10.1. Recent work suggests that the order of

Figure 10.1.

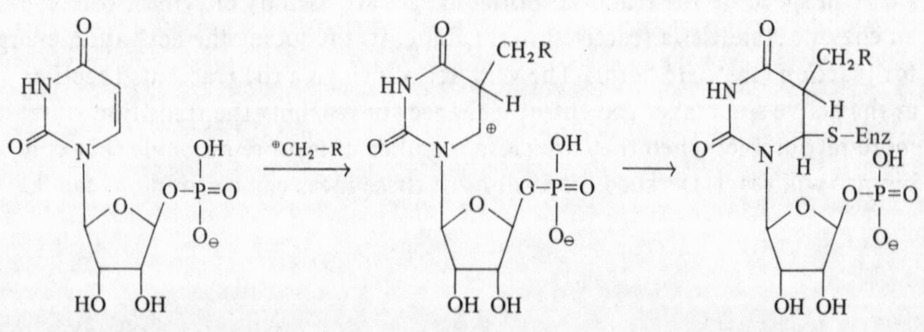

addition to uridylic acid is opposite to that shown in the figure.) The addition of sulphur nucleophiles to cations is often invoked as a rationale for reactions of isoprenoid compounds during biosynthesis (Chapter 5). Such reactions are stereospecific. In general, the sulphur that stabilises the cation can be regarded as preventing inversion of configuration though a planar state (Figure 10.2). In addition to being a powerful nucleophile, thiolates are also leaving groups, and can thus be removed on completion of the required transformation.

Figure 10.2.

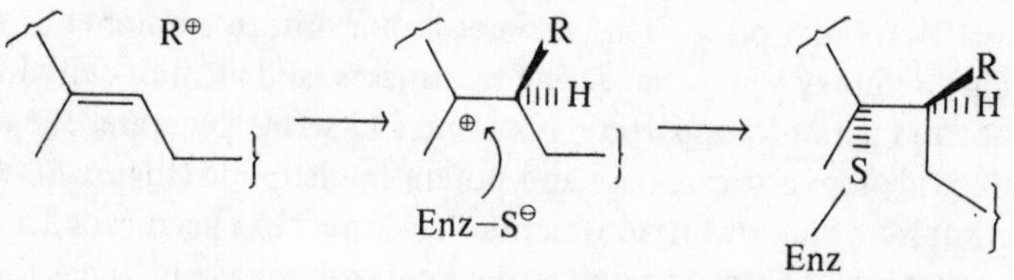

10.1.3. Activation of carbonyl groups towards nucleophilic attack

Although carbonyl groups are inherently susceptible to nucleophilic attack at carbon, amides and many esters are slow to react. One of the ways in which enzymes catalyse addition reactions to such carbonyl groups (e.g. hydrolysis) is to polarise the carbonyl group further. The usual effectors are protons (acid/base catalysis, Chapter 3) or metal cations (Lewis acid catalysis, carboxyl, Chapters 3 and 9). Figure 10.3 summarises the situation. In other cases, the ϵ-amino group of lysine is used to form an iminium cation, which, having a full positive charge, is more susceptible to nucleophilic attack than a bare carbonyl group (porphobilinogen synthesis, Chapter 4). Similar arguments can be applied to phosphate esters' reactions (Chapter 3) with the additional feature that the cation shields the negatively charged oxygen atoms from the incoming nucleophile.

Figure 10.3.

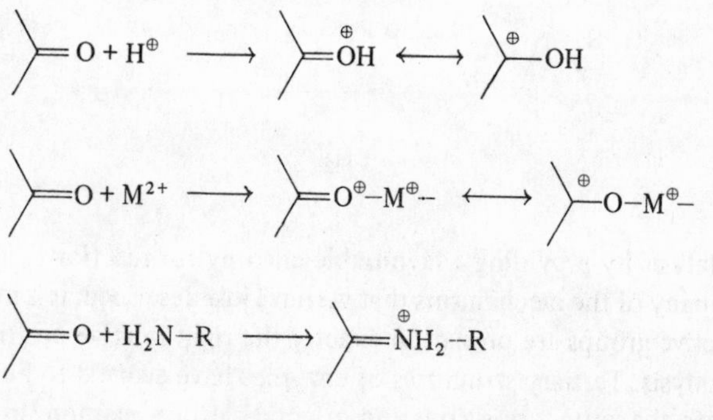

10.1.4. *Generation of reactive nucleophiles*

Although polarisation of a multiple bond provides powerful catalysis, a further acceleration can be obtained if nucleophiles more potent than water can be provided by the enzyme. Anions such as carboxylate or thiolate are readily available from amino acid side-chains (Asp, Glu and Cys); the pK_a of the conjugate acids is sufficiently low to protonate water to a significant extent. On the other hand, an alkoxide ion cannot be formed in aqueous medium except under special circumstances, because alkoxides are stronger bases than water. The ability of enzymes to control the microenvironment in which reaction takes place then becomes important. In a number of enzymes, including chymotrypsin, other serine hydrolases, and alcohol dehydrogenase, the concerted action of aspartate, histidine and serine generates the equivalent of an alkoxide ion, a strong base and potent nucleophile (Figure 10.4). The mechanism by which this triad of amino acids acts has been called a **charge-relay** mechanism. The pivot point is the imidazole ring, which has the ability to act as both a donor and an acceptor in hydrogen bonding. It is thus able to relay the negative charge of the carboxylate to deprotonate the hydroxyl group of serine. Once again, a precise relative orientation of the three amino acids is required for this effect to occur.

Figure 10.4.

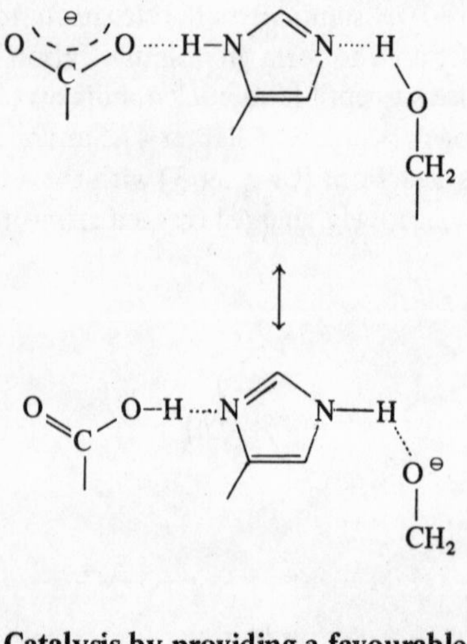

10.2.   **Catalysis by providing a favourable entropy for reaction**

In many of the mechanisms that we have just described, it is important that the reactive groups are present in exactly the right relative positions to optimise catalysis. Tertiary structures of enzymes have evolved to bind their substrates into the correct positions and orientations for reaction. In this way, they can be regarded as providing a favourable entropy of activation.

10.2.1. Intramolecular catalysis

Any reaction in which the participant groups are within the same molecule can be regarded as models of enzyme-catalysed reactions as far as the entropy term is concerned. Hundreds of examples have been discovered, and a simple case is the hydrolysis of carboxyphenyl-β-D-glucosides (Figure 10.5). Only in the *ortho* case can the carboxylic acid donate a proton *intra*-molecularly to the oxygen atom that joins the sugar and the benzene ring. In this case, intramolecular acid catalysis gives rise to a 10^4 rate enhancement over the *para* case. There are similarities between this reaction and the mechanism of action of lysozyme (Chapters 1 and 6). Sometimes, in both enzyme-catalysed and chemical reactions, a metal ion chelates with groups on reactants, thereby bringing them together to the right position for reaction. Several examples have been cited in earlier chapters with phosphates (Figure 3.28) and nitriles (Problem 9.4). The simplest chemical description of intramolecular catalysis is as a subclass of neighbouring group participation. It is a characteristic of neighbouring group effects that only one stereochemical pathway is possible because of the precise geometry of the reactants. This is exemplified by nucleophilic aliphatic substitution reactions at glycosidic carbon atoms (Chapter 6) and is common wherever rigid ring structures are found. The bound substrate to an enzyme's active site is the natural counterpart of the rigid ring.

Figure 10.5.

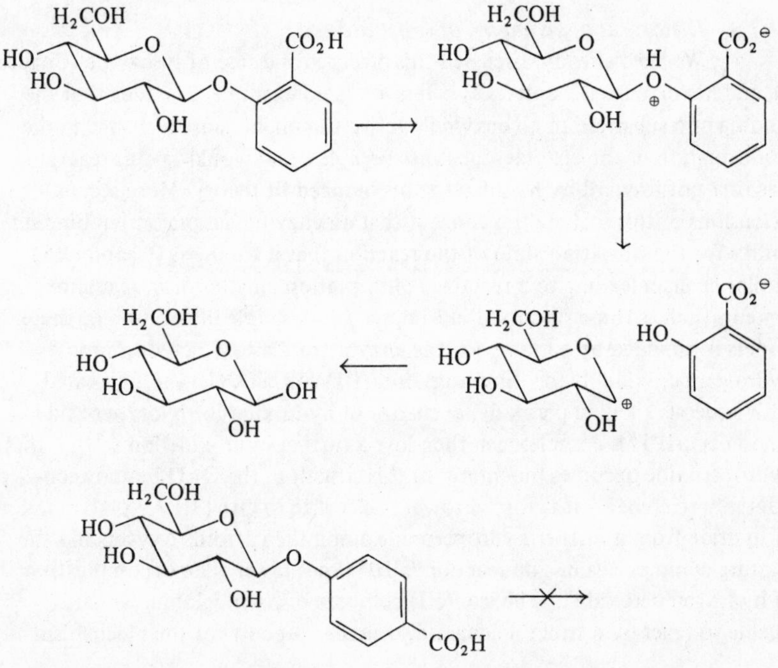

10.2.2. Geometry of reactants

Not only within ring systems is a preferred geometry for reaction found. Even simple substitution and addition reactions proceed by a preferred pathway, which can be studied theoretically or with the aid of X-ray methods (Bürgi, Dunitz, Lehn & Wipff, 1974). It is found that even a small departure from the optimum pathway leads to a substantial decrease in reaction rate. Looking at it the other way round, if the reaction path can be optimised by suitably arranging the reactants intramolecularly or at an active site, a large rate enhancement will be expected. One of the most striking chemical examples of this concerns the hydroxy-substituted phenyl propanoic acids (Figure 10.6). The acid on the left is free to rotate about the aliphatic C–C bonds, but steric interactions between the methyl groups in the second constrict the acid to adopt a conformation close to optimal for lactonisation. The rate enhancement is 10^{11}.

Figure 10.6.

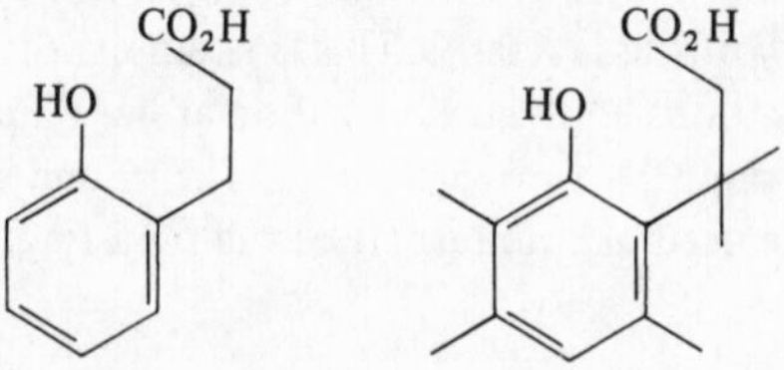

10.2.3. Distortion of geometry of substrate

We have already discussed this effect in the case of lysozyme (Chapter 1), and hexokinase (Chapter 12) offers a related example. The idea that the binding of a substrate to an enzyme's active site might cause a change in the conformation of the enzyme–substrate complex that would favour reaction was first put forward by Koshland in his **induced-fit** theory. More recent extensions of this include the concept that an enzyme has maximum binding ability for the transition state of the reaction that it catalyses (Chapter 15); similar changes leading to a reactive conformation may occur in regulatory systems such as those discussed in Chapter 13. A simple illustration of these effects is provided by a model for the enzyme catalase, which decomposes hydrogen peroxide. If the dihydroxy-iron(III) complex (10.7.1) is treated with aqueous alkaline peroxide, exchange of hydroxide for hydroperoxide anion occurs. This complex can then lose a further hydroxide ion so that the hydroperoxide becomes bidentate. In this situation, the O–O bond is considerably stretched – it is forced towards cleavage to give (10.7.2). Abstraction of hydride from a further hydroperoxide anion then affords oxygen and the starting complex again. The reaction is 10^4 times faster than decomposition of hydrogen peroxide by the iron(III) complex of haemoglobin, which is unable to react by a strain mechanism. Further support for the mechanism is provided by the observation that molecules or ions that occupy a further ligand site inhibit the reaction because hydroperoxide can no longer act as a strained bidentate ligand.

Figure 10.7.

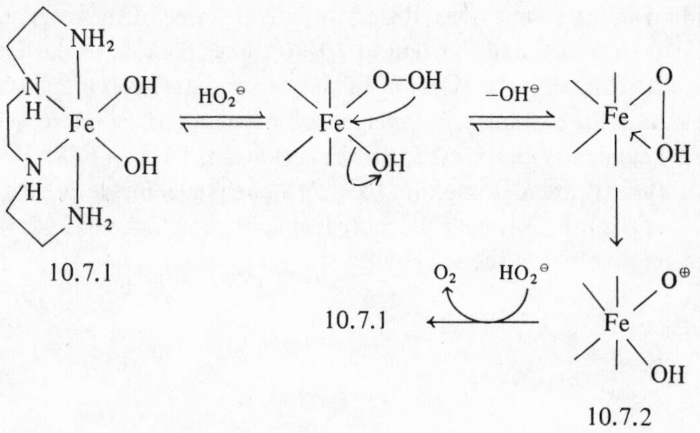

10.7.1

10.7.1 ←

10.7.2

10.2.4. *Binding forces available to enzymes*

Much of the preceding argument depends upon the enzyme's ability to bind its substrate into the correct orientation for reaction. Proper binding is responsible not only for catalysis but also for selectivity of reaction, because the enzyme can bind the substrate so that only one functional group is presented to the active site. The enzyme thus acts as its own protecting group. Almost all known chemical bonds are thought to participate in binding. Covalent, ionic and chelate bonds have already been mentioned and, in addition, hydrogen bonding and the hydrophobic bond should be considered.

Hydrogen bonding is necessary to hold an enzyme in its active tertiary structure: the components of the polypeptide chain hydrogen bond extensively via their amide bonds, often giving rise to the ordered structures known as secondary structure such as the α-helix or β-sheet. It is not surprising to find from X-ray structures that enzymes like chymotrypsin, carboxypeptidase and lysozyme bind their substrates at least in part by hydrogen bonding. Indeed it could be said that if a substrate can hydrogen bond with an enzyme, it will.

Hydrophobic bonding, on the other hand, occurs only when the enzyme's tertiary structure brings together a number of hydrophobic amino acid side-chains (Phe, Tyr, Val, Leu, Ileu) which cannot be solvated by water. Water is therefore excluded from their vicinity. Thermodynamically the enthalpy change for the solution of a hydrophobic molecule in aqueous solution is unfavourable, but the association of several non-polar groups is favourable on entropy grounds. The hydrophobic bond is thereby formed.

Hydrophobic regions in enzymes provide both binding sites for non-polar groups (chmyotrypsin, alcohol dehydrogenase) and also offer an unusual microenvironment in which reaction can take place. The haemoproteins in the previous chapter provide an excellent example of the latter. In non-enzymic chemistry, hydrophobic bonding is the cause of detergency and the formation of micelles, systems which are related to the structure of biological membranes. Micellar chemistry has expanded recently as chemists have sought models for the binding ability of enzymes. Reactions in which one of the reagents forms a micelle have been found to occur at 10–10^6 times the rate of the analogous non-micellar reaction. The effect of binding upon catalysis is emphasised by these results. More strikingly, it is even possible to demonstrate stereoselectivity in micellar catalysis (Figure 10.8; Brown & Bunton, 1974). (10.8.1) catalyses the hydrolysis of the R-isomer of (10.8.2), a phenylalanine derivative, three times faster than the S-isomer. Both hydrophobic bonding and hydrogen bonding are involved in the selectivity.

Figure 10.8.

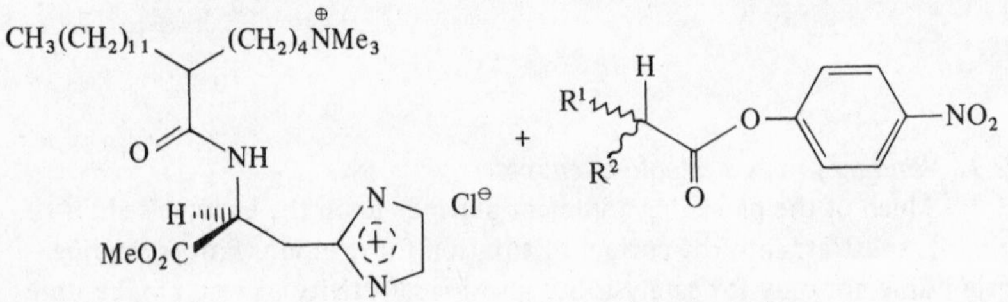

10.8.1

10.8.2

$R^1 = PhCH_2-$
$R^2 = MeCO\,NH-$

Polymeric reagents have also provided clear evidence for the effectiveness of hydrophobic binding. Polymethylene imines in which a proportion of the side-chains are alkylated by a methyl imidazole are true catalysts at pH 7 for the hydrolysis of uncharged nitrophenyl esters (Klotz, Royer & Scarpa, 1971). If a dodecyl substituent is included as a binding site, high reaction rates can be obtained (Table 10.1), approaching those for the analogous enzyme-catalysed reaction.

Table 10.1. *Rates of hydrolysis of nitrophenyl esters by various catalysts*

Catalyst	Rate (k) ($l\,mol^{-1}\,min^{-1}$)
Imidazole	10
Polymer with binding site	2700
α-chymotrypsin	10000

10.2.5. Binding and the transition state

We argued earlier in this chapter that stabilisation of an intermediate would have the effect of lowering the energy of the transition state for attaining that intermediate. It has been suggested that the best that the enzyme can possibly do is to stabilise the transition state directly. In other words, the enzyme should bind most strongly to the transition state itself. The application of this concept to the design of enzyme inhibitors is discussed in Chapter 15, but here it is instructive to consider the enzyme triose phosphate isomerase in this regard; this has been studied in detail by Knowles & Alberry (1977).

Knowles & Alberry carried out a detailed examination of the rates of all the steps catalysed by this enzyme (see Appendix 1 for the reaction) including the bond-making and bond-breaking steps. This analysis showed that the rates of all the steps had been maximised on the enzyme to a limiting rate over which the enzyme itself had no control – the rate at which the substrate and product molecules diffuse on and off the enzyme from solution. This apparent evolution of the catalytic mechanism to perfection has implications for many biological systems, and it would appear that in order to achieve higher rates, if this is desirable, nature must find a way to avoid slow diffusion steps. Multi-enzyme complexes (Chapter 13) may provide such a mechanism.

So far we have examined biochemical reactions from the chemist's point of view. The aim has been to understand the chemical nature of biochemical reactions since this is the molecular basis upon which all living systems are organised. But how far and how precisely can we apply these ideas in biology? The remaining chapters of this book attempt to answer this question. Much depends upon the methods we have available and the accessibility of biological systems for molecular study. In the next chapter we outline how chemical and physical methods are adapted in order to approach the molecular questions on which we wish to focus.

11   Biochemical techniques

11.1.   Introduction

Biochemical studies involve the application of chemical techniques
and concepts to the study of living systems. The range of methods that has
been applied is wide and the nature of biological systems frequently makes the
biological application different from its chemical counterpart. Usually the
differences are in emphasis rather than in principle, and because of this, it is
readily possible for anyone with a basic knowledge of chemical techniques to
appreciate their application in biochemistry, provided that he is aware of the
nature of the biological system being studied. In this chapter we describe
briefly some aspects of biochemical methods and show how they are modified
from those used in less complex systems.

The techniques that can be applied include both purely chemical methods
(e.g. chemical modification of proteins, chemical analysis of substrates) and
physical methods (e.g. spectroscopy) and combinations of these, for example,
spectroscopy of proteins chemically labelled with a reporter group. The
modifications in technique that are necessary depend on the characteristics
of the biological system. For example in biochemistry, like polymer chemistry,
we are frequently dealing with molecules of a rather high molecular weight.
Haemoglobin, the oxygen-carrying protein of red blood cells, has a molecular
weight of 65 000; myosin, protein of muscle, is rather larger (470 000),
whereas a single amino acid like alanine has molecular weight of only 89. The
contrast between the biopolymers and the chemical polymers is that most
biopolymers have a precisely defined structure. Thus a protein, a polymer of
amino acids, is composed of a defined sequence of specific amino acids. Any
slight change will give a protein which is distinguishable from the original protein by a number of methods. In contrast, most chemical polymers have a less
defined structure: a given polymer is usually a mixture of a number of different
molecular species with a distribution of molecular weights, but each with the
same general chemical characteristics. The properties of a chemical polymer
then, are those of a mixture of molecules; those of a protein are of a defined
single molecule.

Another difference between many chemical systems and most biological systems is stability. Many biological preparations are only stable within a very closely defined range of conditions, and indeed some biological molecules have only a limited lifetime *in vivo*. This may be significant in relation to their biological function (e.g. cyclic AMP, Chapter 16). Outside these conditions, biomolecules may lose biological activity whilst retaining much of their chemical character. It is sometimes possible to simulate the natural environment and so to stabilise a biological system. Membrane proteins, for example, are frequently studied in an entirely aqueous environment which is completely different from their natural, partly non-aqueous environment in the membrane. A membrane environment can be simulated by the use of detergents (Chapter 14).

A third difficulty which may arise is the very low concentration in which many biological species are found. Very sensitive methods are needed to detect such molecules, and the determination of their structures is a chemical tour-de-force. Concentrations can be as low as 10^{-9} mol l^{-1}, the concentration of the hormone adrenaline in the blood after its release (Chapter 16), and concentrations can be estimated by the most sensitive methods to three orders of magnitude less than this.

The requirements for a useful biochemical method, then, are:

(1) It must be capable of dealing with large molecules and molecular assemblies if required.
(2) It must not, as far as is possible, cause a loss in the biological activity of the system.
(3) It must be sensitive enough to the small changes in the system or to the low concentrations which may be extremely important in understanding the biological function of the system.

Techniques are required for several main purposes. Firstly we need to isolate and purify our biological material. The approach to be used here depends upon the chemical and physical characteristics of the system being studied: a method for the purification of a protein will clearly be different from a method for the purification of a complex lipid. Once we have obtained a system in as well-defined a form as possible we will want to know something about its chemical structure and how this can be related to the biological role of the system. We shall now consider these two requirements in turn. The overall molecular picture of the biological system, which is our goal, depends on a synthesis of the results obtained by all methods. No single technique is sufficient to provide all the necessary information, and a knowledge of the power and limitations of all biological techniques is therefore essential in understanding their relevance to the system being studied. This is well illustrated by the studies on the kinases described in Chaper 12.

11.2.   Isolation and purification

11.2.1.   *Source of material*

A successful purification scheme firstly requires a readily available source of the material to be studied. This is often a limiting factor, and as a result a few systems have achieved enormous popularity with biochemists. In mammalian systems the prize probably goes to rat liver as a tissue with general metabolic activity, and amongst micro-organisms, yeast and the prokaryote *E. coli* have been extremely popular. Such specialisation in biological sources can lead to a rather blinkered view of biology; in any case one of the fascinations of biochemistry is to see how within a common biochemical structure the wide diversity of life can exist.

The homogeneity of a source can be a problem. With micro-organisms, difficulties can occur because of their relatively rapid rate of mutation, and contamination with unwanted micro-organisms can be a problem. In mammalian systems, the problem is that many organs which can often be isolated rather easily contain several distinct types of cell, each with different functions and which are often difficult to separate. Examples of such systems include the different white cells of blood (Chapter 16) and the two parts of the adrenal gland – the cortex, which synthesises steroid hormones, and the medulla, which produces an entirely different range of hormones including adrenaline (Chapter 16). New techniques for the separation of cell types are becoming available. (Hertzenberg, Sweet & Hertzenberg, 1977) and studies of whole animal cell preparations have become increasingly important.

11.2.2.   *Subcellular fractionation*

The next stage in most schemes of purification is to separate out the part of the cell in which the component of interest is located. Firstly, the cell must be broken open to form a **cell-free extract**. This can be difficult to achieve and must be done with care and as gently as possible. Mitochondria, for example, are very fragile cell organelles. They are studied in great detail because in eukaryotic cells (they are not found in prokaryotes) they are the principal site of the synthesis of ATP, the energy currency of the cell, and the terminal oxidation steps of energy-providing metabolic pathways. Mitochondria consist of two membranes, an outer membrane enclosing the inner and the two enclosing a matrix which contains many enzymes. The function of mitochondria is very closely linked to their structural integrity, so if either membrane should become ruptured, perhaps by over-vigorous homogenisation of the cells or by placing the mitochondria in a medium of the wrong osmotic strength, they are likely to be damaged and will be of little use in experimental work.

Many cells have a very strong wall, and require techniques which develop very high hydrostatic pressures to break them. Such techniques, or alternatively grinding with glass beads or freezing and thawing, are necessary to break open the cell wall of many micro-organisms. In contrast, mammalian liver is a very soft tissue and the cells can be broken open by relatively gentle homogenisation in a tight-fitting mortar and pestle.

Once a cell-free extract has been obtained, and it has been confirmed that the biological system of interest is present in it, the mixture can be further fractionated. This is usually done by differential centrifugation, which depends upon the fact that cell components differ in size and density and so will sediment at different rates in the centrifuge. The rate of sedimentation of a spherical particle is given by:

$$\text{rate} = r^2(\rho_p - \rho)\cdot G/\eta \tag{11.1}$$

where r is the radius of the particle, ρ_p the density of the particle, ρ the density of the medium, η the viscosity of the medium and G the gravitational force provided by the centrifuge. Thus, larger and denser cell components will centrifuge down more rapidly than the others, or alternatively they can be sedimented at lower centrifuge speeds than the less dense and smaller particles. This can be seen from Table 11.1, which shows a typical scheme for the differential centrifugation of a rat liver cell-free extract.

Table 11.1. *Differential centrifugation of a rat liver cell*

Centrifugation conditions	Cell fraction
600 g, 10 min	Cell debris and nuclei (large particles and dense particles, nuclei contain DNA, density 1.75 at 4°C)
15 000 g, 10 min	Mitochondria, lysosomes (moderate-sized particles containing protein and lipid)
100 000 g, 60 min	Microsomes (small particles formed by fragmentation of the endoplasmic reticulum, Golgi apparatus and, to some extent, the plasma membrane)
Supernatant fraction	'Soluble' fraction containing proteins (density about 1.19) small molecules and any free lipid (floats on top of aqueous phase)

Fractions obtained in this way are still not homogeneous as can often be shown by electron microscopy or by the presence of enzymes which are known to be located in another part of the cell. Further purification may often be required, perhaps by using more refined centrifugation methods such as centrifugation through a density gradient of sucrose or caesium chloride (Griffith, 1976).

11.2.3. Isolation of biological molecules

The procedure chosen following differential centrifugation will depend entirely on the chemical and physical properties of the substance that is being purified. Small molecules, substrates and coenzymes for example, can usually be isolated from the supernatant fraction of the cell. Since this fraction also contains protein, it may be necessary to remove it, perhaps by denaturation and precipitation. The small molecules can then be isolated by standard chemical techniques. Care is often required to find suitable conditions for isolation of labile molecules, and some may be so unstable that isolation is very difficult. An extreme example of instability is given by the thromboxanes and prostacyclin (Figure 11.1). These are derivatives of arachidonic acid, and are important in regulating the processes which take place at the interior wall of blood vessels. They are chemically so unstable that they have a half-life when isolated of seconds. This lability is however essential to their biological function. When small molecules are stable enough, final purification can be carried out by chromatography and crystallisation and identity confirmed by chemical synthesis.

Proteins are found either in solution (in the supernatant fraction or within larger organelles such as mitochondria or lysosomes) or they are bound to membranes. In the case of the membrane-bound proteins, present methods require that they should first be 'solubilised', that is released from the membrane. This is frequently done using detergents (Chapter 14). Once the protein is freed from the membrane and is in a smaller aggregate, the techniques applied are basically the same as those used for soluble proteins. Many methods are available and there is a great art in choosing the right ones. The skill and patience required for protein purification is very similar to that required by the synthetic organic chemist when developing a multi-stage synthesis. A feel for the chemical and physical properties of the protein under purification is developed by the biochemist which can, rather like synthesis in organic chemistry, only be learnt by experience.

Figure 11.1.

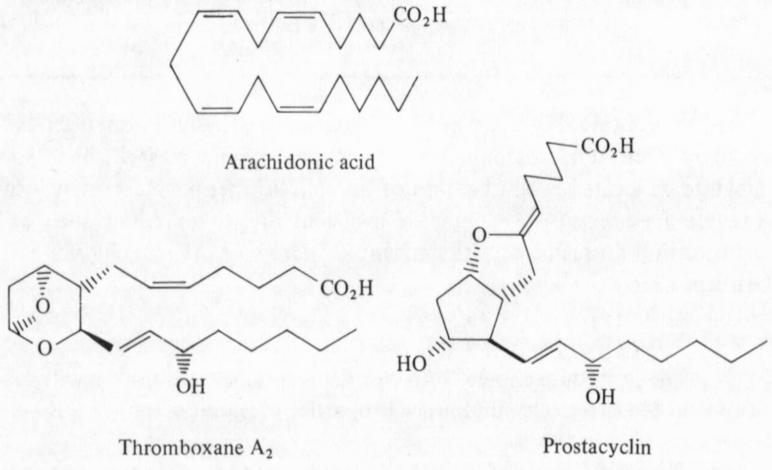

Arachidonic acid

Thromboxane A$_2$

Prostacyclin

The differences in physical properties between components of the protein mixture is then exploited to achieve further purification. One might fractionate a mixture according to molecular weight by methods such as dialysis, ultrafiltration or molecular exclusion chromatography. One could take advantage of the fact that proteins differ in their solubility in solutions of defined ionic strength and pH. Thus, protein fractions can be precipitated by the addition of salts such as ammonium sulphate or by adjusting the pH to a higher or lower value. The electrostatic charge on the proteins caused by ionisable amino acid side chains on their surface can be used to separate them by ion-exchange chromatography or electrophoresis. At each stage the biological activity of the fractions is monitored to ensure that the desired protein is present, and purification is shown by the steady increase of biological activity per milligram of protein as the purification proceeds.

The importance of chromatographic methods in protein purification cannot be overemphasised. We have already met affinity chromatography (Chapter 7); Table 11.2 summarises important chromatographic methods in biochemistry.

Table 11.2. *Chromatographic methods in biochemistry*

Method	Basis of separation	Phases		Types of molecule separated
		stationary	moving	
Absorption	Polarity of organic compounds	Alumina	Organic solvent	Organic compounds
Partition	Polarity of organic compounds	Silica gel	Organic solvent	Organic compounds, lipids
Gas–liquid	Polarity and volatility of organic compounds	Silicone waxes	Inert gas stream	Organic compounds
Ion-exchange	Charge of molecule or macromolecule at pH of experiment	Ion-exchange resin	Buffered aqueous organic solvent	Compounds with ionisable groups, amino acids, peptides and proteins
Exclusion (gel filtration)	Size of molecule	Molecular sieve, e.g. Sephadex	Aqueous solvent	Molecules and macromolecules of various sizes
Affinity	Specific binding of desired molecule to stationary phase	Designed to interact specifically with one component of the mixture	Aqueous solvent	Biological molecules with specific binding properties for stationary phase

A number of steps may be necessary before a protein can be said to have been purified to homogeneity. Proof of homogeneity requires the demonstration that no other proteins are present. This is not quite as easy as determining a melting point. Criteria include the absence of any other biological activity and the demonstration that only one protein band is present on gel electrophoresis under a variety of conditions. This latter method is the protein chemist's equivalent of thin-layer chromatography in organic chemistry. As an illustration of the work involved in a protein purification Table 11.3 summarises the procedures used in the purification of adenylate kinase, an enzyme which we shall discuss in more detail later.

Table 11.3. *Purification of adenylate kinase (from Heil et al., 1974)*

Procedure	Total protein (mg)	Specific activity (units mg^{-1})	Purification per step (-fold)
Extraction of minced pig muscle with 0.01 M KCl	435 000	5.7	
pH fractionation	112 000	19.5	3.42
Chromatography on phosphocellulose	1 716	781	40.0
Gel filtration on Sephadex G-75	211	2590	3.32
Crystallisation	344	2790	1.08

Overall purification 490-fold with a 38.7 % yield of enzyme activity.

We should briefly consider two other important classes of biological compound. **Nucleic acids** are among the largest molecules known, with molecular weights from 10^6 to 10^{12}. They are, despite the great variety of detailed information held within their sequences, chemically relatively more homogeneous than the proteins for which they carry the genetic instructions. Nucleic acids are strongly acidic, and at physiological pH carry a high negative charge. This is usually compensated in the cell by complexation of the nucleic acids with metal ions and, in the case of eukaryotic DNA, with basic proteins such as the histones. Purification of nucleic acids requires removal of protein from the mixture, and one way in which this can be done is to treat a buffered solution of the mixture at pH 7 with an aqueous solution of phenol. A protein-denaturing agent such as sodium dodecyl sulphate may be added. The phenol phase dissolves the denatured protein and the nucleic acid can be precipitated from the aqueous phase by the addition of ethanol in the cold. A formidable array of modern techniques is now available for the isolation and purification of specific nucleic acids; these include chromatographic and electrophoretic methods, similar to those used to purify proteins (Gould & Mathews, 1976). Finally, **lipids**, the biological molecules that are characteristically soluble in organic solvents, include triacyl glycerol, phospholipids and sterols (Fig. 11.2). All contain highly apolar residues and are therefore not soluble in aqueous media. Phospholipids are a major component of membranes, as is cholesterol, the major mammalian sterol (Chapter 14). Triacyl glycerol is an important store of energy in the body (Chapter 2 and Chapter 13). These molecules are commonly extracted from biological tissue with mixtures of chloroform and methanol. Further purification is usually effected by chromatography on silica gel.

Once a reasonably pure preparation has been obtained, the process of structure determination can begin. There is no point in applying some of the structural methods (for example, X-ray crystallographic studies) unless the preparation to be examined is homogeneous, but in many other cases progress can be made with a partially purified system, provided care is taken in the interpretation of results.

Figure 11.2.

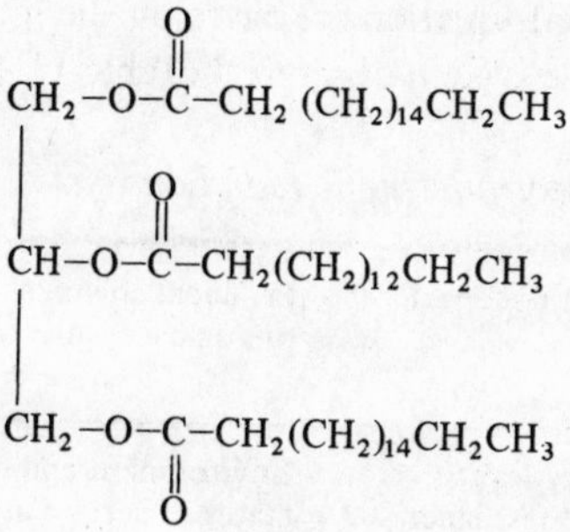

$$CH_2-O-\overset{\overset{O}{\|}}{C}-CH_2\,(CH_2)_{14}CH_2CH_3$$

$$CH-O-\overset{\overset{O}{\|}}{C}-CH_2(CH_2)_{12}CH_2CH_3$$

$$CH_2-O-\underset{\underset{O}{\|}}{C}-CH_2(CH_2)_{14}CH_2CH_3$$

Triacyl glycerol

$$Me_3\overset{\oplus}{N}-CH_2-CH_2-O-\overset{\overset{O}{\|}}{\underset{\underset{O_\ominus}{|}}{P}}-O-CH_2$$

$$CH\rightarrow O-\overset{\overset{O}{\|}}{C}-CH_2\,(CH_2)_{14}\,CH_2CH_3$$

$$CH_2O-\underset{\underset{O}{\|}}{C}-CH_2(CH_2)_{14}CH_2CH_3$$

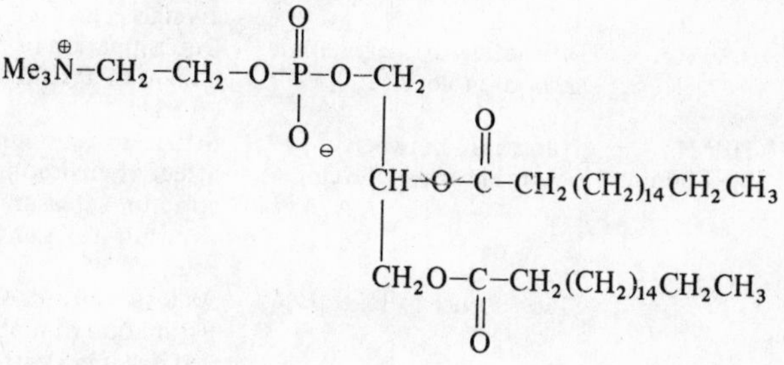

A phosphatidyl choline

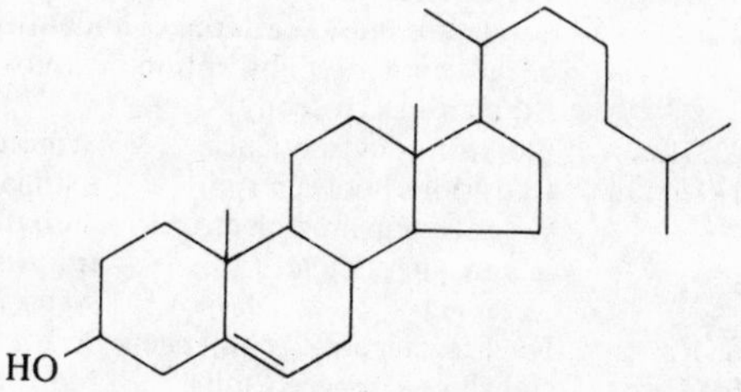

Cholesterol

11.3. Physical methods of structure determination

The application of physical methods of structure determination to both chemical and biological systems has provided many important insights. Physical methods have particular advantages in biochemistry, because they are the mildest techniques that can be applied and cause the least perturbation of the natural structure. The energy applied to a sample is frequently very low, and physical probes can often be so specific that a full purification of the system of interest may not be necessary. The physical methods that are most commonly applied to give molecular information are based on the interaction of radiation of different types with the system of interest (Table 11.4). In this

Table 11.4. *Physical methods using electromagnetic radiation*

Radiation	Energy, wavelength, frequency	Physical event observed	Principal chemical and biological applications
γ-rays	100–1 meV	Nuclear energy-level transitions (Moessbauer effect)	Environment and oxidation states of certain metal ions, especially iron
X-rays	100–1 keV 0.01–1 nm	Diffraction by a crystalline array of molecules	Determination of three-dimensional structure in crystals
(Neutrons)		Diffraction by a crystalline array of molecules	Determination of three-dimensional structure in crystals
Ultraviolet–visible	10^{15} Hz 200–800 nm	Transitions between electronic energy levels	Structure determination, effect of environment on structure. Measurement of concentrations and reaction rates
		Fluorescence (= relaxation)	As for electronic spectra. Estimation of mobility and distances within macromolecules
		Polarised radiation: optical rotatory dispersion, circular dichroism	Asymmetry in structures. Effects of environment on asymmetry
Infrared	10^{14} Hz 2–20 μm	Vibrational transitions of covalent bonds. Scattering of radiation after absorption (Raman spectroscopy)	Structure determination, identification of functional groups
Microwave	10^{11} Hz 1–10 mm	Electron spin resonance: transitions between spin states of unpaired electrons in a magnetic field	Structure determination. Estimation of distances and mobilities. Studies of certain transition metal ion oxidation states and environment
Radio	10^{6} Hz 0.1–1 m	Nuclear magnetic resonance: transitions between spin states of nuclei with a resultant magnetic moment in a magnetic field	As e.s.r., but not studies of most metal ions (see text)

section we shall outline some of the most important physical methods, noting their common features. We shall find examples of their application in the following chapters.

11.3.1. Microscopy

Everyone is familiar with the light microscope, and its use in many branches of biology is very important. One can examine tissues and resolve the types of cells that make them up, especially with the aid of chemical techniques that stain particular types of cell and cell component specifically (Craigmyle, 1975). However, it is not possible to obtain molecular resolution using light of visible wavelengths. Typically, proteins are about 2 nm in their shortest dimension but the wavelength of visible light ranges from 400 to 800 nm, about the size of a bacterium. Since the resolution of a microscope depends on the wavelength of the light used, it is clear that radiation of much lower wavelength is required. Electrons of energy about 60 kV have a wavelength of about 0.005 nm, and so are obviously suitable for microscopy at molecular resolution. In fact, imperfections in the design of electron optics make the resolution of the electron microscope about 1 nm, compared with the optimum for a light microscope of 250 nm.

Electron microscopy, then, can reveal many aspects of the ultrastructure of cells and large molecules: one can observe the lipid bilayer, characteristic of biological membranes (Chapter 14); the overall shape of a large enzyme system can be observed (pyruvate dehydrogenase, Chapter 13) and the shapes of many viruses have been described (Butler & Klug, 1978). Recent techniques have allowed the determination of the orientation of a protein within a membrane (Chapter 14) and the arrangement of ribosomes on the endoplasmic reticulum (Unwin, 1977).

These pictures offer us an overall view of a biochemical system and, like an aerial view of a chemical works, they show the plan but not the details of the processes which take place within the units which they reveal. Other more specific methods are required for this.

The preparation of samples for electron microscopy, however, is perhaps the most drastic of all physical methods. Electron optics operate in a vacuum and, as a result, samples must be prepared in a dehydrated state, often after fixing with a chemical reagent and staining with a heavy metal to provide contrast. One must be aware of the possible consequences of such treatment on a fragile biological sample, which may also decompose under the high energy of the electron beam (Weakley, 1972). There are many specialised methods for preparing samples. In membrane studies, metallic replicas of the membrane are often prepared by a technique known as freeze-fracture (Chapter 14). It is also possible to locate specifically a protein within a cell membrane by labelling it with an electron-dense macromolecule so that it appears as a dark spot on the electron micrograph. This technique allows the labelled protein to be observed on the cell membrane during various cellular events such as secretion of the contents of an intracellular vesicle into the extracellular space (Raff, 1976). This the first example we meet of the important biochemical technique of introducing a probe molecule into a specific site in a complex system.

11.3.2. Methods using electromagnetic radiation

Practically the whole electromagnetic spectrum can be used to study biological systems. Energies and wavelengths cover a very wide range, as can be seen from Table 11.4, which summarises the principal techniques currently in use.

X-ray diffraction

Radiation of any wavelength can be diffracted by an ordered network that has a period of the same magnitude as the wavelength of the radiation. (see van Holde, 1971, for a basic discussion of diffraction). X-rays have a wavelength of 0.01 to 1 nm and so are ideally suited to resolve details of molecular structure at the level of covalent bonds (for example a C–C single bond is about 0.154 nm long). In order to observe a diffraction pattern that can be interpreted in terms of the three-dimensional structure of the sample, an ordered form of the sample is required to act as the molecular diffraction grating. This is usually provided by a small crystal, although other non-crystalline systems such as a membrane may be ordered enough for work at lower resolution. Highly purified material is required in order to obtain crystals, and problems in obtaining such crystals can severely limit the applicability of this technique. Most often the technique is applied to proteins, but nucleic acids (e.g. tRNA) have also been examined in detail. Analysis of the diffraction pattern produced by the crystal leads to a map of the electron density of the molecules within the crystal in three dimensions, and this usually provides enough information for the building of a three-dimensional scale model of the molecule. Initially, this may only indicate the overall shape of the molecule, but structures are usually further refined to reach a resolution of about 0.2 nm for macromolecules, which is sufficient in a protein to define much of the detailed orientation of the functional groups of amino acid side-chains which are so important in determining the structure and specific function of a protein. Examples of protein structure determined by X-ray diffraction will be found in Chapters 12 and 16. For the present, we need to consider how far the structures derived by X-ray diffraction studies relate to the molecules in the intact cell.

There is clearly one major difference. The picture derived from X-ray diffraction studies is static: the molecule is frozen in one conformation, that of the crystal. Thus any changes in the conformation that are important in the catalytic or regulatory mechanism of an enzyme, for example, will not be observed unless the enzyme can be crystallised in forms corresponding to these other states. Such studies have been made and examples will be found in the next two chapters. Recently, attempts have been made to add a dynamic component to X-ray structures (McCammon, Gelin & Karplus, 1977; Huber, 1979). In principle, an X-ray diffraction study should solve the structure of a cr crystal of biological material without the need to perturb the system in any way other than crystallising it. However, difficulties in analysing the diffraction data associated with the so-called 'phase problem' often require that crystals be obtained 'labelled' with a heavy atom. The presence of the heavy atom allows very precise measurements of the intensity of the diffracted X-rays, since the heavy atom scatters the radiation more effectively than the light C, N and O atoms which form most of the molecule. In this type of experiment the heavy atom can be regarded as an **extrinsic probe**, and its use must be controlled by the precautions which we shall apply in the following sections for other extrinsic probes. When a protein is crystallised as a heavy atom derivative it is essential that the heavy atom should not alter the protein structure to any significant degree, and that the 'labelled' protein should crystallise as nearly as possible in the same crystalline form as the pure protein itself.

Electronic spectra

In principle, all atoms can give rise to scattering of X-rays, but only certain atoms or groups absorb electromagnetic radiation of longer wavelengths. This is by no means a disadvantage, since it allows one to focus on the behaviour of a specific component of a system which may closely reflect important aspects of its structure. Electronic absorption bands arise when ultraviolet or visible radiation causes the excitation of an electron from a ground state energy level to an excited state, or higher electronic energy level. The chemical groups which absorb energy of this range of wavelengths include all conjugated systems of double bonds, aromatic systems and also the transition metal ions, which we discussed in Chapter 9. Several important groups which have characteristic electronic spectra are shown in Figure 11.3. The energy of light absorbed is equal to the difference in energy between the ground state and the excited state. This depends upon two things. Firstly, the basic energy levels are decided by the chemical structure of the chromophore. The energy difference between them can be modified by the environment of the group, so that if the environment alters the energy of the ground state but not the excited state, or vice versa, the position of the absorption maximum will be changed on moving the chromophore into that environment. The dependence of the spectra of a substance not only on the substance itself but also on its environment is of special importance in biological studies, where the environment of a group can be very important in the function of a system as a whole.

Figure 11.3.

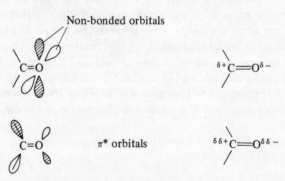

Tyrosine λ_{max} 270 nm   Flavin λ_{max} 280, 370 nm

Haem λ_{max} *bis* aquo Fe^{II} 580 nm

For example, the $n \rightarrow \pi^*$ transition observed in carbonyl groups shows a marked solvent (or environmental) effect (Figure 11.4). In the ground state the n-electrons are not involved in any covalent bond (n = non-bonding), and tend to interact strongly with a polar solvent. As solvent polarity is increased this interaction becomes stronger and the ground state becomes progressively more stabilised. This means that the energy difference between the ground state

Figure 11.4.

Non-bonded orbitals

C=O

π^* orbitals

C=O

$^{\delta+}C=O^{\delta-}$

$^{\delta\delta+}C=O^{\delta\delta-}$

and the excited state increases with increasing solvent polarity since the excited state is not affected by the changes (the distribution of the π^*-electrons being such that they do not solvate as readily as the ground-state n-electrons). The effect of this on the spectrum is the shift of the absorption maximum to a higher energy, a blue shift.

Reasoning such as this can be useful in examining the environment of a particular absorbing group in a biological system, and such studies have a wide application in protein chemistry. For example, the absorption maximum of a tyrosine residue buried in the centre of a protein will change as the protein is denatured and the tyrosine becomes exposed to a more polar aqueous environment. Following these changes can be valuable in studying the way in which polypeptide chains fold and unfold (d'Albis & Gratzer, 1974).

One of the simplest applications of electronic spectra is the assay of biochemicals. Since by Beer's law the intensity of an absorption under defined conditions is proportional to the concentration of the chromophore, the concentration of an unknown sample can readily be determined from its absorbance under the defined conditions. By measuring the rate of change of absorbance with time one can measure the rate of change of concentration of a substance and hence the rate of a reaction by which it is produced or destroyed.

Pyruvate kinase, which is discussed in some detail in the next chapter, can be assayed using absorbance measurements (Figure 11.5). Pyruvate is produced from phosphoenol pyruvate in the presence of ADP, the reaction is catalysed by pyruvate kinase. The pyruvate so produced is reduced to lactate by lactate dehydrogenase and NADH. The conditions are arranged so that the rate-limiting step in the two coupled enzyme-catalysed reactions is the formation of pyruvate. Under these conditions the rate of loss of NADH is equivalent to the rate of formation of pyruvate. This can be followed by measuring the decrease in absorbance of the mixture at 340 nm, the absorption maximum of NADH.

In all spectroscopic methods the absorbed energy must be dissipated in some way as the system *relaxes* from its excited state to its ground state. Energy is often lost in the form of heat by vibrations within a molecule or by collisions between molecules. In some biological systems, the energy released by relaxation is channelled so that it is used for the synthesis of organic material. This is the process of photosynthesis, which takes place in green plants and photosynthetic micro-organisms. Synthesis of carbohydrate using light energy by these organisms is the biggest chemical industry on earth. (Hall, 1977).

Another way in which relaxation can take place is by emission of a photon of light, the process known as fluorescence. If the rate of relaxation by other mechanisms is slow, fluorescence can be observed. This is an example of a spectroscopic technique which has an inherent time-scale that can be put to use in biological experiments. The excited electron is initially promoted into a higher vibrational energy level of the excited state on absorbing a photon of light of the required energy. It decays very rapidly into the ground vibrational level of the excited state, and the quantum of light emitted by fluorescence is usually emitted within 10^{-9} s of reaching this state. It is possible to take advantage of this time scale to estimate the rate of motion of biological molecules by using polarised light as the source of exciting radiation, since the rates of tumbling of large biomolecules are of the same order as the time scale of the fluorescence event.

More commonly, fluorescence measurements are used in a similar way to absorption measurements, that is for assays or for structural studies, since for similar reasons to absorption maxima, emission maxima are also sensitive to the environment of the emitting group. Fluorescence spectroscopy is a more sensitive technique than electronic absorption spectroscopy and thus allows studies to be made at lower concentrations. A further way in which the excitation energy can be lost from an excited state is by transfer to another

Figure 11.5.

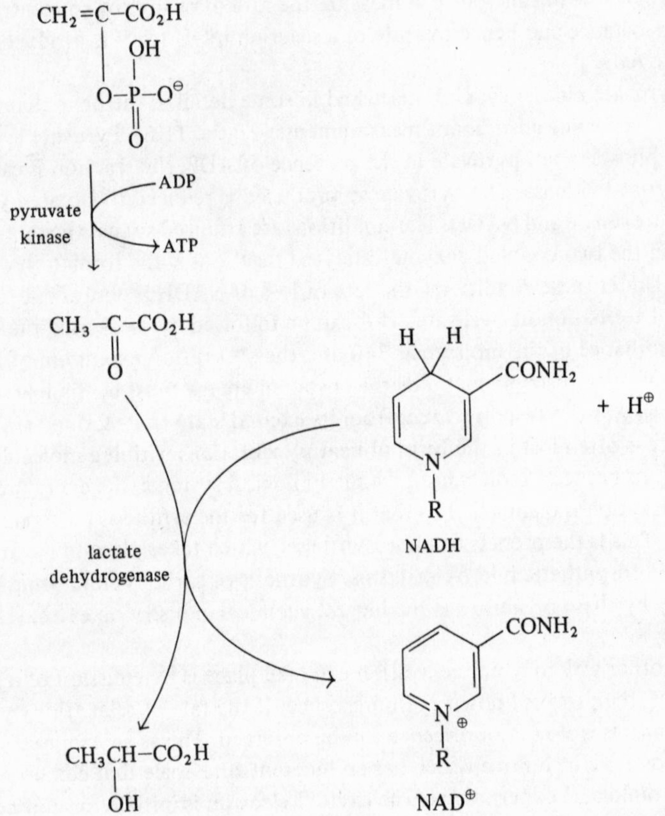

chromophore without emission of light, a process known as resonance tranfer. Since the efficiency of resonance transfer depends upon the distance between the two chromophores, this phenomenon provides a method by which distances can be estimated within a macromolecule in solution (if the macromolecule contains suitably placed groups). Fluorescence and magnetic resonance methods provide complementary approaches for measuring distances and rates of molecular motion in solution.

Often a system of interest does not contain a chromophore; in these circumstances extrinsic probes are often used. Extrinsic probes are used in all types of spectroscopy; they are atoms or molecules not normally present in the biological system which give information on their environment from their spectra. Two examples of extrinsic probes with useful electronic spectra are shown in Figure 11.6. Probes such as 8-anilinonaphthalene-1-sulphonic acid (ANS) are often used to investigate the polarity of their environment: in this case it is the fluorescence spectrum that responds to the polarity of the molecule's environment. In contrast, *p*-nitro-ω-bromo-acetanilide (11.6.2) is used as a covalent label of proteins. Nucleophilic groups such as amino and thiol groups on amino acid side-chains can displace the bromide, thus covalently labelling the protein with a chromophore with a characteristic u.v. spectrum. Changes in the environment of the probe due to changes in the protein structure may be observed in the spectrum of the probe, and hence the nature of the change in protein structure may be inferred. Some of these probes are rather large molecules, and they may cause some perturbation of their immediate environment. Results obtained with such probes are most useful when supported by data obtained from independent methods.

Electronic spectra can be readily obtained on partially purified systems. The electron-transporting haemoproteins, the cytochromes, were originally detected using a very simple spectroscopic equipment by examining subcellular fractions (by MacMunn, 1886 and by Keilin, 1925). It was many years later

Figure 11.6.

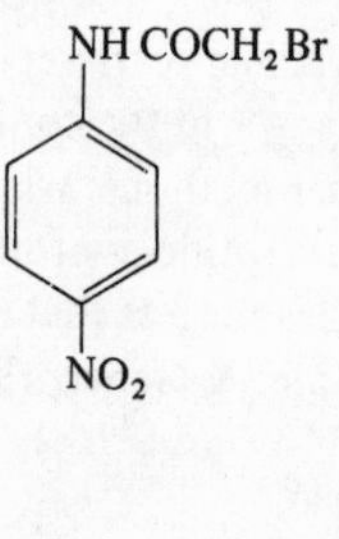

11.6.2

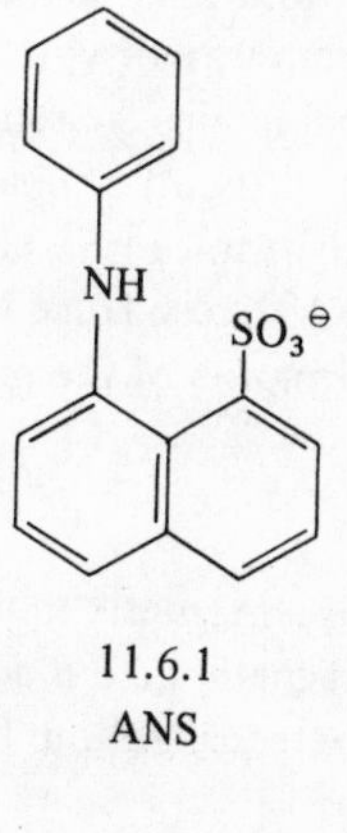

11.6.1

ANS

that the crystal structure of cytochrome c was determined by X-ray diffraction. Even today, many pigments, particularly those associated with membranes, have not yet been obtained in pure form, and much has been learnt about their behaviour by electronic spectroscopy. A recent example of this has been work on the cytochrome P 450 system, which is important in the metabolism of drugs (Chapter 15). Until very recently this cytochrome could not be obtained in pure form from mammalian systems. Nevertheless, spectroscopic studies allowed much to be learnt about its molecular characteristics and its catalytic mechanism, and these have provided a sound basis for current work with the purified protein.

To complete our survey of methods used in biochemistry which take advantage of electronic transitions we should mention the two techniques which use polarised light: optical rotatory dispersion and circular dichroism. These methods have been especially used in studies of protein conformation. They allow, for example, an estimate to be made of the ordered structure within a protein (d'Albis & Gratzer, 1974).

Vibrational spectra

Radiation of infrared wavelengths is absorbed by transitions between vibrational energy levels of covalent bonds. The frequencies at which the radiation is absorbed depends upon the type of bond, and so this technique is of very great value in organic chemistry in the identification of functional groups in molecules. In more complex biological molecules, however, the large number of functional groups with similar infrared absorption bands makes interpretation of spectra difficult. Also water absorbs strongly in the infrared, masking many vibrational transitions of interest.

A relatively new technique which also depends upon vibrational transitions is resonance Raman spectroscopy. In this method, a laser is tuned to an electronic absorption band of a system of interest, for example the porphyrin in haemoglobin. The scattering of the incident light is observed (the Raman effect) and from this the vibrational modes of the porphyrin, free from those of the peptide and also from water (which is a poor Raman scatterer) can be observed. These vibrational modes are sensitive to changes in the conformation of the porphyrin, which is difficult to observe by other methods. Whereas binding of oxygen is thought to cause a change in the peptide conformation in haemoglobin (Chapter 13), resonance Raman studies suggest that there is little effect on the conformation of the porphyrin (Spiro & Gaber, 1977).

Magnetic resonance

When a particle which has a magnetic moment, such as a proton or an unpaired electron, is placed in a magnetic field it adopts an orientation with respect to the applied field, its axis precessing about the direction of the field. If radiation of a frequency corresponding to the frequency of the precession is applied to the particle it will absorb the energy and move to a higher magnetic energy level by changing its orientation with respect to the field. This is the phenomenon of magnetic resonance, which has become increasingly applied to biological systems in recent years (Knowles *et al.*, 1976), particularly with the development of equipment which allows the resolution of the many signals that can be observed from large biomolecules and also from intact cells and organs.

One can extract useful information from the magnetic resonance spectrum of a system in a number of ways.

(1) The chemical environment of the particle affects the frequency of its resonance (measured as the chemical shift in nuclear magnetic resonance, n.m.r., or the g-value in electron spin resonance, e.s.r.). This can allow specific sites in a molecule to be observed and gives detailed structural information.

(2) Resonances are often split into multiple lines. This is because of the effect of a neighbouring species with a magnetic moment, a nucleus or an electron. In n.m.r. the splitting depends upon the spatial relationship of the interacting nuclei and thus conveys stereochemical information.

(3) The width of the lines depends upon the rate of molecular motion and on the presence of a neighbouring species with a magnetic moment. Many biological molecules are large and tumble slowly in solution. A consequence of this slow tumbling is that the magnetised nuclei or electrons relax very rapidly to their lower magnetic energy level. This causes the lines to be broadened, and the line-broadening effect can be turned to advantage to estimate the mobility of a biological molecule (Chapter 14).

(4) A neighbouring species with a magnetic moment can also increase the line width by increasing the relaxation rate. This effect depends upon the distances between the interacting particles, allowing distances to be estimated very accurately within macromolecules, for example within the active site of an enzyme (Chapter 12).

The range of nuclei which can be studied by n.m.r. is quite large and includes many of biochemical importance, for example ^1H, ^2H, ^{13}C, ^{19}F, ^{23}Na, and ^{31}P.

Free electrons are found in the reactions of some biochemical systems such as the flavins and unpaired electrons in metal ions also give rise to signals which can be observed by e.s.r. (Chapter 9).

Despite the wide range of sources of magnetic resonance signals available from biological systems, probe methods have also proved valuable, especially in the case of e.s.r. Non-paramagnetic metal ions can often be replaced by paramagnetic ions with little loss of biological activity. For example, non-paramagnetic zinc can be replaced by paramagnetic cobalt in carbonic anhydrase, and we shall see in the next chapter the value of Mn^{2+} as a paramagnetic probe substituting for Mg^{2+} in the kinases. The range of systems that can be studied by e.s.r. has been vastly increased by the use of stable free radicals, or spin labels, which can be incorporated into many types of biological system. These compounds are nitroxides (11.7.1); the free radical is stabilised by the steric hindrance of the four alkyl groups which shield it. These probes can be used to report on the mobility and polarity of their environment. Some, such as the bromoacetamide (11.7.2) can (like (11.6.2)) be covalently attached to proteins and their behaviour used to infer the behaviour of the protein in solution. A non-covalent probe, 12-doxyl stearic acid (11.7.3), is used to assess the fluidity and order of biological membranes (Chapter 14).

Spin labels have been criticised as probes because of the perturbation which may be caused by their relatively bulky structure. This may be significant in some cases, but the sensitivity and wide applicability of spin labels makes them a valuable addition to magnetic resonance methods.

Figure 11.7.

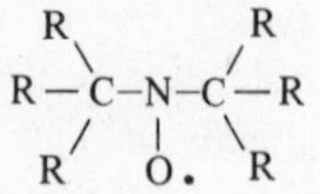

R = an alkyl group
11.7.1

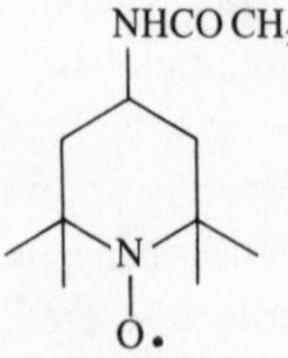

11.7.2

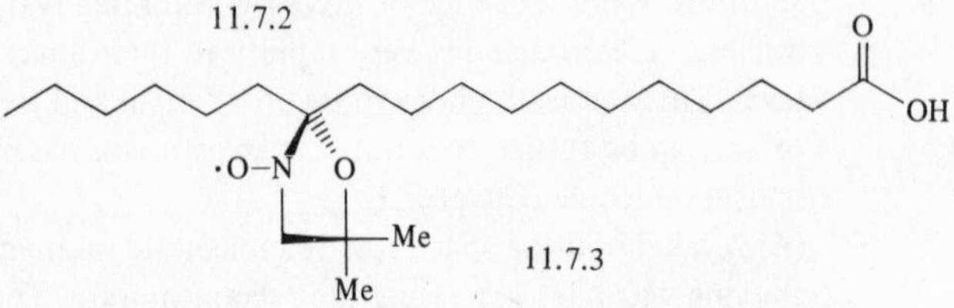

11.7.3

11.4. Chemical and enzymatic methods of structure determination

We have seen earlier (Chapter 5) examples of the classic chemical approach to the structure determination of small molecules. The approach depends upon being able to carry out controlled reactions to degrade the unknown into smaller, simpler fragments and then identifying the products. The structure of the unknown is then deduced from the structure of its degradation products and the known course of the reactions which produced them. This approach has been supplemented with the use of physical methods and, for small molecules, physical methods have to a large extent superseded the classical chemical method. The size and complexity of biological macromolecules however, requires that chemical degradation is still necessary for a complete structure determination.

Very powerful and specific techniques have been developed which depend on a combination of chemical degradation methods with physical methods for separation, purification and analysis of the degradation products. The specificity of enzymes is also exploited to degrade molecules in a defined way. More recently, enzymes have also been used to synthesise molecules as part of a structure determination (Section 11.4.2).

Two examples are discussed in this section, one to illustrate the methods used in determining the amino acid sequence of an enzyme and a second to illustrate the corresponding activity in studying DNA. In both these cases we are dealing with linear polymeric systems whose basic structural unit is known – amino acids and deoxyribonucleotides respectively. The methods aim to determine the linear sequence of these components upon which all their other properties depend. The sequence is usually known as the primary structure of the biopolymer. Other orders of structure which describe the three-dimensional shape of the molecule can also be determined (Chapters 12 and 16).

11.4.1. Protein sequencing

The strategy of protein sequencing is very much that of the classical organic chemist in determining the structure of a smaller molecule. Firstly the molecular weight of the pure protein and its amino acid composition is determined. This can be done by complete hydrolysis of the polypeptide followed by separation of the resulting amino acids by ion-exchange chromatography in an automated amino acid analyser. Knowing the numbers of residues of each of the individual amino acids present in the protein we are in a position to find out how they are joined together and we can approach this by carrying out controlled partial hydrolyses of the polypeptide. These hydrolyses are carried out at known sites in the protein and give rise to fragments whose structures can be investigated further, after purification by gel filtration or ion exchange chromatography and characterisation for amino acid composition.

There are a number of ways in which peptide chains can be hydrolysed at specific sites. Some depend upon the specificity of an enzyme such as trypsin, whilst others are entirely chemical (Table 11.5). The chemistry of these methods used in protein sequencing is worth further study (see Thomas, 1974). If these primary peptides are short enough it is probably possible to determine their sequence directly by the chemical method known as the Edman degradation. If not, further specific hydrolyses of the primary peptides will be necessary to yield manageable-sized fragments. The sequence of small peptides (up to eight amino acids) can be determined by mass spectroscopy.

Once the sequence of all the fragments has been determined, it remains to fit them together to give the complete primary structure. Since there could obviously be several ways in which the fragments might be joined, it is necessary to resolve the ambiguity by using different initial hydrolysis methods to give different primary fragments which will overlap the first set. Combination of these fragments with the first set will often lead to the complete sequence.

Table 11.5. *Some degradation methods used in protein sequencing*

Chemical methods

1-fluoro-2,4-dinitrobenzene   Used for determining the N-terminal amino acid

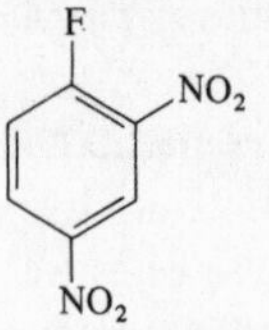

Dansyl chloride   Used for determining the N-terminal amino acid

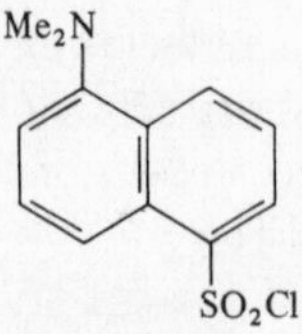	
Cyanogen bromide (CNBr)	Cleaves peptides at methionine residues
Phenyl isothiocyanate (Edman degradation)	Used to selectively remove the N-terminal amino acid. Can be used to sequence small peptides and can be used in automated equipment

Enzymic methods

Carboxypeptidase	Hydrolyses selectively the C-terminal amino acid.
Trypsin	Hydrolyses at marked peptide bond where R = arginyl or lysyl side-chains
Chymotrypsin	Hydrolyses at marked peptide bond where R = tryptophanyl, phenylalanyl or tyrosyl side-chains 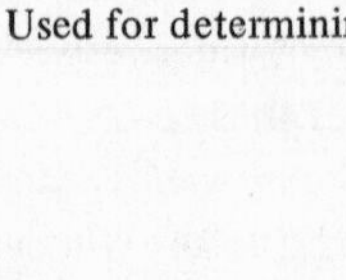

In such an outline we cannot cover many of the aspects of protein sequencing which complicate matters, for example the localisation of amide derivatives of acidic amino acids or the positions of disulphide bridges between cysteine residues, all of which will have important consequences in the tertiary structure of the protein and in its biochemical properties. Figure 11.8 shows a simplified scheme of the sequencing of adenylate kinase, which was first purified by the methods already outlined in Table 11.3 (Heil *et al.*, 1974). More details about sequencing methods can be found in Thomas (1974).

11.4.2. DNA Sequencing

Unlike proteins, DNA is a polymer with only four different monomeric units, the four nucleotides derived from adenine, guanine, cytosine and thymine. Partial hydrolysis of the phosphate ester bonds between these bases

Figure 11.8.

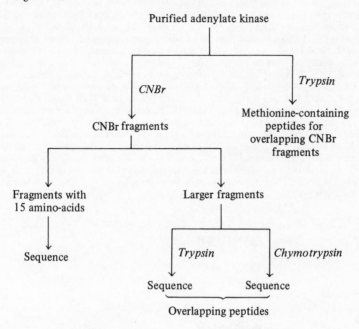

may be expected to give many similar fragments, whereas hydrolysis of a protein usually leads to a few unique peptides. Several ways have recently been
developed to sequence DNA molecules and the one we shall describe differs
in many respects from the methods of protein sequencing. This is the 'plus–
minus' technique of Sanger and his colleagues, which has recently provided
some very important results for molecular biology (Fiddes, 1977). Progress in
this field is at present very rapid, so several newer methods are likely to gain
importance.

In the plus–minus method enzymes are used to degrade the DNA and they
are also used to synthesise fresh DNA in a controlled way so that the sequence
of bases is revealed. Firstly, a circular DNA molecule, such as may be found
in a bacterium or virus, is specifically hydrolysed into fragments by enzymes
known as restriction endonucleases. These enzymes hydrolyse DNA at specific
base sequences, so that the terminal sequences of the fragments produced are
known. The fragments are separated by gel electrophoresis and one of the
purified fragments is allowed to bind to its complementary single strand of the
DNA molecule from which it was produced, so that the complementary strand
can act as a template for further synthesis of DNA. The fragment binds at the
specific site complementary to its own sequence and from this site new DNA
is synthesised by the addition of the enzymes and nucleotides required. The
nucleotides are added in three ways:

Figure 11.9. Primer is extended under three different conditions after being annealed to the template (whose nucleotides are here arbitrarily numbered from 1 through 18). In the 'zero' system the primer and template are incubated with DNA polymerase and all four nucleotide building blocks under conditions such that successive nucleotides are added, according to the base-pairing rules, to produce extension products (*open boxes*) of every possible chain length. A subsample of these extension molecules is separated according to size by electrophoresis on a polyacrylamide gel (*vertical slab at right*). Other subsamples are subjected to 'minus' or 'plus' treatment. In the minus-A system, for example, the A building block is withheld from the mixture; the extension products are further extended (*heavy boxes*), but only up to the position just before an A. Now electrophoresis reveals only three bands, each band representing the position in the sequence one nucleotide before the next A. The same procedure is followed with each of the other three nucleotides withheld from the synthesis process to establish the positions one nucleotide before each appearance of a G, T or C. In the plus-A system only As are provided, and the polymerase is one that degrades DNA except in the presence of an excess of a nucleotide building block. Extension products are therefore degraded (*broken boxes*) until position of an A is reached. Electrophoresis again shows three bands, this time at position of As. From Fiddes, 1977.

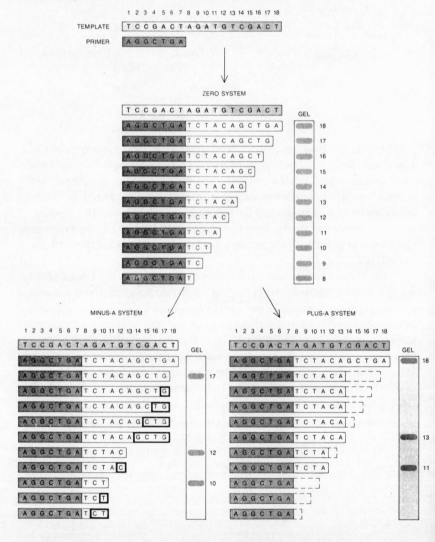

(1) In the 'zero system' all four nucleotides are added, so that new strands of DNA are produced of all possible lengths (Figure 11.9). The new strands are separated from the template strand of DNA by use of the same restriction enzyme as before and then are separated from each other by gel electrophoresis which gives a pattern of many bands, one for each new strand separated according to their molecular weight.

(2) In the 'minus system' one of the four nucleotides is omitted, and a sample of the products of the zero system is incubated with the new mixture so that synthesis takes place only up to points where the omitted nucleotide is required. Electrophoresis of this mixture will lead to fewer bands and all of these must have as their next base in the sequence the nucleotide which was omitted. This experiment can be repeated for the other three nucleotides.

(3) In the 'plus system' only one nucleotide is provided and enzymes are added such that the extended DNA molecule produced by the zero system is degraded until the position of the added nucleotide is reached. Fragments identified by electrophoresis from the plus system will have the added base as their terminus.

If plus and minus experiments are carried out from all four nucleotides and the strands produced compared with those from the zero system on gel electrophoresis it is possible to read off the base sequence of the strand of the DNA which had initially been cleaved by the restriction enzyme, and the sequence extends for as far as the resolution of the gel electrophoresis will allow.

Recently, techniques for sequencing DNA have become so rapid and convenient that, provided material is available, it is now easier and faster to determine the sequence of bases in DNA than it is to determine the primary structure of the protein for which the DNA carries the information (Wu, 1978).

12   Enzyme mechanisms of phosphate transfer

12.1.   Introduction

There is no doubt that an understanding of enzyme mechanisms is essential to the molecular approach to biochemistry. Practically all chemical reactions in cells are catalysed by enzymes, and many other processes such as transport, in which a chemical reaction does not occur, have features in common with enzyme mechanisms (Chapter 15). In addition, the forces which operate when substrate binds to the enzyme active site are the same as those found in many other specific binding processes found in biology, for example the binding of a hormone to its receptor and an antigen to an antibody (Chapter 16).

There are both chemical and biological reasons for studying the mechanisms of enzymes. The chemical goal is to understand the basis of the unequalled catalytic efficiency and specificity of enzymes (see Chapter 10). From a biological viewpoint, knowledge of enzyme mechanisms is essential if we are to understand how individual enzymes function in relation to the overall chemical activity of the cell. In addition to questions of efficiency and specificity we want to understand how the activity of a given enzyme is adapted to the requirements of the cell in which it is found (Chapters 13 and 16).

There are sound practical reasons, too, for studying enzyme mechanisms. The chemist may be able to design more efficient catalysts using the example of nature as a model. Knowledge of the mechanism of an enzyme may help a pharmaceutical chemist to design a drug which can interact specifically with that enzyme (Chapter 15).

Pure enzymes are one of the simplest biological systems that we can study (Chapter 2) but they are by no means easy to obtain and in many cases compromises have to be made. For example, in the kinetic studies described below, the use of a pure enzyme may not be necessary, although it would be desirable, as long as the preparation used has no catalytic activity which might compete or interfere with the reaction being studied. In contrast, a relatively large amount of pure crystalline material is required for X-ray crystallography.

To illustrate the ways in which enzymes can be studied and the molecular concepts that can be derived from these studies we have chosen to discuss the group of enzymes that catalyses phosphate-transfer reactions, the phosphotransferases or kinases. Since one of the most important chemical components in a cell is a phosphate, ATP (Chapter 3), which has a central role in the transfer of phosphate between many important metabolites (Appendix 1), and many control steps depend upon phosphorylations (Chapter 13), phosphate transfer could be regarded with hydrogen or electron transfer (Chapter 8) as being one of the most important chemical reactions in biology.

Kinases catalyse the general reaction shown in Figure 12.1. The reaction can take place in both directions. An example of Figure 12.1 in a forward direction is the transfer of a γ-phosphate from ATP to the nucleophilic group X on the substrate R–X (such as glucose). In the reverse direction the phosphate donor might be phosphoenolpyruvate, which donates its phosphate to the β-phosphate of ADP, thus forming ATP. Figure 12.2 shows the reactions catalysed by the kinases which we shall discuss in this chapter.

Hexokinase catalyses the conversion of glucose to glucose 6-phosphate with the phosphate donor, ATP. This is the first step in the metabolism of exogenous glucose in many cells. Such early steps in biochemical pathways are often found to limit the rate of the whole pathway.

Glucose 6-phosphate can be utilised in a number of ways (Appendix 1). Commonly, if the cell is oxidising glucose to provide energy the glucose will

Figure 12.1.

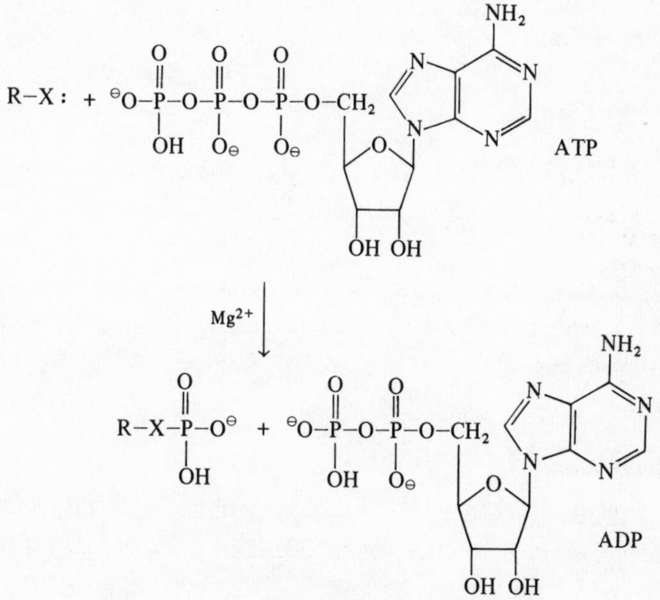

enter the glycolytic (or glucose-breakdown) pathway, which is catalysed by a series of enzymes present in the cytoplasm of the cell. Towards the end of this pathway phosphoenolpyruvate is formed.

Pyruvate kinase catalyses the reaction in which phosphoenolpyruvate donates its phosphate to ADP, forming ATP and pyruvate. This is an example of the generation of ATP by direct transfer of a phosphate from an intermediate in a metabolic pathway to ADP, a process known as **substrate-level phosphorylation**. This differs from the main way in which aerobic cells generate ATP which is by **oxidative phosphorylation** (Chapter 2). Substrate-level phosphorylation takes place in almost every type of cell, and is particularly important when the oxygen supply to the cell is very low, for example in fermenting yeast and in muscle under strenuous exercise. It allows the muscle to continue to contract (a process which requires ATP) in the absence of a sufficient oxygen supply for the complete oxidation of glucose to carbon dioxide and water.

Another kinase that is important in muscle is **adenylate kinase**, which in the form found in muscle is also known as myokinase. This enzyme is important in many cells as a means by which AMP, which is the product of many acti-

Figure 12.2.

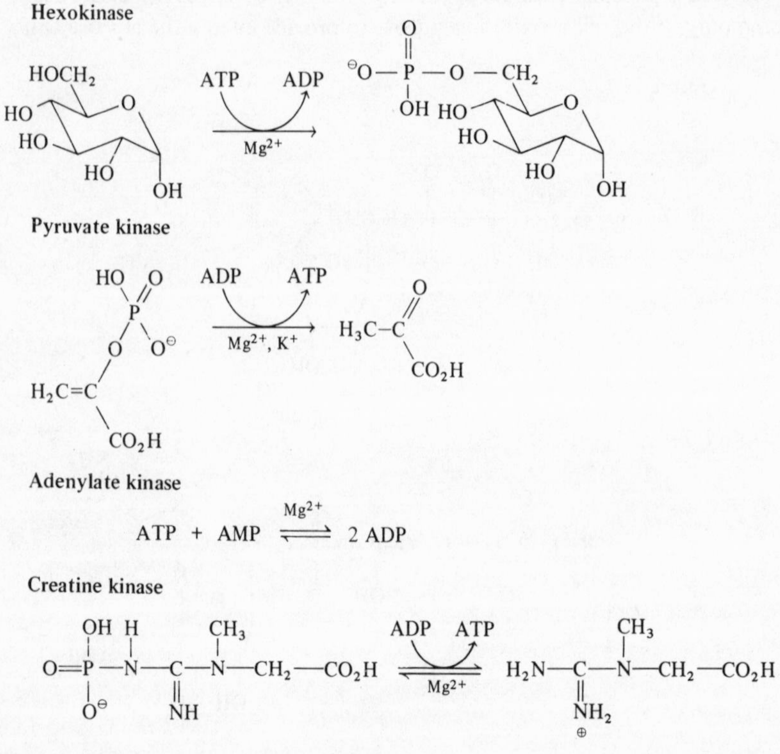

Hexokinase

Pyruvate kinase

Adenylate kinase

$$ATP + AMP \underset{}{\overset{Mg^{2+}}{\rightleftharpoons}} 2\ ADP$$

Creatine kinase

vation reactions (Chapters 3 and 7), can be converted to ADP for rephos-
phorylation to ATP. In muscle it provides an additional means by which ATP
can be generated (from two molecules of ADP) and the AMP also produced
acts as an activator of glycogen phosphorylase, the enzyme which catalyses
the hydrolysis of the energy store glycogen. In this way, adenylate kinase is
thought to promote the supply of energy to a contracting muscle cell.

Adenylate kinase is found in the space between the two membranes of the
mitochondria of eukaryotic cells. **Creatine kinase**, another muscle enzyme, is
found in the sarcoplasm, the cytoplasm of the muscle cell. Creatine phosphate
is stored by the muscle cell, and provides a buffer for the ATP levels within
the cell. As soon as ADP is formed by contraction, the phosphate from creatine
phosphate is transferred to ADP, catalysed by creatine kinase, thus maintaining
the ATP concentration until the supply of creatine phosphate runs out.

In all chemical reactions, biological as well as laboratory, it is impossible to
define every atomic and electronic movement. The best that can be done is
to eliminate some of the possible mechanisms by well-chosen experiments,
and the ideal approach is to use a wide range of techniques, including many of
those described in Chapter 11. In the pages that follow, we shall see examples
of physical methods, including X-ray diffraction and n.m.r. spectroscopy, and
chemical methods, including chemical modification studies and studies of the
rates of the enzyme-catalysed reactions. Information from all these sources
can be brought to bear on the catalytic mechanism to achieve a coherent
picture of the process.

12.2. The chemical mechanism of phosphate transfer

The mechanism of reaction of phosphate esters depends greatly on
the structure of the organic component and upon the nature of the catalysis
available, as we saw in Chapter 3. Unless there are good reasons to favour C–O
cleavage, such as allylic activation of the phosphate ester, alkylation reactions do not occur. More commonly, phosphate esters undergo P–O cleavage,
behaving as phosphorylating agents; this is the reaction catalysed by the kinases.
In order to transfer a phosphate group, the negative charge cloud on the oxygen
atoms must be shielded from the attacking nucleophile. Metal ions fulfil this
role in the kinases.

As with nucleophilic aliphatic substitution reactions (Chapter 6), we can
envisage two extreme mechanisms for nucleophilic substitution at phosphorus
in a phosphorylation. If the leaving group departs before the nucleophile
attacks we will have a unimolecular or S_N1 route, but if attack of the nucleophile and the departure of the leaving group are concerted we have a bimolecular S_N2 case. Phosphates are tetrahedral in structure, and the stereochemical
consequences of S_N1 and S_N2 substitutions are similar to those for carbon. It
is, however, more difficult to follow stereochemistry at phosphorus because

of the difficulty in obtaining chiral phosphates, although these have now been obtained, and the stereochemistry of the reaction catalysed by the enzyme alkaline phosphatase has been determined (Jones, Kindman & Knowles, 1978). Chirality was achieved by labelling the oxygen atoms in the phosphate with ^{16}O, ^{17}O and ^{18}O. The overall reaction proceeded with retention of the stereochemistry of the phosphate, and since a phospho-enzyme intermediate is well established in this system, unlike the kinases (as we shall see shortly), this result is consistent with two S_N2-type displacements taking place at phosphorus.

12.3. Kinetic studies of mechanisms

Many of the chemical mechanisms we have discussed earlier are based on studies of the rates of the reactions and how this varies with the concentrations of reactants. Such experiments allow the chemist to distinguish between bimolecular and unimolecular processes (Chapter 6). In chemical reactions the sequence of events is usually as follows: (1) collision of reactants; (2) conformational change to reacting conformation; (3) reorganisation of electrons to form intermediate (if present in reaction) by way of transition state; (4) breakdown of intermediate again via a transition state with electronic movement; (5) diffusion of products away from each other. This sequence of events is central to all reactions in which bonds are made or broken and so it also applies to enzyme-catalysed reactions. There are, however, additional steps in enzymic reactions which may have a very great significance on the rate of the reaction and its dependence on the concentrations of substrates and other molecules. The sequence of events in a two-substrate enzyme-catalysed reaction such as a phosphate transfer could be: (1) binding of each substrate to enzyme; (2) conformational change of enzyme or substrate or both; (3) bond-making and bond-breaking steps (as above); (4) further conformational change on formation of products; (5) release of products.

It is the initial and final events in this sequence that can be most directly studied by kinetic methods, the substrate-binding and product-release steps. In the case of the kinases, as two-substrate reactions, there are clearly several ways in which these steps could take place. For example, ATP might bind first to the active site of hexokinase and donate its phosphate to a group at the active site. The resulting ADP might diffuse away before the glucose binds and the phosphate is transferred to the glucose. In such a case, direct transfer of phosphate from ATP to glucose is not possible. Alternatively, the glucose and the ATP might bind to the enzyme, forming a ternary enzyme–substrate complex. Now a direct phosphate transfer is possible. Clearly these two binding mechanisms restrict the kind of bond-making and -breaking mechanism, so kinetic studies are important in defining the system on which electronic shifts take place.

Three possible kinetic models for creatine kinase are illustrated in Figure 12.3. The first two both involve a ternary complex between creatine phosphate and ADP and enzyme. The third model involves binary enzyme–substrate complexes. The two ternary-complex models can be distinguished by the order in which the two substrates bind to the active site of the enzyme. In the first, equilibrium binding and release of either substrate or either product takes place in any order (a random-order mechanism), whereas in the second the enzyme requires that one substrate binds before the other and that the products too leave the enzyme in a defined sequence (an ordered sequential mechanism).

By assuming that the reaction is taking place at a steady state, a common assumption in many chemical kinetic studies, one can derive rate equations for these three mechanisms. Their exact form need not concern us here: details can be found in Roberts (1977) or Ferdinand (1976). These kinetic models allow us to make predictions about how the rate of the reaction will vary under defined conditions. For example, we may investigate the dependence of the rate on substrate concentration or the pattern of inhibition produced by products or other inhibitors of the reaction. We can compare the predictions with experimental results and so provide evidence for or against the possible kinetic models.

It is also possible to study kinetics on the phase of the reaction before a steady state is reached. Such studies provide a powerful method which may uncover mechanistic features (e.g. a conformational change of the enzyme) which are not detectable by steady-state kinetics (Halford, 1974).

Steady-state kinetic studies led to the conclusion that the random-order mechanism fitted the data for creatine kinase best. Note that we have not said

Figure 12.3. Possible kinetic models for creatine kinase.

Cr = creatine
CrP = creatine phosphate
E = creatine kinase
P = phosphate

Random-order mechanism

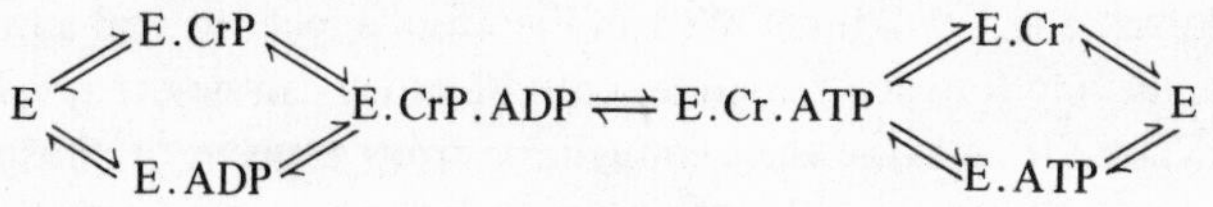

Ordered sequential mechanism

$$E \rightleftharpoons E.CrP \rightleftharpoons E.CrP.ADP \rightleftharpoons E.Cr.ATP \rightleftharpoons E.Cr + ATP$$
$$E + Cr$$

Binary enzyme–substrate complex

$$E + CrP \rightleftharpoons E{-}P + Cr \rightleftharpoons E{-}P.ADP \rightleftharpoons E.ATP \rightleftharpoons E + ATP$$

that creatine kinase operates by a random-order mechanism; we are only entitled to say that this kinetic model is the best description of the experimental data. We can envisage during the enzyme-catalysed reaction either substrate binding to the active site by a rapid equilibrium process. Phosphate transfer, by a mechanism which must be investigated by other methods, takes place once the ternary complex has formed, and the products diffuse away from the enzyme, again in random order. A direct transfer of the phosphate from creatine phosphate to ADP is, therefore, possible in the ternary complex, and this step in creatine kinase is probably rate determining. However, this need not always to be the case in kinases. The mechanism of hexokinase, determined by similar methods to creatine kinase, is also found to be random order, but it can further be shown that the rate at which the product, glucose 6-phosphate, is released from the enzyme is slower than the rate of phosphate transfer (Cleland, 1975). This situation is similar to the one we met in Chapter 10, where we saw that measurements of all the steps involved in the catalysis of the bond-making and -breaking steps in an enzyme-catalysed reaction showed that they were all extremely rapid, and that the rate-limiting step was the diffusion of substrates to and from the enzyme. Such 'evolution to perfection' provides a strong advantage for systems which eliminate slow diffusion steps, such as the multi-enzyme complexes we consider in the next chapter.

Kinetic studies of the effects of activators and inhibitors of enzymes are often used to provide evidence for possible ways in which the activity of the enzyme is controlled *in vivo*. We shall meet examples of such control mechanisms in the next chapter, but we should remind ourselves that before such experiments can be regarded as relevant to the situation in the living cell we should be satisfied that the experimental conditions used are not unacceptably different. As we mentioned in Chapter 2, particular attention needs to be given to the concentrations of enzyme, substrate and activators or inhibitors. Reliable values for these concentrations *in vivo* can be very difficult to determine.

Returning to the kinases, we can conclude that the random-order mechanism appears to be a good model for many kinases, including creatine kinase, pyruvate kinase and hexokinase. We can therefore expect to find a binding site on the enzyme for each substrate, both of which can be occupied at the same time. The sites may be close enough for direct transfer of phosphate between the two substrates. The kinetic model does not tell us anything about the mechanism of the phosphate transfer. To elucidate this we must focus on the events which take place once the substrates have bound to the active site of the enzyme and are close enough to react.

12.4. The bond-making and bond-breaking steps
12.4.1. Introduction: protein structure
In order to describe the catalysis of the chemical reaction by any enzyme we need to known something about the structure of the active site,

for example what amino acid side chains are present and how they and the substrate are arranged in three dimensions. We can then extrapolate this to propose a scheme by which electron shifts could take place to carry out the changes in chemical bonding. This model could be expressed in the formalism of curly arrows (Appendix 2).

The most direct way of defining the three-dimensional structure of a protein is by X-ray crystallography. This can give us a static picture of the way in which the participating chemical groups are aligned and provide the basis for a more dynamic interpretation. To carry out such an analysis we need to have a pure crystalline protein, and it is also a great help if we know the amino-acid sequence of the polypeptide chains in the protein (Chapter 11). We can distinguish in the three-dimensional structures derived from X-ray crystallographic studies four orders of protein structure. These will be familiar to many biochemists and we outline them only briefly here. More details can be found in one of the recommended biochemistry texts.

The amino acid sequence of the polypeptide chains is known as the primary structure of the protein. Ultimately it is the primary structure which defines all the other orders of structure, because within it are arranged in a precise sequence all the amino acids in the protein with their associated side chains. The further orders of structure depends on the bonding which is possible between the side chains and between the atoms of the peptide bonds. These interactions, together with the fact that a peptide bond can only have a limited number of conformations, are now making it possible to predict the three-dimensional structure of a protein from a knowledge of its primary sequence alone, using a computer to carry out the necessary calculations (Sternberg & Thornton, 1978).

If one examines the three-dimensional model of a protein derived from X-ray crystallography, one can often discern areas of the folded polypeptide chain which appear regular in structure. These are secondary structures, and the most common are the α-helix, in which the helix is stabilised by hydrogen bonds between residues three amino acids apart on the same chain, and the β-sheet, which is also stabilised by hydrogen bonding but this time the bonds are formed between different chains (Figure 12.4). Examples of secondary structure will be found in later sections, particularly hexokinase (Section 12.4.2) and immunoglobulins (Chapter 16).

Tertiary structure, observed as non-regular folding of the polypeptide chain, arises from any type of intramolecular non-covalent interaction (hydrogen bonding, hydrophobic bonding, ionic bonding) involving amino acid side-chains.

It is important to realise that, apart from the peptide bonds of the polypeptide chain and the occasional disulphide bond between suitably placed cysteinyl residues, all the forces which hold the protein in a defined conformation are non-covalent in nature. The effects of these interactions, hydrogen bonds, ionic bonds and hydrophobic bonds, are cumulative and give the protein structural stability in the cell and also, by slight changes in the relationship of groups, flexibility to respond to the influences of substrates or other small molecules (Chapter 13). Non-covalent interactions are of the greatest importance in many biochemical systems. Examples to be discussed later include the interactions between a hormone and its receptor and an antigen and an antibody (Chapter 16).

Many proteins show a fourth order of structure, known as quaternary structure. These proteins contain more than one polypeptide chain or subunit, which can often be dissociated from each other. Subunits can be identical, each having the same function, or they may have different functions, for example one may be involved in catalysis of a chemical reaction whilst others may be concerned with controlling the rate of the reaction. Kinases show quaternary structure; further examples will be found in the next chapter.

Figure 12.4. Types of secondary structure in proteins. (From Yudkin, M. and Offord, R. 1973 'Comprehensible Biochemistry', Longman, London.)

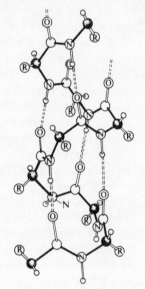

The α-helix

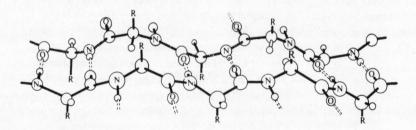

The β-pleated sheet

for example one may be involved in catalysis of a chemical reaction whilst others may be concerned with controlling the rate of the reaction. Kinases show quaternary structure; further examples will be found in the next chapter.

12.4.2. X-ray diffraction studies of kinases

In recent years the resolution achieved by X-ray crystallographic analyses of the kinases has improved sufficiently to provide us with a framework on which detailed mechanistic models can be built. Hexokinase, a dimer of molecular weight 104 000 in yeast, has been described at a resolution of 0.35 nm, and the monomer structure has been resolved to 0.23 nm, (Steitz *et al.*, 1976). Comparing this with a carbon–carbon single bond, which extends to 0.154 nm, we can see that this resolution is nearly sufficient to identify the positions of specific covalent bonds in the protein, and with a knowledge of the amino-acid sequence, a three-dimensional structure can be derived which shows how the protein is folded and which can give some indication how substrates bind. This is shown in Figure 12.5 for yeast hexokinase. Such a diagram

Figure 12.5. The structure of yeast hexokinase, as deduced from X-ray crystallography studies. α-Helices are represented by tubes and β-sheets by arrows; bound substrate molecules are represented by the bond structures; the numbers refer to amino acid residues. (From Steitz, Anderson, Bennett, MacDonald & Stenkamp (1977), *Biochemical Society Transactions*, **5**, 622.)

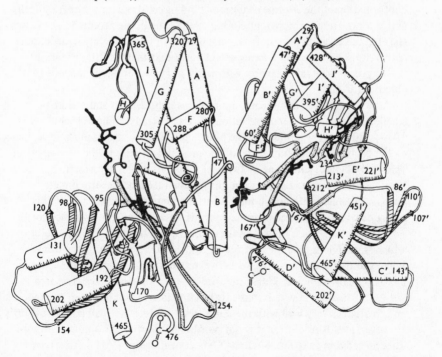

makes clear the basic structure of the hexokinase dimer. The two subunits are not identical (an unusual situation, most oligomeric proteins show some symmetry in their quaternary structure) and each consists of two lobes separated by a deep cleft. One lobe consists of mainly α-helix, whilst the other contains both helix and β-sheet structures. This model was built without a knowledge of the amino-acid sequence of the protein, so the identity of many of the amino acid residues in the structure is unclear.

The diagram also shows the binding sites for glucose and ATP, which have been deduced from the enzyme crystallised in the presence of glucose. Unfortunately, attempts to bind ATP to the active site in the crystal leads to breaking up of the crystal structure. Similarities have been observed in the nucleotide binding sites of a number of enzymes, which has led to the suggestion that all nucleotide binding sites have evolved from a common ancestor. The proposed ATP binding site (deduced from the binding of adenosine rather than ATP) in yeast hexokinase is clearly different from that in other nucleotide binding sites (for example that for NAD in lactate dehydrogenase), so the idea of a common ancestor may be regarded as unlikely for these enzymes.

The X-ray work further suggests that a marked change takes place in the structure of the enzyme on bonding of glucose. A large part of the β-structure lobe moves about 0.6 nm nearer the mainly α-helical lobe. Such a change is consistent with Koshland's ideas of induced fit (Chapter 10), in which the conformation of the enzyme changes on binding substrate into the catalytically active form. Here the apparent closing of the cleft in the protein might have the effect of excluding water from what is thought to be a mainly hydrophobic active site. The absence of water would mean that there is little chance of hydrolysis of ATP taking place, with water as a competing nucleophile (with glucose) for ATP.

A model of adenylate kinase, molecular weight 22 000 and one of the smallest kinases, at 0.3 nm resolution has also been determined (Schulz *et al.*, 1974). This study was aided by the prior knowledge of the amino-acid sequence of the protein. The overall shape of the protein is similar to the monomers of yeast hexokinase, consisting of two lobes separated by a deep cleft. It was not possible to determine the location of the substrate binding sites directly because crystals containing bound substrate could not be prepared. Other methods which we consider in the next two sections provide evidence for this.

Examination of cat muscle pyruvate kinase has reached 0.6 nm resolution, which is sufficient to show that the basic structure is similar to the other kinases in that a deep cleft exists between two lobes, but unlike the lobes in hexokinase or adenylate kinase they are of very different sizes. The area within the protein in which substrate binding takes place has been located, and this agrees very well with the n.m.r. results which we shall consider shortly. However, at this resolution it is not possible to identify the amino acid residues involved in substrate binding. X-ray crystallography of creatine kinase has been hampered by the lack of suitable crystals.

Although we know of the overall shape of several kinases, and in the case of hexokinase we can observe a conformational change which may have bearing on the mechanism of the enzyme, X-ray diffraction studies have not given any useful information about the mechanism of phosphate transfer. This is in marked contrast to the serine proteases, for example chymotrypsin (Chapter 3), where X-ray crystallographic structures could be easily interpreted in terms of a chemical mechanism for peptide bond hydrolysis. Before the three-dimensional structure of enzymes like chymotrypsin was known, much was known about the mechanism by chemical studies of the enzymes. We shall see in the next section how far chemical modification experiments and other chemical probes can help us to describe the way in which phosphate transfer takes place in the kinases.

12.4.3. Chemical methods

One of the fascinating things about the chemistry of enzymes and enzyme active sites is that although we know what kind of reactivity might arise from those amino acids whose side chains contain reactive functional groups, we do not know how reactive these groups will be. They may be more reactive than would be expected for the amino acid in solution because of the effect of neighbouring groups, or they may be so hindered by the protein structure that they show very little activity. The existence of a very reactive functional group in an enzyme does not necessarily imply that this group is essential for the catalytic mechanism (Thomas, 1974).

The chemical approach to the elucidation of an enzyme mechanism is to attempt to identify the functional groups of amino acid side chains which are essential for catalysis by a specific chemical modification of the enzyme. This is done by exposing the enzyme to a reagent which is specific under defined conditions for only one type of functional group. If the enzyme is inhibited by such a reagent, it is reasonable to conclude that the reagent has modified a functional group which is necessary for the activity of the enzyme. The functional group could be important in carrying out the chemical reaction itself, or in binding substrate or in effecting a change in the conformation of the protein to one which is needed for the reaction to take place. You will find a series of reagents for the various functional groups in Table 12.1.

Problem 12.1.

We shall discuss the applications of some of the reagents in Table 12.1 to kinases in the following paragraphs, but it would be valuable at this stage to consider how you would expect these reagents to react with amino acid functional groups stated.

The combination of a knowledge of the amino acid sequence and X-ray crystallographic studies on adenylate kinase suggested that a cysteinyl residue is present at the active site. The thiol group which is present in a cysteinyl

Table 12.1.

Amino acid	Functional group	Reactivity	Reagent
Lys	$-NH_3^{\oplus}$	Ionic bonding, $H^{\oplus}$ donor, acid–base catalysis, nucleophile in unprotonated form.	Anhydrides, HNO_2,
Cys	$-SH$	Nucleophile	ICH_2CONH_2,
Ser	$-OH$	Nucleophile	$[(CH_3)_2CHO]_2PF$ (with P=O)
Asp } Glu }	$-CO_2^{\ominus}$	Ionic bonding, $H^{\oplus}$ acceptor	CH_2N_2
His		H-bonding, $H^{\oplus}$ donor, acid–base catalysis	XCH_2CO_2H, $(EtO)_2CO$,
Trp		Hydrophobic bonding	
Tyr		Hydrophobic bonding, proton donor (phenol)	, HOI
Phe		Hydrophobic bonding	
Met	$-S-CH_3$		XCH_2CO_2H
Arg		Ionic bonding	$CH_3-\underset{\underset{O}{\|}}{C}-\underset{\underset{O}{\|}}{C}-CH_3$, $Ph-\underset{\underset{O}{\|}}{C}-\underset{\underset{O}{\|}}{C}-H$

X = leaving group such as I

Discussions of the reactivity of amino acid functional groups can be found in these chapters: lysine, Chapter 3; cysteine, Chapter 12; serine, Chapter 3; histidine, Chapter 6; tryptophan, Chapter 6; tyrosine, Chapter 7; methionine, Chapter 6; arginine, Chapter 12.

residue is a nucleophilic group, and it readily reacts with reagents which carry a suitable leaving group. Examples of such reactions for reagents illustrated in Table 12.1 are given in Figure 12.6. The thiol group is one of the most nucleophilic groups found on amino acid side-chains and often takes part in catalytic mechanisms, for example in some peptidases such as papain and in glyceraldehyde-3-phosphate dehydrogenase. The seryl hydroxyl group is an intrinsically weaker nucleophile, and the ϵ-amino group of lysine is usually protonated at physiological pH and is therefore not a good nucleophile, although it is important in ionic bonding and in binding carbonyl groups as imines (Chapter 4).

A good example of a reagent which reacts with a thiol group is 7-chloro-4-nitrobenzo-2,1,3-oxadiazole (NBD-Cl) and this has been used in studies on adenylate kinase. There are two cysteinyl residues in adenylate kinase, one at position 25 and one at position 187 in the amino acid chain. One or both of these residues may be involved in catalysis or binding and we need to know how they can be distinguished. The special environment of a functional group at the active site of an enzyme often makes it more reactive than it would be in isolation, particularly if it is directly involved in catalysis, as, for example, is serine-195 in chymotrypsin.

Figure 12.6.

$$R-SH + ICH_2CONH_2 \longrightarrow R-S-CH_2CONH_2$$
Iodoacetamide

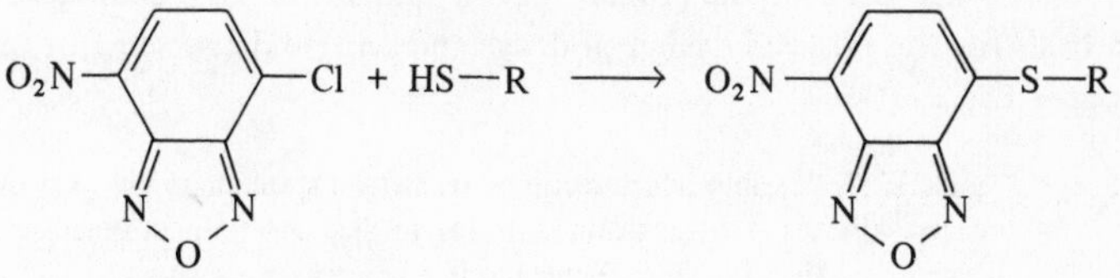

NBD−Cl

If adenylate kinase is treated with NBD-Cl, the thiol group on residue 25 is found to react 40 times faster than that at position 187. The interpretation of these results is by no means as precise as one might expect from a chemical experiment. Two possibilities might apply. Firstly the group at residue 25 might be a component of the active site; alternatively it is possible that this thiol group is more reactive because it is more accessible to the reagent than the group on residue 187, for example by being on the surface of the protein rather than deeply buried in its interior. It is, however, clear that modification of one of the two thiol groups is relevant to the catalytic mechanism because the enzyme becomes catalytically inactive when the thiol group on cysteine-25 is completely blocked. Now that we know that cysteine-25 is necessary for catalysis we can ask what role it plays. Some idea can be obtained from the following experiment. If adenylate kinase is first incubated with Mg-ATP, so that the ATP binding site is occupied, and the enzyme treated with NBD-Cl the reactivity of the two thiol groups is found to be reversed. Cysteine-25 now reacts more slowly than cysteine-187, whose reactivity is hardly affected. A reasonable interpretation of this result is that the Mg-ATP binds to the enzyme close to cysteine-25 so that the NBD-Cl is sterically prevented from reaching the reactive thiol group. These experiments indicate that cysteine-25 has a role to play in the reaction catalysed by adenylate kinase and that it is probably located in the region of the Mg-ATP binding site. The precise role, however, remains undefined (Price, Cohn & Schirmer, 1975). Creatine kinase has also been shown to possess an 'essential' thiol group by similar chemical modification experiments, although it is probably not directly involved in catalysis.

Another amino acid implicated in the creatine kinase and adenylate kinase reactions is arginine. Here the specific reagent is butanedione or phenylglyoxal (Table 12.1). The modified enzyme will no longer bind nucleotides, and it is possible that the positively charged arginyl residue is required in all kinases for the binding of the negatively charged phosphate groups (cf. Figure 8.11).

A hint that the mechanism of phosphate transfer in creatine kinase may be S_N1 was obtained by the observation that planar monoanions such as nitrate or formate protect the enzyme–substrate complex from modification of its reactive thiol group (Milner-White & Watts, 1971). Since nitrate and formate are similar in size, shape, and charge to the PO_3^- anion, the expected intermediate in the S_N1 mechanism, it was proposed that anions such as nitrate and formate behave as transition state analogues which lock the two substrate molecules onto the enzyme (Figure 12.7). Transition state analogues are used as a basis for the rational design of drugs and will be discussed further in Chapter 15.

Figure 12.7. Possible relationship of transition state analogue to substrates in creatine kinase. The transition state for an S_N2 mechanism is shown above the diagram and that for an S_N1 mechanism below.

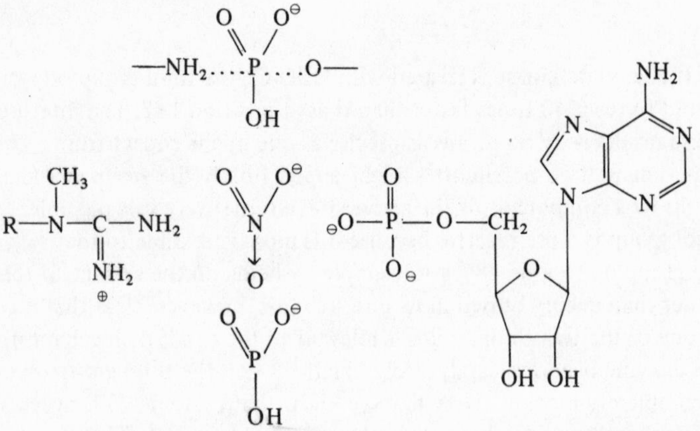

We must consider how strong this evidence is for either mechanism of phosphate transfer. Chemical studies of the hydrolysis of creatine phosphate at pH 1 shows that a PO_3^- ion is formed after an initial protonation of the creatine phosphate, and an S_N1-type process occurs. This evidence, and the fact that the analogue monoanions bind strongly to the enzyme–substrate complex, certainly suggest that an S_N1 process is possible, but we cannot exclude an S_N2 reaction on this basis. The transition state situation simulated in Figure

12.7 could be equally close to the transition state for an S_N2 process at the instant when the five-coordinate phosphorus atom reaches trigonal bipyramidal geometry.

Hexokinase and pyruvate kinase are also thought to possess essential thiol groups, and in addition pyruvate kinase appears to have a site that can be phosphorylated by a protein kinase. This may be important in the regulation of the enzyme activity (Chapter 13). Pyruvate kinase is also inhibited by oxalate, which acts as a substrate analogue and prevents binding of pyruvate to the active site.

The view obtained from these chemical studies does not help us in defining a complete chemical mechansim for phosphate transfer. If we combine the kinetic and chemical results into one picture we could propose a mechanism such as that in Figure 12.8. Here we suggest that a thiol group is required in some ill-defined way together with a positively charged group which may be a lysine or an arginine. Since chemical studies do not indicate that an amino acid residue takes part directly in the phosphate transfer (no phosphorylated groups can be detected) and the kinetic model is consistent with this idea, a direct transfer of phosphte between the substrates may occur. The enzyme probably orients the substrates in the optimum way for phosphate transfer, at the same time excluding water from the active site. We are still unable to distinguish between the possible intermediates required by the S_N1 or S_N2 mechanisms.

Figure 12.8. Mechanistic summary of studies on the mechanism of pyruvate kinase.

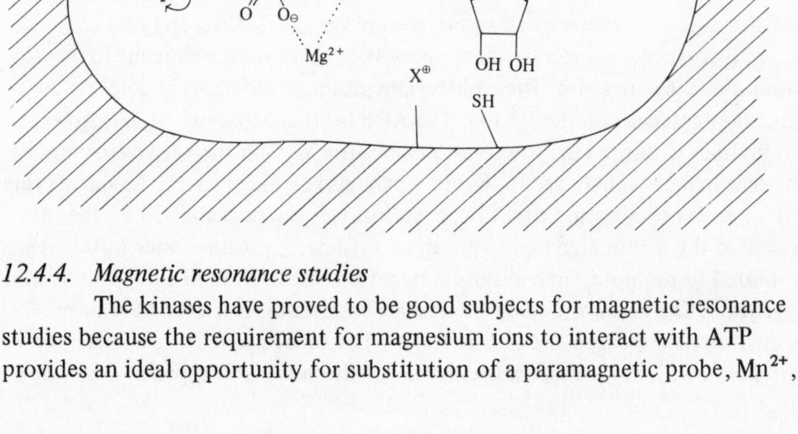

12.4.4. Magnetic resonance studies

The kinases have proved to be good subjects for magnetic resonance studies because the requirement for magnesium ions to interact with ATP provides an ideal opportunity for substitution of a paramagnetic probe, Mn^{2+}, which has very similar bonding characteristics to Mg^{2+}. As we saw in Chapters 9 and 11, substitution with an ion like Mn^{2+} allows very accurate determination of angles and distances within the enzyme from the paramagnetic enhancement of nuclear relaxation caused by the Mn^{2+} ion. Such methods have helped to confirm the ideas about the active site of adenylate kinase which were derived from X-ray diffraction studies, and the arrangement of the planar monoanions in the transition state analogue complex of creatine kinase (Figure 12.7) has also been supported.

Pyruvate kinase requires two metal ions for activity, Mg^{2+} and K^+, and each of these can be replaced with ions that are useful in n.m.r. experiments. As we have just seen Mn^{2+} can replace Mg^{2+}; Tl^+, which can be observed by n.m.r. spectroscopy, can replace K^+. Using these two probes it was possible to determine the spatial relationships between various substrate nuclei when bound to the active site with much greater accuracy than can be obtained with X-ray diffraction studies of molecules of the size of enzymes: distances can be estimated accurate to 0.01 nm. Resonances from ^1H, ^{13}C and ^{31}P were used (Mildvan, 1977). Direct measurement of the orientation of the substrate at the active site is not possible by other methods, so any agreement between the results of such magnetic resonance experiments with those of X-ray diffraction (which are obtained with crystals) helps to increase our confidence in the value of both techniques.

The role of the two metal ions in pyruvate kinase has been clarified by the n.m.r. experiments. The potassium ion appears to interact with the carboxylate group of the phosphoenolpyruvate, anchoring this part of the molecule. The manganese ion is positioned to shield some of the negative charge of the phosphate, and this also serves to orient the phosphoenolpyruvate and the ADP in such a way that an S_N2 transfer of phosphate seems likely.

We are now in a position to use all the evidence from different approaches to propose a model for the chemical mechanism of pyruvate kinase. The scheme which follows is based on all these data, and is best regarded as a summary of the experimental evidence to date in one diagram which will doubtless be refined as more experimental results become available (Figure 12.9).

In this scheme we can firstly see how the substrates are thought to be bound to the active site. The carboxylate group of the phosphenolpyruvate is anchored ionically to the K^+ ion. The ADP is bound to a site which allows it to interact with the Mn^{2+} (and presumably the Mg^{2+} in the native enzyme) by its terminal phosphate group. Binding may also be aided by the imidazole ring of a histidyl residue and the ATP, once formed, can be stabilised by the presence of the protonated amino group of a lysine. A proton donor (−B−H) is required to promote the enolisation or reverse enolisation of the phosphoenolpyruvate. The enolisation and the reverse enolisation reactions have been shown to be stereospecific (Rose, 1970). The other ligands of the metal ions are unknown , but the overall geometry of the active site appears to be suited to an S_N2 type of phosphate transfer in which the manganese (or magnesium) ion guides the phosphate from one substrate to the next. This interpretation may, however, be incorrect in the light of recent work with pyruvate kinase using ^{18}O-labelled ATP and ^{31}P n.m.r. to analyse the products. The authors concluded that an S_N1 mechanism probably operates (Lowe & Sproat, 1978).

Figure 12.9. Suggested mechanism for pyruvate kinase based on n.m.r. data. (From Dunaway-Mariano, Benovic, Cleland, Gupta and Mildvan (1979), *Biochemistry* Vol. 18, 4347. ($M^+ = K^+$)

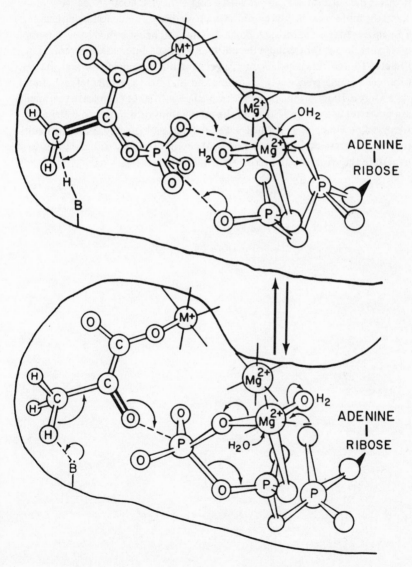

Other recent work with ^{18}O-labelled substrates suggests that hexokinases and pyruvate kinase have the same stereochemistry (Wimmer & Rose, 1978).

There are clearly many parts of this mechansim that are not sufficiently precisely defined. This is often the case in the study of biochemical mechanisms. As with many mechanisms in organic and inorganic chemistry, summaries such as the one presented here must be treated critically and be based on as wide a variety of experimental approaches as possible, since no single method is sensitive to every significant interaction in the system. The chemist may wonder if this is the furthest we can get with a single enzyme how far can we understand the more complicated systems which we discuss in the following chapters? The answer may be surprising. Some systems can be described in even more molecular detail, for example the antibody-antigen interaction described in Chapter 16, but this depends upon a very fortunate coincidence that the system of interest is relatively easily accessible and that the experimental techniques, here X-ray diffraction and magnetic resonance with some chemical experiments, are powerful enough to resolve all the interactions of interest. In other areas we may know much less because a well-defined biological system is difficult to obtain and we do not have a powerful enough armoury of techniques to penetrate the structure and resolve the important interactions.

13 Enzyme systems and intracellular control

13.1. Introduction

The overall chemical activity of a cell is a very complex network of related reactions. Some idea of its complexity can be gained from the metabolic pathways presented in Appendix 1. Here we can see how one simple compound, pyruvate for example, can be derived from several sources (from glucose, lactate and phosphoenolpyruvate). Pyruvate can also be converted into many compounds, including those above and also acetyl CoA, oxaloacetate and, very desirably, in yeasts to ethanol. The metabolic activity of a cell can be divided into several subsystems which usually have an identifiable metabolic function. Some pathways such as glycolysis are primarily degradative or **catabolic** pathways. In the case of the breakdown of an energy source such as glucose or a fatty acid, a catabolic pathway is also an energy-producing pathway. Other pathways are synthetic or **anabolic**, for example fatty acid synthesis, discussed later in this chapter. ATP or reduced nicotinamide coenzymes are often required for anabolic processes, and other pathways such as the citric acid cycle provide a link between anabolic and catabolic processes.

Pathways are spatially organised within the cell. A part or a whole pathway can be located in a specific cell compartment, as we saw for fatty acid synthesis in Chapter 2. Since access to a compartment is selective, owing to the presence of membranes containing transport systems, compartmentation is a form of control of the metabolism of the cell. This is additionally favourable when one pathway requires a concentration of a common metabolite (for example NAD^+) different from the concentration required in another. If these two pathways are present in different cell compartments there is no difficulty in achieving the optimum concentration for each pathway. The multi-enzyme systems discussed shortly are examples of a smaller-scale compartmentation.

The cell aims to achieve within itself a balance between energy production and energy utilisation; it can do this by regulating the flux of metabolites through the network of pathways. A cell must also respond to its environment. Separate control mechanisms can be identified here, and some of these are discussed in Chapter 16.

The most economical place to control a sequence of enzyme-catalysed steps is at the step which commits the molecules irrevocably to that route. Thus control points are frequently found early in pathways and at reactions which lie well displaced from equilibrium. Examples of such committing steps are the reactions catalysed by phosphofructokinase in the glycolytic pathway and acetyl CoA carboxylase in fatty acid synthesis. Porphobilinogen synthase, discussed in Chapter 4, is also a control enzyme situated early in the biosynthetic route to tetrapyrrolic compounds. Current thinking, however, tends to place emphasis on several enzymes in a pathway, rather than on a single enzyme, as being the sites of regulation.

One way in which we can attempt to understand how pathways are controlled in the cell is to identify which enzyme-catalysed steps are rate limiting and then to analyse how the activity of these enzymes can be modulated. The rate-limiting step in the utilisation of a given substrate may not be a chemical reaction. In the case of glucose metabolism in muscle, the rate limiting step is the permeation of glucose into the muscle cell, which depends on the action of the hormone insulin.

If the rate limiting enzymes are to be sensitive to the requirements of the cell and its environment for the products of the pathway which they control, they need some way of sensing the demands made on them. One way in which this could be done is for products of the pathway to inhibit the rate limiting step if they are produced in excessive amounts. This is known as feed-back inhibition. Feed-forward effects are also known.

Under aerobic conditions, citrate can be regarded as a product of glycolysis. It is able to leave the mitochondria, where it is produced, by way of a carrier system which allows it to cross the otherwise impermeable inner mitochondrial membrane. Citrate inhibits phosphofructokinase which, as we mentioned, is rate limiting in glycolysis.

Compounds other than the products of a metabolic pathway also influence the activity of control enzymes. Molecules such as ATP and NADH are produced during catabolic processes and are consumed by anabolic processes. A high concentration of either of these suggests that the cell has a large amount of metabolic energy available for anabolic pathways or other energy-requiring functions such as active transport. In these circumstances it is not economical to produce more metabolic energy, so we might expect the high concentrations of these substances which link energy-producing pathways with energy-requiring processes to inhibit the processes of energy production. In this respect, metabolites such as ATP and also AMP, ADP and others are both reactants in pathways and carriers of information about the metabolic energy available to the cell. When food supplies are plentiful, cells will also store sources of energy, and a similar control mechanism applies to the energy-storage pathways. The same signals that inhibit energy production should promote the accumulation of energy stores, which include glycogen, a store of glucose, in liver and muscle and triacylglycerol or fat in adipose tissue. We shall see this idea in action in Chapter 16,

So by considering the cell as a series of subsystems, some of which produce energy and some of which consume it, we can identify molecules which link these systems as carriers of both metabolic energy and of information about the energy immediately available to the cell.

We now need to consider what the possible molecular mechanisms are for these and other control systems to work. This study requires further investigation of enzyme structure and function, and is a direct development of the study of enzyme mechanisms.

Table 13.1 shows several ways in which the metabolic activities of enzymes

Table 13.1. *Control of enzyme activity*

---

(1) Alter total enzyme concentration
    by increasing rate of synthesis
    by reducing rate of degradation
(2) Alter activity of enzyme already present
    by limited proteolysis (activation only)
    by covalent modification (usually reversible, requires an enzyme for addition and removal of regulatory group)
    by allosteric modification (non-covalent and reversible, no enzyme required)

---

may be controlled *in vivo*. The first method of increasing metabolic activity is to increase the concentration of enzyme by promoting its synthesis. This is most effective when the enzyme is saturated with substrate. This is a very common mechanism in micro-organisms; a mammalian example is the induction of drug-metabolising enzymes in the liver (Chapter 15). A complementary way of altering the enzyme concentration in the cell is to alter its rate of degradation. This may be a less specific mechanism but it is known that enzymes have different lifetimes within cells; their half-lives range from seconds to many days.

The mechanisms for activating enzymes which we shall concentrate on in this chapter form a second group. In these cases the activity of existing enzyme is modified without changing the concentration of enzyme present. There are three different mechanisms of this group, and all have a common effect – namely to alter the conformation of the enzyme to give a more active or inactive catalyst as appropriate.

In some tissues it is particularly advantageous to have a specific form of an enzyme present so that reaction in one direction is favoured over its reverse as the tissue requires. This is found in the case of the isoenzymes of lactate dehydrogenase. The muscle form favours the production of lactate; the forms in heart and liver, both tissues which can utilise lactate, favour the oxidation of lactate to pyruvate.

However, before we examine how systems of separable enzymes might be controlled at the rate limiting step, we shall look at systems in which physical control is exerted on a whole system of enzymes in one complex. These multi-enzyme complexes will allow us to introduce examples of some of the control mechanisms in Table 13.1 which we shall discuss in more detail later in the chapter.

13.2. Multi-enzyme complexes
13.2.1. Introduction

We suggested in the last chapter that the rate limiting step in many enzyme-catalysed reactions is the diffusion of the product from the enzyme. In order for a sequence of catalysed reactions to take place with maximum efficiency it is possible to avoid some slow diffusion steps by having all the enzyme activities associated together in one complex. Multi-enzyme complexes in which different enzyme-catalysed reactions take place in sequence on one isolable unit occur in many parts of cells. For example, assemblies of membrane-bound enzymes may be regarded as multi-enzyme complexes if the components are precisely arranged with respect to each other. The ribosome, the protein-synthesising machinery of the cell, can also be regarded as a multi-enzyme complex. We shall examine two different complexes in this section. They are biotin-dependent carboxylases, for example acetyl CoA carboxylase (Figure 13.1a), and pyruvate dehydrogenase (Figure 13.1b).

These are both typical multi-enzyme complexes, consisting of aggregates of two or more enzyme activities in one well-defined physical system. The activities may be associated with one or more polypeptide chains, and the molecular weight of the complex may range from 100 000 to several million. In addition to the avoidance of slow binding and diffusion steps, multi-enzyme complexes segregate any enzymes which might compete for the same substrate,

Figure 13.1.

(a) Acetyl CoA carboxylase

$$CH_3COSCoA \xrightarrow{\quad biotin \quad} HO_2CCH_2COSCoA$$

$$HCO_3^{\ominus} \quad ATP \quad ADP + P_i$$

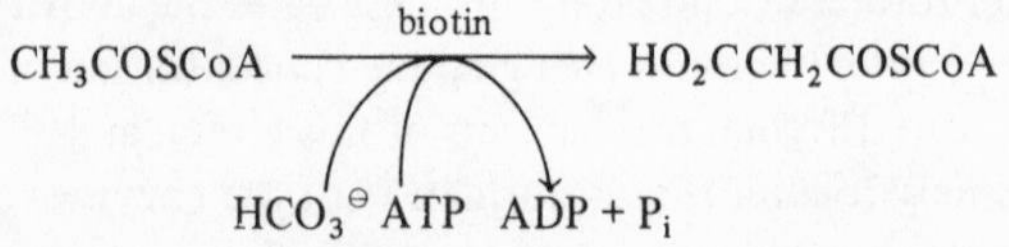

(b) Pyruvate dehydrogenase

$$CH_3COCO_2H \xrightarrow{\quad CoASH \quad CO_2 \quad} CH_3COSCoA$$

$$NADH \quad NAD^{\oplus}$$

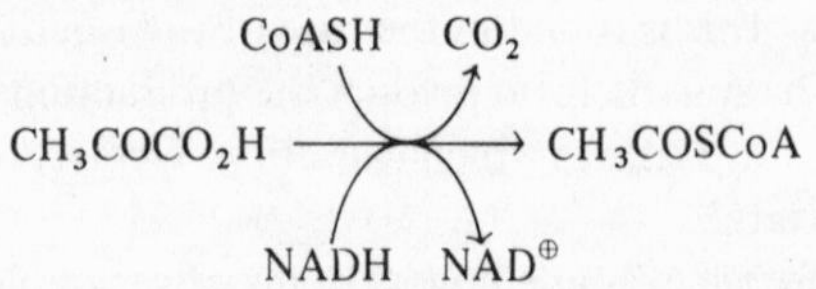

and intermediates which are not stable or soluble in the aqueous environment can be transferred from one active site to another without decomposition or aggregation. The whole sequence takes place in one defined compartment and can be controlled as if it were one step.

The advantage of associating enzymes closely together has been demonstrated in a model by Mosbach. He showed that a system containing the three enzymes citrate synthase, lactate dehydrogenase and malate dehydrogenase chemically immobilised on the same support gave eight times the catalytic activity of the same amount of the same enzymes in free solution (Srere, Mattiason & Mosbach, 1973).

13.2.2. Biotin-dependent enzymes

Biotin (Figure 13.2a) is an essential cofactor in enzymes which cata-lyse the addition of carboxyl groups to saturated carbon atoms (13.2b) or in enzymes which catalyse the transfer of carboxyl groups (13.2c). In these enzymes, the biotin is covalently bound as an amide to a lysyl amino group through its aliphatic chain. We shall concentrate on the enzyme that catalyses reaction 13.1a, the carboxylation of acetyl CoA to give malonyl CoA. This is the key step in the biosynthesis of fatty acids, which is carried out on another multienzyme complex discussed at the end of this chapter.

Figure 13.2.

(a)

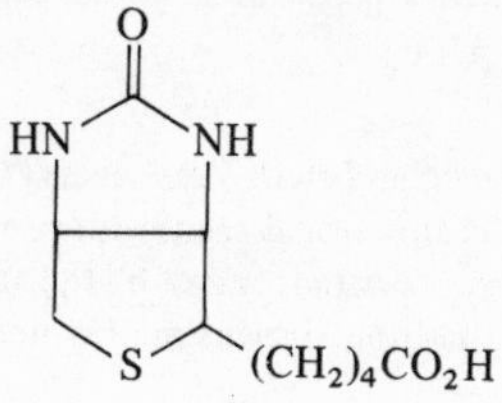

Biotin

(b)

$$CH_3COCO_2H \longrightarrow HO_2CCH_2COCO_2H$$

Pyruvate   Oxaloacetate

(c)

$$CH_3CHCOSCoA + CH_3COSCoA \rightleftharpoons CH_3CH_2COSCoA$$
$$\overset{|}{CO_2H} \qquad\qquad\qquad\quad + HO_2CCH_2COCO_2H$$

(d)

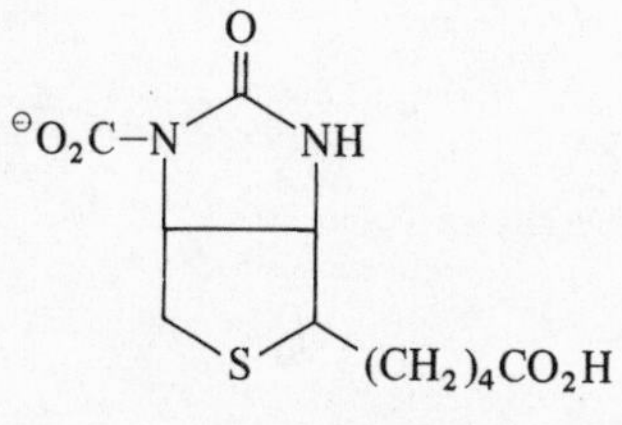

Biotin was shown to react with bicarbonate, the source of the carboxyl group, to form a carboxylated derivative (Figure 13.2*d*). This step is coupled to the conversion of ATP to ADP and inorganic phosphate. The second catalysed step in the complex is the transfer of the carboxyl group from carboxybiotin to the receiving molecule, acetyl CoA. Thus there are two enzyme-catalysed reactions in this system, and carboxybiotin is transferred from the site at which it is formed to the site at which it releases its carboxyl group to acetyl CoA. Acetyl CoA carboxylase from yeast has been shown to possess all these functions on one polypeptide chain of molecular weight 190 000. This is illustrated schematically in Figure 13.3.

There are two further directions in which we could examine this system. We could investigate the chemical mechanism of each step, or we might consider if there is a control mechanism which can modulate the rate of the carboxylation. Since the carboxylation of acetyl CoA occurs at the start of the pathway to fatty acids (Appendix 1) and the product of carboxylation, malonyl CoA, is not required in any other pathway, the acetyl CoA carboxylase is obviously a candidate for the rate determining step of the overall fatty acid synthesising process.

Firstly let us consider the mechanism of each step. Questions which arise include: what is the role of ATP in the carboxylation of biotin? Why is biotin carboxylated on nitrogen? Many approaches to these questions have involved the use of models (Wood, 1976). You may like to try your hand at a mechanistic interpretation by considering the data given below, which provide some evidence for the role of ATP.

Figure 13.3. Schematic mechanism of yeast acetyl CoA carboxylase. Note that this scheme illustrates the functional relationships within the enzyme and does not illustrate the precise arrangement of the sites in space. This would require a knowledge of the three-dimensional structure of the protein.

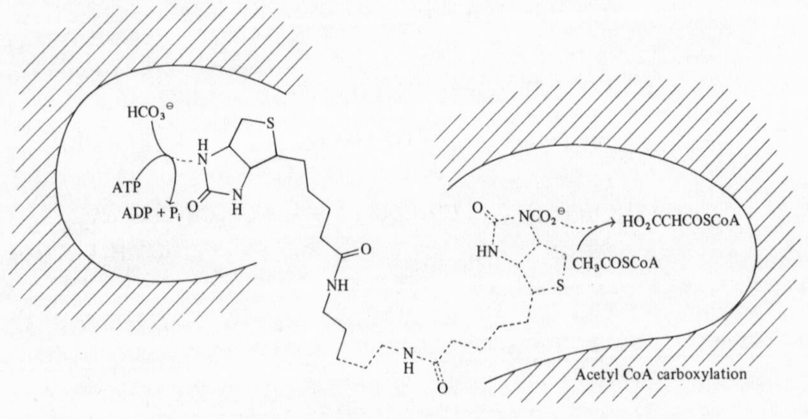

Problem 13.1.

Propionyl CoA carboxylase catalyses the reaction shown in Figure 13.4*a*. This reaction is important because it allows three-carbon acids to be utilised for energy production. It is particularly significant in ruminants, which absorb large quantities of propionic acid, produced by fermentation in the rumen, as their main source of carbon for the synthesis of glucose. The following experimental observations show us something of the way in which the reactants, biotin, bicarbonate, and ATP, interact:

> (1) Incubation of the enzyme with $HC^{18}O_3^-$, ATP and propionate leads to the labelling pattern in Figure 13.4*b*. There is clearly some specificity in the transfer of oxygen during the formation of carboxy-biotin. Suggest a mechanism consistent with these results.
>
> (2) If $C^{18}O_2$ were the substrate what labelling pattern would you expect?

Figure 13.4.

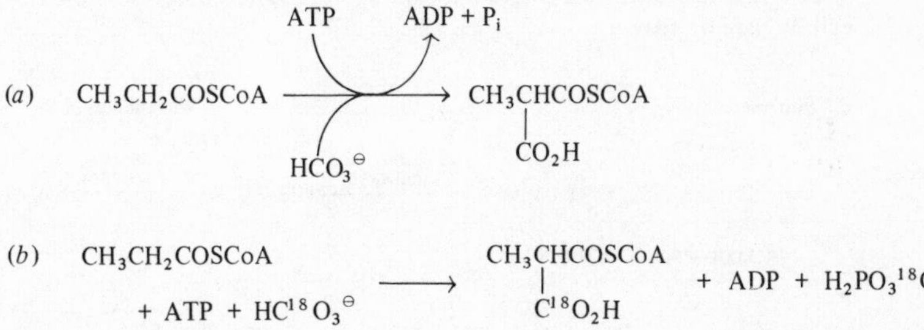

(*a*)
$$CH_3CH_2COSCoA \xrightarrow[\substack{HCO_3^{\ominus}}]{\substack{ATP \quad ADP + P_i}} CH_3\underset{\underset{CO_2H}{|}}{C}HCOSCoA$$

(*b*) $CH_3CH_2COSCoA$
$\quad + ATP + HC^{18}O_3^{\ominus} \longrightarrow CH_3\underset{\underset{C^{18}O_2H}{|}}{C}HCOSCoA \quad + ADP + H_2PO_3{}^{18}O^{\ominus}$

Returning to the control of enzymic and metabolic processes, it is interesting to compare the control mechanisms that operate in different organisms. It is usually found that the catalytic mechanisms, which presumably evolved early, are the same in many species. Evolutionary modification has concentrated on the control of the metabolic system to best meet the needs of the evolving organisms, and this is demonstrated by biotin-dependent carboxylases. Biotin functions as a carboxylate carrier in many organisms, but the control mechanism has evolved in several ways. In *E. coli* (a prokaryote) acetyl CoA carboxylase is composed of three separate polypeptides, one for each function. In yeast (a eukaryote) the three functions of the carboxylase are all associated with one polypeptide chain (Obermayer & Lynen, 1976). Thus the organisation of the complex has become tighter.

In an animal system, for example rat liver, another change has taken place. A specific regulatory site has been added which allows the activity of the complex to be altered in the following way: Acetyl CoA carboxylase of rat liver exists in a monomeric form in which it is catalytically inactive. If citrate is added to the purified monomer, the enzyme polymerises to give a polymer containing about 20 units of enzyme, each unit with a molecule of bound citrate. (Evidence is growing that many so-called 'soluble' enzymes may be physically associated in the cell with related enzymes from the same pathway.) The polymerisation increases the rate of transfer of carboxylate from carboxybiotin to acetyl-CoA. Why should citrate be the activator of this system? We mentioned earlier that citrate is produced in the mitochondrion (from acetyl CoA and oxaloacetate) and that if sufficient citrate is present it can permeate out of the mitochondrion into the cytoplasm, where it serves as a source of acetyl CoA. Thus a high concentration of citrate in the cytoplasm is a signal that plenty of acetyl CoA is being produced in the mitochondrion for use elsewhere. A major pathway for the utilisation of acetyl CoA is the synthesis of fatty acids, which begins by converting acetyl CoA into malonyl CoA. Citrate is an appropriate activator for this enzyme because it signals that supplies of starting materials are available for fatty acid synthesis (Figure 13.5).

Figure 13.5. Outline of fatty acid metabolism and control of fatty acid synthesis. The broken line and asterisk shows the activation of acetyl CoA carboxylase by citrate.

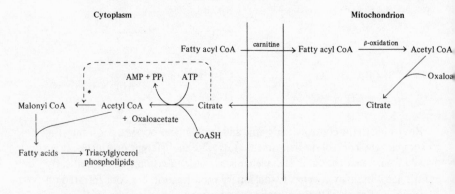

In the rat liver, then, a fourth function is added to the multi-enzyme complex, a regulatory site to which the activator citrate can bind. Such a site, distinct from the catalytic site, at which the enzyme activity is modified by the binding of a small molecule, is known as an **allosteric** site. This type of modification is a very precise way of tuning the activity of an enzyme to the system to which it belongs. We shall meet further examples and examine the nature of allosteric effects later in the chapter. Recent evidence suggests that acetyl CoA carboxylase can also be regulated by another mechanism, phosphorylation–dephosphorylation, which we shall discuss in Section 13.2.3.

The next stage in the synthesis of fatty acids is the polymerisation of the malonyl CoA on the fatty acid synthase multi-enzyme complex. Some data on this system are presented as Problem 13.3 at the end of this chapter.

13.2.3. Pyruvate dehydrogenase

Our final example of a multi-enzyme complex illustrates some important chemistry, and introduces phosphorylation as a means of modulating the activity of an enzyme. The oxidative decarboxylation of 2-oxoacids (pyruvate and 2-oxoglutarate) is carried out by multi-enzyme complexes which comprise several subunits (Reed, 1974). Coenzymes (thiamine pyrophosphate, NAD and FAD) are required for many of the steps; the coenzymes have been discussed from a chemical point of view earlier (thiamine pyrophosphate in Chapter 4 and NAD and FAD in Chapter 8). The overall reaction is catalysed in several steps by distinct subunits of the enzyme. In several cases the individual subunits have been purified and reconsitituted into an active system. The reactions catalysed by each of the subunits are shown in Figure 13.6.

Figure 13.6. The reactions catalysed by the pyruvate dehydrogenase complex.

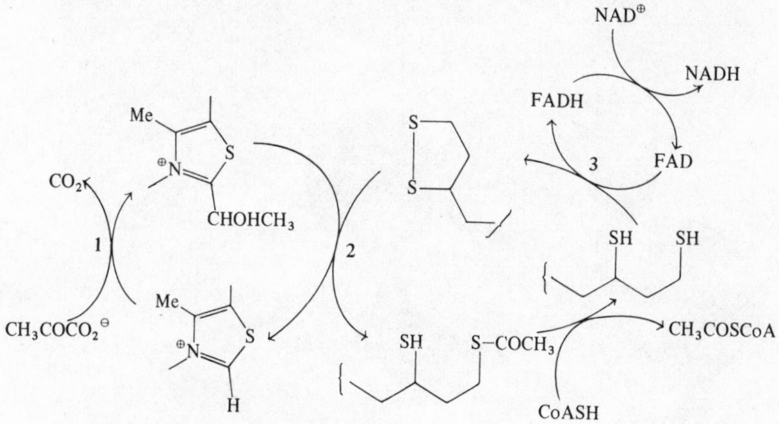

Enzymes: 1 pyruvate decarboxylase
2 dihydrolipoate transacetylase
3 dihydrolipoate dehydrogenase

In the active enzyme complex, the subunits are combined in a fixed relationship to each other, as illustrated in Figure 13.7. The complex in *E. coli* is of a similar size to a ribosome, about 18 nm across, and it allows the decarboxylation and oxidation of pyruvate to occur in a single place and in a controlled way. The highly reactive acetyl intermediate (the product of reaction (1)) is never released and so cannot be lost to any other system.

We can examine each step of the pyruvate dehydrogenase complex in mechanistic detail. The role of thiamine pyrophosphate in stabilising carbanions has been considered previously (Chapter 4, see also Chapter 15). Lipoic acid is an oxidisable acetyl group carrier. During the sequence of reactions shown

Figure 13.7. A diagram of the pyruvate dehydrogenase multi-enzyme complex of *E. coli*. 1, pyruvate decarboxylase; 2, dihydrolipoate transacetylase; 3, dihydrolipoate dehydrogenase. This figure was kindly supplied by Dr G. Hale, University of Cambridge Department of Biochemistry.

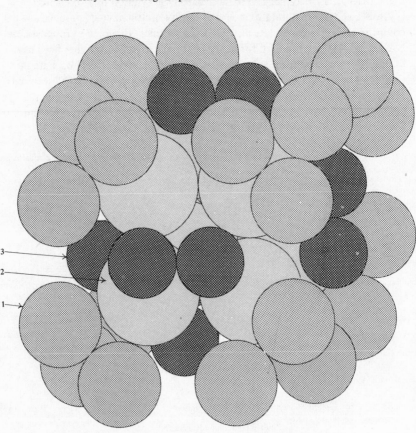

in mechanistic terms in Figure 13.8 it is oxidised and acts as an acceptor for the acetyl group from the acetylated thiamine pyrophosphate. Acetyl lipoic acid then donates the acetyl group to CoA, forming acetyl CoA and is itself reduced by the remaining subunit of the enzyme which depends on the co-enzymes NAD and FAD for its activity.

The work of Bates *et al.*, (1977) shows that each lipoate acetyltransferase subunit contains two active lipoyl residues. It is thought that the transfer of acetate from one active site to another takes place by a series of internal trans-acetylations involving both lipoyl residues, rather than the movement of a single acetyl lipoate from one subunit to another.

Figure 13.8. The mechanism of pyruvate dehydrogenase.

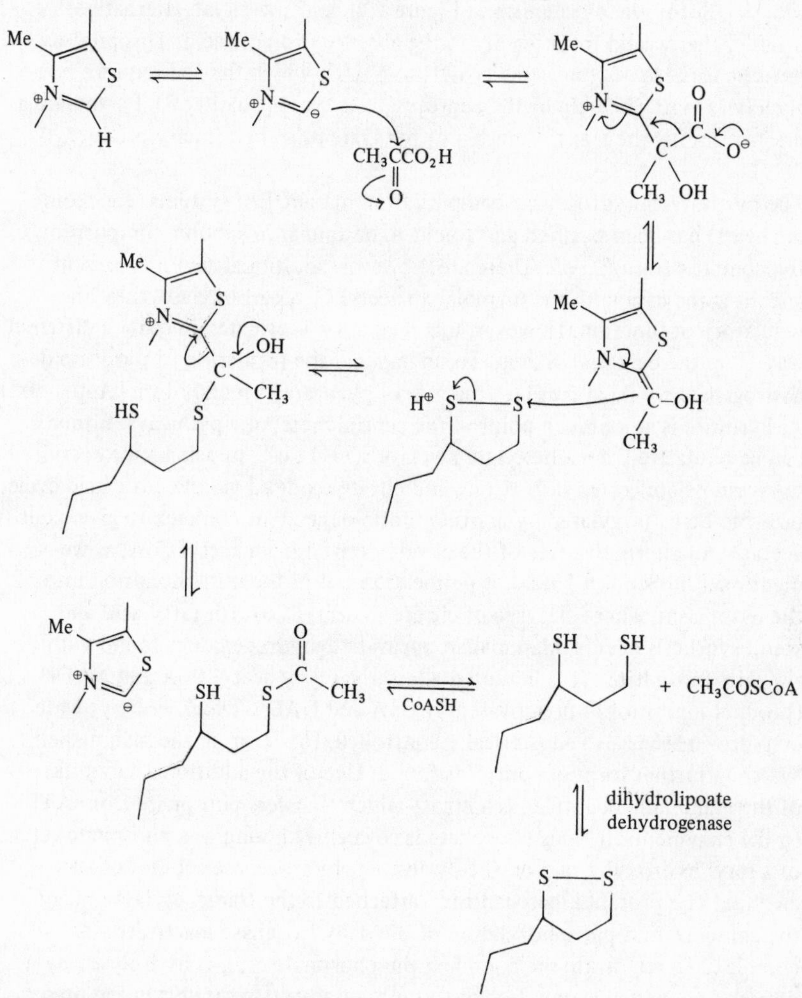

Alternative fates of pyruvate

Oxidation and decarboxylation are not the only fates of pyruvate. Under anaerobic conditions, the absence of oxygen prevents the reoxidation of the NADH that is produced by glycolysis. Since NAD^+ is required for pyruvate dehydrogenase, this step cannot take place in the absence of oxygen. In muscle, NAD^+ is regenerated by the reduction of pyruvate to lactate by lactate de-hydrogenase, a reaction which requires NADH, and glycolysis can then proceed.

Yeast is an organism that can grow under either aerobic or anaerobic con-ditions: such organisms are called facultative anaerobes. By using part of the pyruvate dehydrogenase system and another enzyme it can remove the pyruvate produced by glycolysis, regenerate NAD^+ from NADH and yield a commer-cially desirable product.

Problem 13.2.

Study the mechanism in Figure 13.8 and suggest an alternative product that would form readily in the absence of lipoic acid. This product must be capable of converting NADH to NAD^+, which the cell requires for glycolysis, with the help of the appropriate enzyme (Chapter 8). Formulate a mechanism for the transformation of pyruvate into the primary product.

The pyruvate dehydrogenase complex from mammalian systems (e.g. from pig heart) has been purified and found to be similar in subunit composition to the complex from *E. coli*. There are, however, additional subunits present and, as is the case with the mammalian acetyl CoA carboxylase, they are regulatory in function. However, this regulation is effected in quite a different way from the bacterial system. To understand the regulation of pyruvate dehydrogenase we must consider the role of pyruvate in metabolism (Appendix 1).

Pyruvate is at a branch point in the central metabolic pathways. Either it can be oxidatively decarboxylated to give acetyl CoA, in which case acetyl CoA will be converted into citrate and citrate oxidised via the citric acid cycle, or it can be carboxylated by another biotin-dependent complex to give oxalo-acetate. An alternative fate of the citrate derived from acetyl CoA, as we mentioned in Section 13.2.2, is permeation out of the mitochondrion into the cytoplasm, where cleavage of citrate to acetyl CoA for fatty acid and sterol synthesis occurs. Mammalian pyruvate dehydrogenase is found within the mitochrondrion. It is inhibited allosterically by acetyl CoA and NADH (product inhibition) but activated by CoA and NAD^+. The *E. coli* pyruvate dehydrogenase is also allosterically controlled. However, in the mammalian system a further form of control is found. One of the additional subunits of the mammalian complex is a kinase which transfers phosphate from ATP to the enzyme itself. This phosphate is covalently bound as a phosphate ester of a seryl hydroxyl group on the pyruvate dehydrogenase subunit of the enzyme. The protein kinase is firmly attached to the transacetylase unit of the complex, and phosphorylation of the dehydrogenase inactivates the whole complex. To reactivate the complex, the phosphate ester is hydrolysed by a phosphatase which is only bound to the transacetylase subunit in the presence of Ca^{2+}.

So far we have control of pyruvate dehydrogenase by an allosteric mechanism and by phosphorylation–dephosphorylation, but the phosphorylating enzyme itself is subject to a further set of controls which, therefore, indirectly act upon pyruvate dehydrogenase too. The protein kinase is inhibited by pyruvate and by ADP. This control by small molecules is easily rationalised (cf. Section 13.1). Overall it seems likely that the intramitochondrial ATP/ADP concentration ratio (which reflects the level of metabolic energy immediately available to the cell) regulates the activity of the protein kinase, and the concentrations of free Mg^{2+} and Ca^{2+} regulate the binding and the activity of the phosphatase. It is possible that these effects may be further controlled from outside the cell by hormones.

The whole complex of enzymes, including the regulatory ones, are made sensitive to all the relevant information within the cell and outside it, so that their activity can be best suited to the needs of the cell. The activity which would be observed in a cell *in vivo* is a result of an integration by the enzyme system of all this information. In this way the enzyme system behaves not only as a catalyst of a complex set of chemical reactions but also as a stage in the transfer of information about the level of activity required from one system in a cell to another. The two mechanisms we have introduced that effect this integration are **allosteric modification** and **protein phosphorylation**. Both are very common, and we shall return to them and to their molecular basis a little later in this chapter.

13.3. Activation by limited proteolysis

In the control mechanisms we have met so far, specific interactions, either covalent (phosphorylation) or non-covalent (allosteric), cause a change in the structure of a protein which renders it more or less efficient as a catalyst. These changes are reversible: the phosphorylation can be reversed by hydrolysis of the phosphate ester and the allosteric effect by release of the allosteric modulator from its binding site. There are also a number of systems in which enzymes are activated irreversibly. These include the digestive enzymes, the triggering of the system which leads to blood clot formation and the complement system. We shall mention some aspects of the complement system, which assists in the destruction of foreign cells in animals, in Chapter 16.

In all these systems the enzymes are synthesised in inactive forms, which are often known as zymogens, and the active protein is formed by specific hydrolysis of a fragment of the polypeptide chain of the zymogen. The protein is then free to assume its active conformation. We shall illustrate the mechanism in action in the digestive system.

In Chapter 1 we saw how chymotrypsin, a digestive enzyme which hydrolyses protein, is able to catalyse the hydrolysis of a peptide bond using the amino acid side chains present at the active site. Chymotrypsin is synthesised as chymotrypsinogen, with many other digestive enzymes (also as zymogens) in the pancreas, and is secreted from there into the duodenum. The hydrolytic removal of two small fragments of chymotrypsinogen is catalysed by the peptidase trypsin, which arises itself by hydrolysis of its zymogen, trypsinogen (Figure 13.9). In the digestive system, trypsinogen is hydrolysed by the enzyme enteropeptidase, which is present in an active form in the intestine. The active trypsin molecules are able to catalyse the activation of further molecules of trypsinogen as well as the hydrolysis of chymotrypsinogen and other digestive zymogens. Thus, once one molecule of active trypsin has been formed the remaining activation steps take place very quickly.

Figure 13.9. Activation of chymotrypsinogen.

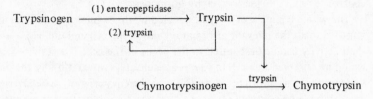

The activation of digestive peptidases illustrates an important point in many control systems, namely the amplification of an initial stimulus. Amplification is possible because a single molecular event, for example activation of trypsinogen, gives rise to many further events, each of these being catalysed by the enzyme which was activated by the first hydrolysis reaction. A sequence of such steps can give rise to amplification by a factor of several million. Another example is the action of adrenaline (Chapter 16). Having outlined the biochemical structure of this control mechanism we are in a position to examine some of its details at a molecular level. Firstly, let us consider the mechanism of hydrolysis of the small peptides from the zymogens. The action of chymotrypsin depends upon a seryl hydroxyl group at the active site which forms a covalent intermediate with the acyl fragment of the peptide. This mechanism of hydrolysis of peptide bonds appears to have proved very satisfactory in evolution, because many of the proteolytic enzymes involved in these amplification or 'cascade' systems have active sites which are related to chymotrypsin, and they also show similarities in their tertiary structure. Probably these 'serine proteases', as they are called, have evolved from a common ancestral enzyme. A second important question concerns the way in which the structure of the protein is altered by the activation process so that it assumes its active conformation. In the case of chymotrypsin the activation produces an increase of activity of about 10^4–10^6 times. (Chymotrypsinogen has weak hydrolytic activity.) The increase in activity is the result of the release of two dipeptides from a total of 245 amino acids. In order to answer this question we need to know the structure of the precursor and of the active enzyme; X-ray crystallography has supplied the information.

The overall tertiary structure of both chymotrypsin and chymotrypsinogen is very similar. This indicates that only relatively small changes in conformation take place during the activation. In fact the relative positions of the catalytically important groups – the serine, the histidine and the aspartate – change little. The changes which take place appear to affect the way in which the substrate is oriented at the active site (Chapter 10), allowing tight binding of the substrate without which catalysis could not occur (Figure 13.10). A similar mechanism has been observed in the activation of trypsinogen (Huber & Bode, 1977). The amino acid residues which contribute to the substrate binding site are in a very flexible part of the trypsinogen molecule. On activation, 85 % of the structure is unchanged, but the flexible regions rigidify to form the binding site, thus giving the enzyme activity.

Figure 13.10. Changes in protein structure on activation of chymotrypsinogen (*left*) to chymotrypsin (*right*). The chymotrypsin is labelled with a tosyl group, shown with solid bonds to distinguish it from the amino acids forming a specific cavity in the enzyme. (From Freer et al., 1970, Biochemistry, 9, 2006)

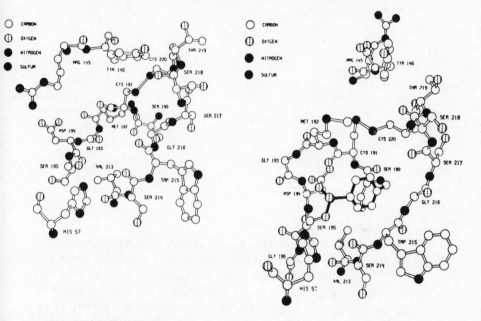

13.4. Covalent modification

Several types of control mechanisms which operate by covalent modification of a protein are known. Of these the most common is phosphorylation–dephosphorylation, which we met in the mammalian pyruvate dehydrogenase system. Other modes of covalent modification include ADP-ribosylation, part of the mechanism of action of the diphtheria toxin, and adenylation, which is found in the regulation of glutamine synthase in *E. coli* (Figure 13.11). The pyruvate dehydrogenase system already described illustrates the principles of the phosphorylation–dephosphorylation system well. Phosphorylation is catalysed by a protein kinase, which may be a specific enzyme (as in glycogen phosphorylase, Chapter 16) or which may have more general specificity. Since we are dealing with a control mechanism the protein kinase must be activated in some way, and in the case of pyruvate dehydrogenase activation appears to be mediated by an allosteric mechanism. Phosphorylation, which may activate or inhibit the system, is reversed by the action of a phosphatase whose activity must also be controlled in some way. Perhaps the best-understood systems of this type are those involved in the activation of glycogen synthesis and breakdown in muscle. This system, which is sensitive to hormone action, is examined in Chapter 16.

Figure 13.11. Covalent activation of glutamine synthase.

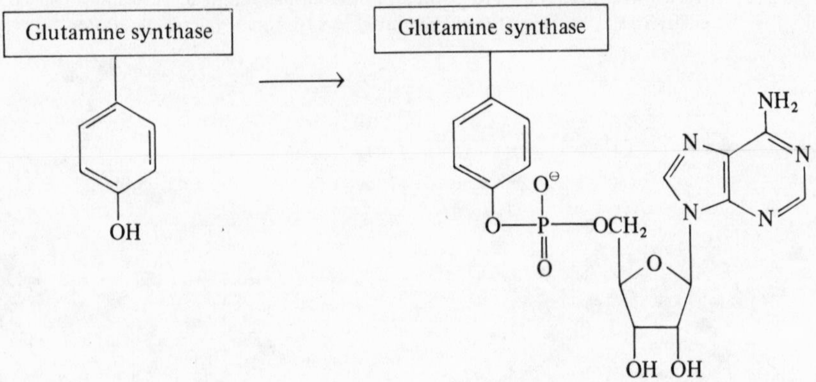

Repeated on tyrosine residues of each of the twelve subunits of the enzyme

13.5. Allosteric effects

Allosteric modification of enzyme activity is, with covalent modification, probably the most important mechanism for regulation at the protein level within a cell. We have already discussed examples of this type of control in the multi-enzyme systems acetyl CoA carboxylase and pyruvate dehydrogenase. As with phosphorylation, the binding of the allosteric effector to a specific site which is remote from the catalytic site causes a change in the tertiary structure of the enzyme. This change may be very subtle but it is sufficient to alter the catalytic efficiency of the enzyme. The allosteric site can be on a separate subunit of the enzyme (as in aspartate transcarbamylase or the protein kinase discussed in Chapter 16) or, more commonly, it is part of the same subunit.

Allosteric systems can often be recognised by their kinetic behaviour. For a non-allosteric system the dependence of the rate of the reaction on the concentration of substrate for a fixed concentration of enzyme follows a rectangular hyperbola (Figure 13.12a). This curve implies that at high concentrations of substrate all the active sites on the enzyme are filled, so that a maximum rate is achieved. The enzyme is said to be saturated with respect to that substrate. The corresponding curve for an allosteric enzyme is commonly sigmoidal (Figure 13.12b). This curve suggests that at low substrate concentrations the rate of the reaction is low but as the substrate concentration increases the rate increases. In other words, increase in rate seems to depend on the concentration of substrate; binding of one molecule of substrate causes the binding of the next to take place more readily, and so the rate increases more rapidly. The observation of such a sigmoid curve is indicative of the phenomenon known as cooperativity. In this case the cooperativity is positive, since the rate is increased more rapidly than in the absence of cooperativity by increasing substrate concentrations until saturation is reached. Systems are known which exhibit negative cooperativity, and in these the reverse is true; the plot of rate

Figure 13.12. Dependence of the velocity of an enzyme-catalysed reaction on substrate concentration for (a) a non-cooperative, (b) a cooperative and (c) a negatively cooperative system. (Redrawn from Lehninger, 1975.)

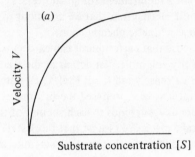

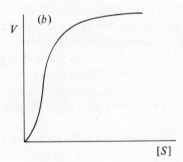

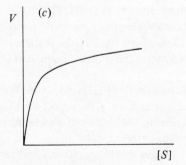

of reaction against substrate concentration would look like Figure 13.12*c* in this case.

Theories which attempt to explain the kinetics observed with allosteric systems were originally developed to describe an 'honorary enzyme', haemoglobin. Oxygen binds to haemoglobin in a cooperative manner. In Figure 13.12(*b*) the axes would correspond to the amount of oxygen bound on the rate axis and the partial pressure of oxygen on the concentration axis. Haemoglobin consists of four subunits each of which contains an oxygen binding site (see Chapter 9). Many apparently allosteric enzymes are similarly constructed although a tetrameric structure is not a requirement for an allosteric system, nor is cooperativity always due to the mechanisms that we are about to consider. Cooperativity is best defined as a kinetic phenomenon.

Two of the current theories propose that each subunit can exist in two conformations. We do not need to specify any molecular details of these conformations, and we can call them the T ('tense') and the R ('relaxed') states following Monod, Wyman and Changeux, who proposed the first model we shall describe. As the substrate, here oxygen, binds to each subunit of the system, changes in conformation take place. We assume that the T state has a lower affinity for oxygen than the R state, and that the four subunits are initially in the T state.

The Monod–Wyman–Changeux model, also known as the symmetry model, envisages that the first two molecules of oxygen bind in turn to the first two subunits, both of which remain in the T or low-affinity state. The binding of the second molecule of oxygen favours a conformational change and as a result all four of the subunits change to the R state. Since the R state binds oxygen more readily, the affinity of the haemoglobin has been increased; in other words the system shows cooperative binding. Throughout the process the haemoglobin remains symmetrical all the subunits having the same conformation (Figure 13.13).

An alternative view due to Koshland, and known as the sequential model, derives from earlier ideas of Adair and from the concept of induced fit which was developed by Koshland himself (Chapter 10). Here, binding of the *first* oxygen molecule is thought to cause a change in the conformation of the first subunit and some of this effect is transmitted to the next. This makes binding of the next molecule easier. As each molecule of oxygen binds, the conformation of the subunit it binds to changes from the T to the R state, and the process continues in a cooperative way until the protein is saturated (Figure 13.14).

These models can be treated mathematically (Ferdinand, 1976; Roberts, 1977) and can be applied to an allosteric enzyme, if we regard the binding of an allosteric activator as being similar to the binding of a molecule of oxygen, that is tending to cause a change in the conformation of the enzyme to favour binding of substrate. An allosteric inhibitor would operate in a contrary sense.

Figure 13.13. The Monod–Wyman–Changeux (symmetry) model of allosteric activation.

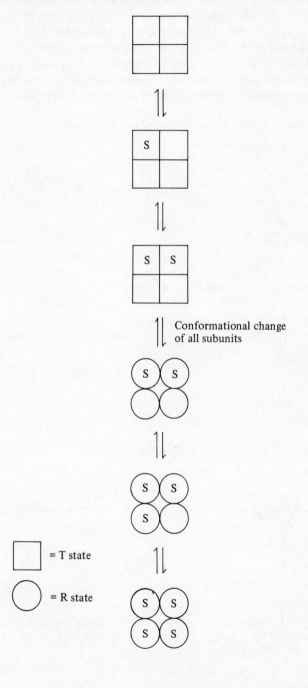

Given these two models we can reasonably ask which applies to the binding of oxygen to haemoglobin; again some answers are available from X-ray crystallographic studies. The complete structures of both oxy- and deoxyhaemoglobin have been determined by Perutz and his colleagues (Perutz, 1978). The four subunits of haemoglobin consist of two pairs of indentical subunits the so-called α- and β-chains. In the tetramer, most of the inter-subunit interactions are between α- and β-chains; like chains do not interact very much with each other. The interactions between the chains are non-covalent in character and

Figure 13.14. The sequential model of allosteric activation.

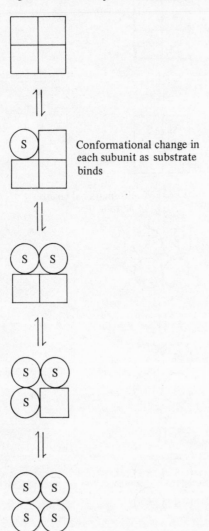

Conformational change in each subunit as substrate binds

ionic bonds play a very important part. On oxygenation, small changes in quaternary structure take place. The α- and β-subunits rotate with respect to each other and there are some substantial changes in the interchain bonding. For example a hydrogen bond between a tyrosine and an aspartate residue in the α- and β-subunits respectively is broken and a new one between an aspartate residue in the α-chain and an asparagine residue in the β-chain is formed. The tyrosine moves out of its original buried position in the deoxyhaemoglobin. A number of ionic bonds are broken as the oxygen binds and the haem iron moves out of the plane of the porphyrin ring. This cooperative change can be modified by the allosteric effector 2,3-diphosphoglycerate. The change attributed to a conformational change by the two models is in fact a change in quaternary structure.

The X-ray studies appear to favour the symmetry model, and this is confirmed by other methods which can demonstrate changes in conformation of proteins, notably magnetic resonance methods.

It is not easy to decide what changes are taking place in enzyme systems which exhibit cooperativity. The idea that the enzyme exists in two conformational states may be an oversimplification. A clearer understanding of enzyme systems depends upon our being able to determine the conformation of the enzyme–substrate and the enzyme–substrate–effector complexes at atomic resolution. However, we can get a feel for what may be taking place in enzyme-catalysed reactions if we take the view of Pauling that the most active conformation of an enzyme–substrate complex is close to the transition state of the reaction (cf. Chapter 15). We can then consider how the structure of the enzyme–substrate complex progresses as the reaction proceeds. Initially, both enzyme and substrate are in their more stable solution conformation, but when binding takes place both may change in conformation and reach a combined state which is close to the transition state. In this state, the activation energy for the reaction to be supplied thermally will be low and the reaction will proceed rapidly. In the case of an enzyme which requires an activator for maximum activity, the binding of enzyme to substrate alone is not sufficient for this near-transition state configuration to be achieved. The activation step, which may be allosteric or covalent, causes a change in the conformation of the ensemble to give a state sufficiently close to the transition state for maximum activity. We can only speculate on the precise nature of such changes, and no doubt many different types of conformational change can lead to the same result. A model for such a process may be the activation of chymotrypsin discussed earlier in the chapter. In this case, catalytic efficiency was not achieved until the full binding capacity for substrate was attained by the enzyme, although the rest of the catalytic apparatus was present. This again emphasises the importance of binding steps in enzyme catalysed reactions. If the catalytic steps are as rapid as possible, control must be exerted at the binding steps if a modulation of rate is to be achieved.

Finally we should make one or two points on the biological significance of the mechanisms discussed in the last few sections. The three types of kinetic behaviour illustrated in Figure 13.12 represent ways that make enzymes sensitive to changes in substrate concentration to different degrees. According to Lehninger (1975) an 81-fold increase in substrate concentration is required to bring the activity of an enzyme with a hyperbolic activity profile (13.12*a*) from 10 % to 90 % maximum activity. A system showing positive cooperativity (13.12*b*) is much more sensitive: a change of substrate concentration of only nine times is necessary to bring about the same change in rate. Such a system is obviously ideal in a control enzyme which has to respond to small changes in concentration of substrate and activators or inhibitors. On the other hand a negatively cooperative system (13.12*c*) is very insensitive to substrate concentration. This would be an advantage in an enzyme which is required to operate at a steady rate in the presence of a fluctuating concentration of substrate. Here an increase of substrate concentration of over 6000 times is necessary to increase the rate from 10 % to 90 % of the maximum.

How relevant are these models of mechanism to the metabolism of the cell *in vivo*? Most of the work we have outlined in this chapter was carried out using purified enzymes *in vitro*. As we pointed out in Chapter 2, there may be some hazards in this, particularly when we attempt to establish whether the control mechanisms which can be demonstrated *in vitro* take place inside the intact cell. This is mainly because it is difficult to determine with confidence the concentrations of substrates, enzymes and activators or inhibitors within a cell at any particular instant. Many of the experiments on which the control mechanisms are based depend on the enzyme being present in small amounts compared with the substrate: it is difficult to do meaningful kinetic studies in any other way. Yet it is entirely possible that the enzyme and substrate are present in the cell in equimolar concentrations, and the kinetic consequences of this would have to be determined separately.

Problem 13.3 Fatty acid synthase

This multi-enzyme complex catalyses the overall reaction shown in Figure 13.15.

(1) What intermediates would you expect to find, and how many catalytic activities would you expect on such a complex. (The individual reactions have already been described in Chapters 4 and 5. Your job is to put them together. The enzyme requires NADPH.)

Figure 13.15.

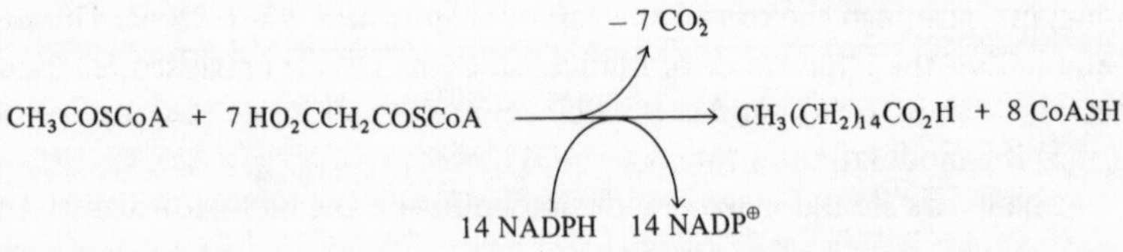

$$CH_3COSCoA + 7\ HO_2CCH_2COSCoA \xrightarrow{\quad\quad} CH_3(CH_2)_{14}CO_2H + 8\ CoASH$$

$-7\ CO_2$

$14\ NADPH \quad 14\ NADP^{\oplus}$

(2) What factors might regulate the activity of the complex both at the level of the enzyme and also with regard to its role *in vivo*?

Figures 13.16–13.18 are on pp. 365–366.

14   Biological membranes

14.1.   Introduction

One of the main targets of current biochemical research is a molecular understanding of the function of biological membranes, one of the most important structures in living systems. Membranes form the boundary between many of the quasi-independent systems which biochemists try to isolate and study (Chapter 2). Cells, for example, are bounded by membranes, nutrients and waste products pass through, and the sites at which cells can be recognised by other cells and by small molecules are situated on the cell membrane. This means that the recognition of a target cell by a hormone or the recognition of a foreign cell by an antibody is a membrane-dependent event (Chapter 16).

The cell organelles of eukaryotic cells, – the nuclei, mitochondria, endoplasmic reticulum and others – are bounded by membranes, each with its characteristic function. This subcellular organisation has advantages, some of which were mentioned in the previous chapter. A cell which is divided into compartments by membranes is able to control its activity in a most sophisticated way. It becomes possible for the different compartments to contain different concentrations of metabolites. The asymmetric distribution of small molecules that can result allows the development of energy-conservation devices such as mitochondria and chloroplasts. Processes within individual compartments can be controlled separately, leading to a greater efficiency in energy production and utilisation. Membranes also provide an apolar matrix within which the synthesis of water-insoluble compounds such as cholesterol can take place.

Another reason why membranes are being studied so intensively at present is technical. It is only relatively recently that techniques of sufficient power have become available for the analysis of the structure of their components. Clearly we would not expect techniques developed for water-soluble systems to be useful in membrane studies directly. As we discussed in Chapter 11, membrane proteins must first be released from their membrane by detergent treatment before more conventional techniques of protein separation can be applied. Recent developments in physical methods such as n.m.r., e.s.r., electron microscopy and neutron scattering now allow membrane systems to be studied more directly, and we shall discuss the application of some of these methods a little later in this chapter.

14.2. Membrane components

The basic components of membranes are well known. Apart from the rigid cell wall structures in plants and many micro-organisms, the main membrane components are protein and lipid. Many types of lipid molecule are known, and they all have one basic chemical property in common: they are all soluble in organic solvent mixtures (commonly a mixture of chloroform and methanol is used) and insoluble in water. This water immiscibility is due to the high hydrocarbon content of all lipid molecules, and means that when suitably organised they are capable of serving as a boundary between aqueous regions.

Two of the most commonly studied lipids in membranes are the phosphatidylcholines, members of the class of phospholipids, and the sterol cholesterol (Figure 14.1). The high hydrocarbon content of these molecules is immediately apparent: one would not expect them to be soluble in water. Table 14.1 shows that these two lipids are very common components of membranes; the structures of the other lipids mentioned in the table are given in Figure 14.2.

Membrane proteins have a large variety of functions. Too few have been sufficiently characterised for a really general picture of their structure to emerge. We shall see how the proteins that have been studied fit into the molecular view of the membrane when we have considered the matrix into which they fit, the lipid bilayer, which provides the basic unit of membrane structure.

Figure 14.1.

Phosphatidylcholine (lecithin)

$$(CH_3)_3\overset{\oplus}{N}-CH_2-CH_2-O-\overset{\overset{\textstyle O}{\|}}{\underset{\underset{\textstyle O_\ominus}{|}}{P}}-O-CH_2 \qquad\qquad 1$$

$$CH-O\,CO\,(CH_2)_n CH_3 \qquad 2$$

$$CH_2-O\,CO\,(CH_2)_n CH_3 \qquad 3$$

$$n = 12, 14, 16$$

Cholesterol

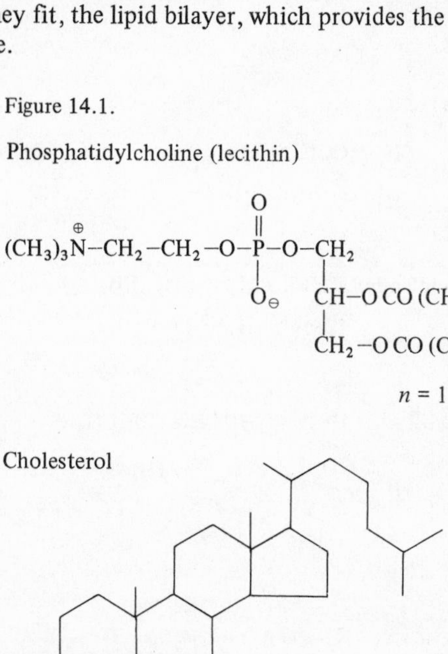

HO

Table 14.1. *The lipid composition of various membranes (Data from Korn, 1966)*

	Lipid composition (%)		
	Myelin	Erythrocyte	Mitochondrion
Cholesterol	25	25	5
Phosphatidyl choline	11	23	48
Phosphatidyl ethanolamine	14	20	28
Sphingomyelin	6	18	0
Phosphatidyl serine	7	11	0
Phosphatidyl glycerol	0	0	1
Remainder	37	3	18

14.3. The phospholipid bilayer

If we examine the structure of phosphatidylcholine more closely (Figure 14.1) we can distinguish two main features. Firstly the long hydro-carbon chain which we commented on previously and, in addition, a polar group, the phosphate ester of choline, which is attached to C-3 of the glycerol. We would expect this polar group to prefer an aqueous environment. One way in which this can be demonstrated is to spread a monomolecular layer (or monolayer) of lipid onto a water surface and examine the way in which the

Figure 14.2.

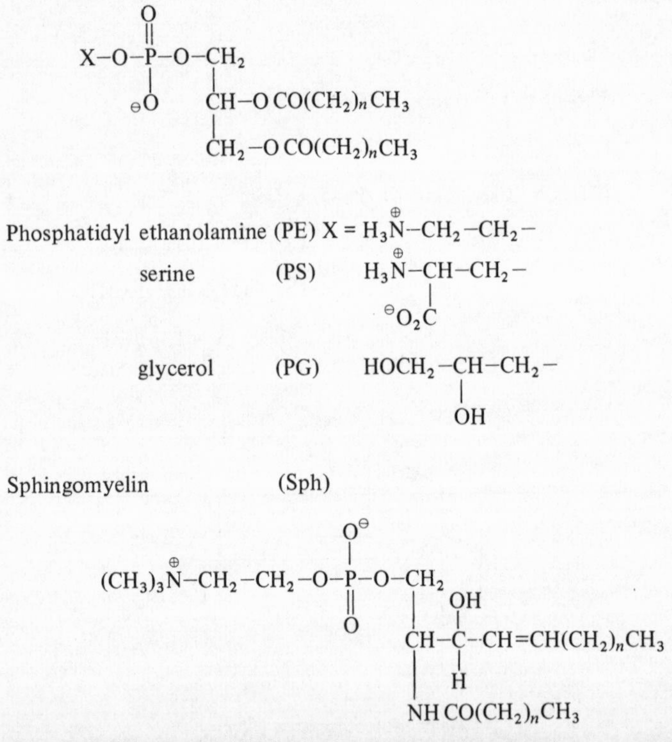

$$X-O-\overset{\overset{\displaystyle O}{\|}}{\underset{\underset{\displaystyle \ominus O}{|}}{P}}-O-CH_2$$

$$CH-O\,CO(CH_2)_n\,CH_3$$

$$CH_2-O\,CO(CH_2)_n\,CH_3$$

Phosphatidyl ethanolamine (PE) $X = H_3\overset{\oplus}{N}-CH_2-CH_2-$

serine (PS) $H_3\overset{\oplus}{N}-CH-CH_2-$
$\qquad\qquad\qquad\quad\ \overset{|}{\underset{}{\ominus O_2 C}}$

glycerol (PG) $HOCH_2-CH-CH_2-$
$\qquad\qquad\qquad\qquad\quad\ \overset{|}{OH}$

Sphingomyelin (Sph)

$$(CH_3)_3\overset{\oplus}{N}-CH_2-CH_2-O-\overset{\overset{\displaystyle O^{\ominus}}{|}}{\underset{\underset{\displaystyle O}{\|}}{P}}-O-CH_2$$

$$CH-\overset{\overset{\displaystyle OH}{|}}{\underset{\underset{\displaystyle H}{|}}{C}}-CH=CH(CH_2)_n CH_3$$

$$NH\,CO(CH_2)_n CH_3$$

lipid molecules orient on the surface. A convenient way of doing this is to use the Langmuir trough (Figure 14.3). Water is placed in the trough and a solution

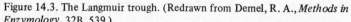

Figure 14.3. The Langmuir trough. (Redrawn from Demel, R. A., *Methods in Enzymology*, 32B, 539.)

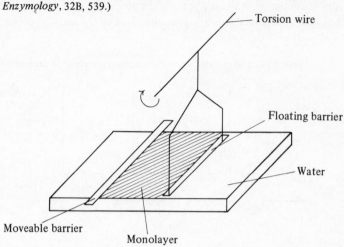

of a known amount of the lipid to be examined is pipetted onto the water between the moving barrier and the floating barrier. The area of the monolayer so formed is varied by moving the barrier, and the change in surface pressure which this causes is measured by its effect on the torsion wire. Since the number of molecules in the film is known and the surface area is also known, the mean area occupied by each lipid molecule can be calculated. It is usual to plot the variation of surface pressure (π) as a function of the mean molecular area. Such a plot for phosphatidylcholine is given in Figure 14.4. At a surface pres-

Figure 14.4. A pressure–area curve for dipalmitoyl phosphatidylcholine. (Data from Shah, D. O. & Shulman, J. H., 1967, *J. Lip. Res.*, 8, 219.)

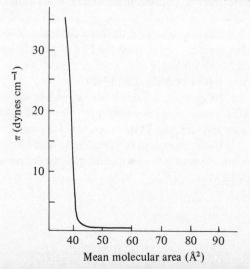

sure of 12 dyn cm^{-1}, about the surface pressure found in a typical membrane, the area per molecule of a phosphatidylcholine molecule is about 80 Å^2. This, from the molecular dimensions of the lipid molecule, is consistent with the orientation of the phospholipid shown in Figure 14.5. The polar headgroup resides in contact with the surface of the water and the hydrophobic hydro-carbon chains extend vertically above the plane of the water.

Figure 14.5. The orientation of phospholipid molecules at an air–water interface.

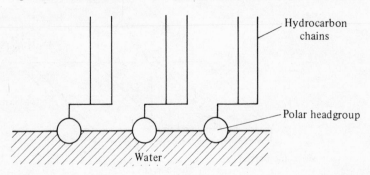

The forces involved in this interaction are clearly those of polar groups being stabilised in a polar environment, and also a strong hydrophobic inter-action between the adjacent hydrocarbon chains of the phospholipid.

A structure corresponding to this bilayer can also be observed in three dimensions. If we coat the inside of a round-bottomed flask with a film of phospholipid (say egg lecithin, a readily available phosphatidylcholine), add a salt solution and agitate the mixture for a few minutes, the lipid film soon hydrates and leaves the side of the flask, forming a milky suspension. If this suspension is examined by electron microscopy it is found to contain many closed vesicles inside each other (Figure 14.6). Under high magnification it is clear that these vesicles are made up of concentric double layers of lipid. Now, this bilayer can also be observed in electron micrographs of cell membranes and it is now accepted that this lipid bilayer is the fundamental unit of all biological membranes. The vesicles formed from the film of phospholipid are known as liposomes, and have been very widely used as a model for the lipid part of biological membranes (Colley & Ryman, 1976).

The concept of the lipid bilayer is now over fifty years old. It was first suggested on the basis of monolayer experiments on red blood cell lipids by two Dutch workers, Gorter and Grendel. Their experiments depended on many uncertain assumptions, but now the concept of the phospholipid bilayer is confirmed by many modern physical methods including X-ray diffraction and electron microscopy (Zwaal *et al.*, 1976).

Figure 14.6. Electron micrograph of liposomes formed from phosphatidyl-choline in an aqueous salt solution. (From Bangham & Horne (1964), *Journal of Molecular Biology*, 8, 660.).

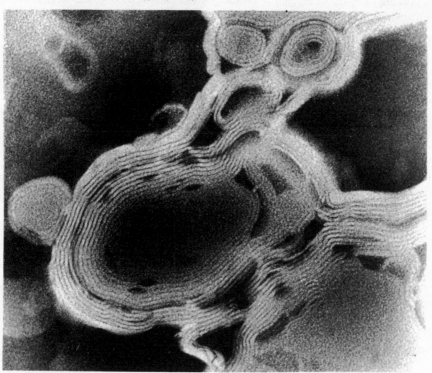

In many ways the phospholipid molecules, in their orientation at an aqueous surface, are behaving like detergent molecules. Detergents are also amphipathic (contain both a polar and an apolar group) and can form micelles (Figure 14.7). Micelles are short-lived aggregates of amphipathic molecules, for example the micelles of bile salts, cholesterol and triacyl glycerols that aid fat absorption in the intestine, whereas liposomes are stable defined structures; the ability to form the closed liposome structure seems to be restricted to certain lipids such as phosphatidylcholine.

Figure 14.7. Some detergent molecules.

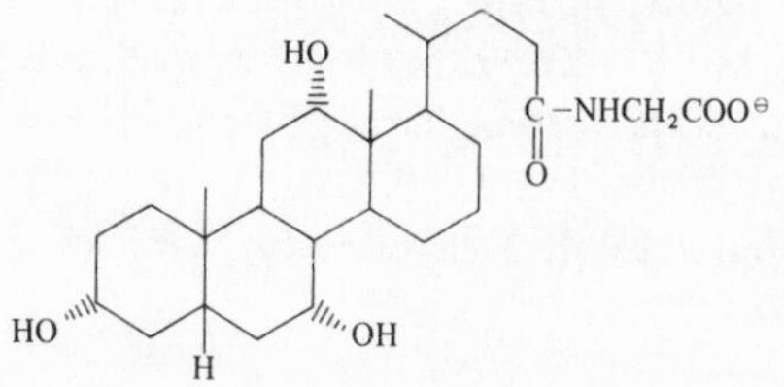

Glycocholic acid

Cetyl trimethylammonium bromide

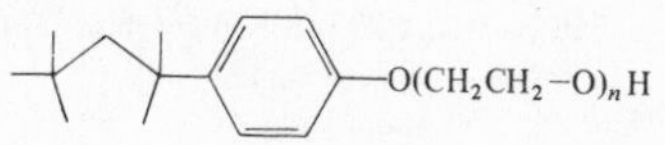

Triton X series

The phospholipid bilayer can be understood as a liquid crystal, that is an ordered fluid phase. One of the most characteristic properties of a liquid is its fluidity, and this is thought to be a very important property of a biological membrane. Too rigid a membrane may prevent the diffusion of small molecules across the bilayer or inhibit membrane-bound proteins. A cell with too fluid a membrane may be over-fragile and unable to maintain its internal environment satisfactorily.

The fluidity of membranes can be regulated in a variety of ways. One of the simplest is to vary the fatty acid components of the phospholipids. Unsaturated fatty acids melt at lower temperatures than saturated fatty acids with the same fatty acyl chain length, and we might expect a similar difference to apply to phospholipids. Egg phosphatidylcholine, for example, contains about one unsaturated fatty acid (about C_{16}) per molecule, and is fluid at temperatures below 0 °C. Dipalmitoyl phosphatidylcholine, a fully saturated phospholipid with fatty acids containing 16 carbon atoms, is fluid only above 41 °C. Proteins and cholesterol may also modify the fluidity of the membrane, as we shall see shortly.

14.4.   Cholesterol–phospholipid interactions

Table 14.1 showed that cholesterol is a major component of certain membranes, notably red blood cells, and in general the plasma membranes of many animal cells. The highest amounts of cholesterol seem to be found not in intracellular membranes such as endoplasmic reticulum and mitochondria, but in external cell membranes such as the plasma membrane. Since there seems to be a specificity in the distribution of cholesterol, it is reasonable to ask what the function of cholesterol is and how it is maintained in one membrane type. Some of the answers are known to the first question but very little is known in answer to the second.

The effect of cholesterol can be studied with liposomes. Small molecules such as glycerol and glucose and ions such as Cl^- will diffuse through phospholipid bilayers, and their rate of diffusion is greatly reduced by the addition of cholesterol or other closely related sterols containing a 3-β-hydroxyl group, a planar ring system and an intact side-chain. Cholesterol appears to be causing an increased rigidity of the membrane. The fatty acyl chain mobility is presumably decreased, and hence it becomes harder for small molecules to penetrate the more rigid bilayer. This is similar to the decrease in motion that one would imagine would take place with a decrease in temperature.

The biological significance of the rigidifying effect of cholesterol on otherwise fluid membranes is still unclear, but it is likely that it gives an advantage of mechanical stability and rigidity to parts of some membranes which require this under physiological conditions.

Magnetic resonance experiments have also demonstrated the effect of cholesterol on the rigidity of phospholipids and their hydrocarbon chains. This can be seen from the use of spin label probes as well as from proton and ^{13}C n.m.r. Here we describe the use of a sterol spin label to demonstrate the effect of cholesterol on the rigidity of the bilayer of egg phosphatidylcholine liposomes. The spin label used is shown in Figure 14.8; it is known for short as 3NC. Liposomes were prepared containing egg lecithin and cholesterol in various molar proportions. They also contained a small quantity of the spin-labelled sterol and their spectra were measured.

Figure 14.8.

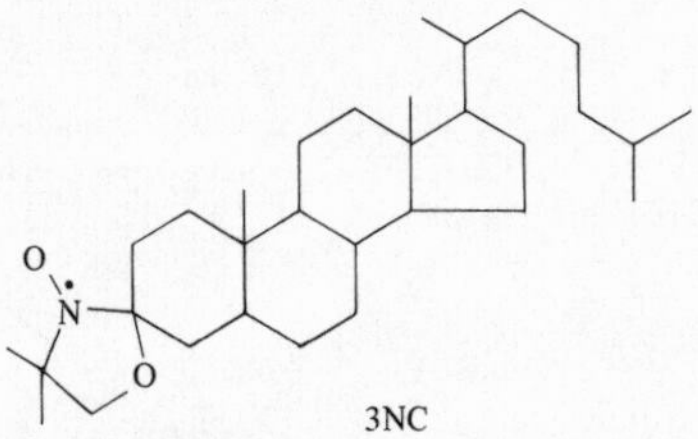

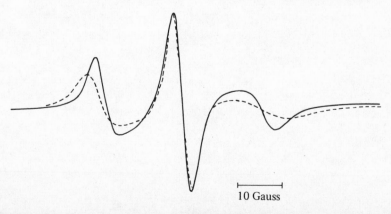

3NC

3-spiro[2'-(*N*-oxyl-4', 4'-dimethyloxazolidine)]cholestane

A spin-label molecule such as 3NC is sensitive to its environment in several ways. For this experiment it is important that the spin label is sensitive to the degree of order in the membrane in which it is incorporated. This is because the signals observed depend on the orientation of the nitroxide with respect to the magnetic field applied in the experiment. The greater the separation of the three lines of the nitroxide spectrum, the greater the degree of anisotropy in the sample, and therefore the more ordered the environment of the spin labelled molecule (Figure 14.9). The interpretation of the order parameters, S, derived from the separation of the lines is purely qualitative in this experiment. The data can also be analysed mathematically and from this analysis one can describe the motion of the spin-labelled molecule in some detail.

Figure 14.9. Spectra of spin label 3NC in phosphatidylcholine liposomes containing 0 mole % cholesterol (solid line) and 40 mole % cholesterol (broken line).

10 Gauss

The results for varying amounts of cholesterol are plotted in Figure 14.10. Clearly the degree of ordering increases as the amount of cholesterol increases, and this is consistent with the ideas that we have developed from other experimental approaches.

Figure 14.10. Order parameter S for 3NC in mixtures of egg phosphatidylcholine and cholesterol. S = 1 for a highly ordered system. Unlike the spectra shown in Figure 14.9 these data were obtained using an oriented planar membrane which allows a more accurate determination of S. (Data from Schreier-Mucillo, S. *et al.*, 1973, *Chem. Phys. Lipids*, **10**, 11.)

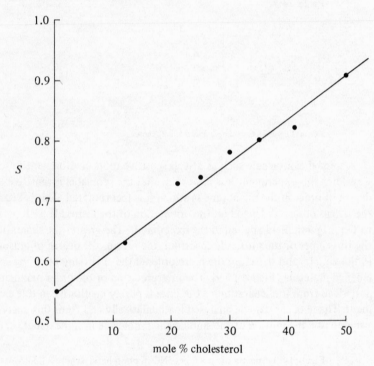

mole % cholesterol

We have suggested in the foregoing paragraphs that the fluidity of the membrane may determine many of its properties, particularly its permeability. We may also ask if the fluidity of the membrane varies at different depths in the bilayer. This is equivalent to asking how the motion of the carbon atoms of the fatty acyl chains of the phospholipid molecules varies as the distance from the polar ends of the molecule increases. This question can also be answered by magnetic resonance. If, for example, we use a series of fatty acid spin labels such as those in Figure 14.11 in which the nitroxide is placed at different positions down the fatty acyl chain, it should be possible, by incorporating them into liposomes and observing their spectra, to see how their ordering varies with the distance from the polar end of the bilayer.

Figure 14.11.

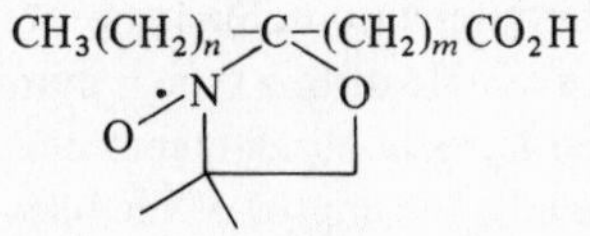

$$CH_3(CH_2)_n\!-\!C\!-\!(CH_2)_m CO_2H$$

Problem 14.1.

The order parameters obtained from such an experiment are plotted against the carbon number in the fatty acid chain in Figure 14.12. How would you interpret these results?

Figure 14.12. Order parameters for fatty acid spin labels (Figure 14.11) incorporated into liposomes. (From Seelig *et al.* (1972), *Journal of the American Chemical Society*, **94**, 6364.).

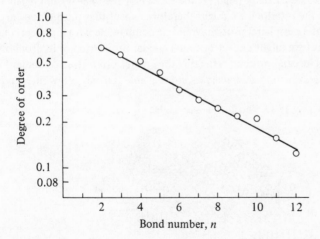

Similar conclusions to those from this experiment can be reached using ^{13}C n.m.r. or deuterium n.m.r. although there are some differences in the details of the interpretation. Recently, ^{31}P n.m.r. has become important in studying the phospholipids in model and biological membranes, since the ^{31}P resonance is sensitive to both the motion and phase of the lipids (Cullis & McLaughlin, 1977).

14.5. Membrane proteins

Having described the matrix in which all membrane events occur, we must now consider how these ideas can accommodate the various kinds of protein that are found associated with membranes. The most popular current picture is that due to Singer & Nicolson (1972), and is known as the fluid mosaic model (Figure 14.13).

In this view the proteins can be associated with the lipid bilayer in various ways. Protein *A*, for example, is in contact with the surface of the bilayer only and is presumably held there by polar interactions. Such a protein is called by Singer a peripheral protein. An example of this type is cytochrome c, which is found associated with the inner mitochondrial membrane of many tissues. Because it is not very tightly bound to the depths of the bilayer, cytochome c can be relatively easily released from the membrane and purified by conventional techniques.

Other proteins such as *B* and *C* are strongly embedded in the bilayer, emerging at one or both sides. These proteins are known as integral proteins and are relatively difficult to release from the membrane – it usually requires the use of detergents, which have the function of breaking up the lipid bilayer and providing a partially hydrophobic environment for the integral proteins. The interactions between integral proteins and the lipid bilayer would be expected to be partially polar (at the surface of the bilayer) and partially apolar (at the interior). We might therefore expect that the amino-acid composition of a peripheral protein would be comparable with a typical soluble protein, but we might expect to find a greater proportion of hydrophobic residues in an integral protein, reflecting the greater proportion of apolar binding.

At present, amino acid analyses are available for only a few integral pro-

Figure 14.13. The fluid mosaic model.

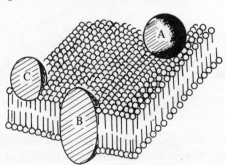

teins. One example is glycophorin, a glycoprotein of the human erythrocyte. This protein has two polar ends and a central region that is composed of predominantly hydrophobic amino acid residues. The central apolar region appears to be big enough to span the membrane. Another example is cytochrome b_5. This cytochrome, which is found in the endoplasmic reticulum of liver (rat is usually studied), is held in position on the membrane by a hydrophobic peptide tail. The tail can be removed by enzymic hydrolysis with trypsin, leaving a protein whose composition is mainly characteristic of a peripheral protein like cytochrome c.

The requirement of an integral protein for a hydrophobic environment is now easily understood. It must be stabilised in a conformation that is enzymatically active. This requires a degree of hydrophobic bonding, and if this is to be achieved in a disrupted membrane a detergent may have to take the place of the natural phospholipid. It is not difficult to imagine that since the lipid environment appears to stabilise the active conformation of a membrane protein, the protein will be sensitive to the nature of the surrounding lipid. An example of this will be discussed in the next section.

Some of the strongest evidence for the fluid mosaic model of membrane structure has come from the electron microscopic technique of freeze-fracture. In this method a membrane fragment is frozen rapidly and then fractured along the middle of the bilayer. The surfaces of the two halves of the bilayer can be replicated by platinum-carbon shadowing and the two replicas can be examined in the electron microscope. The micrographs show globular particles which either appear to protrude from the surface or appear as depressions (Figure 14.14). These are thought to represent the sites occupied by the integral proteins. During the fracture process only the lipid bilayer is cleaved; the integral proteins remain with either of the two halves, appearing as a depression or a raised particle on the replicas. Pictures such as this are a visual representation of the fluid mosaic model.

14.6.   Lipid and protein systems in membranes

Perhaps the most active area of research in the whole field of membrane biochemistry is the study of the interaction of lipids and proteins. This study is thought to hold the key to a molecular understanding of many aspects of membrane function. At present only a few well-defined studies have been carried out, and these give some indication of what may be expected from further systems. It is obviously too complicated to attempt to analyse the lipid-protein interactions in a heterogeneous natural membrane, so approaches have tended to use purified proteins where possible. This means that at present results are biased towards proteins that are easily purified, and this might be an unrepresentative sample of the whole population of membrane proteins.

Several types of question can be asked. The first, as usual, is structural. How are the protein molecules arranged with respect to the lipid bilayer? Are

Figure 14.14. The inner face of the membrane of a red blood cell as revealed by freeze–fracture. (From Verkleij & Ververgaert (1978), *Biochimica et Biophysica Acta,* **515**, 310.)

they found in areas of the membrane which may be composed of one particular type of lipid? One can then ask how the lipid environment of the protein affects its function. Another important question concerns the mobility of the membrane components. Unlike a molecule in free solution, a membrane component is constrained within the lipid bilayer and as a result two types of motion are available to it. The first is lateral motion within the plane of the bilayer. This type of motion can be very rapid, and both protein and lipid molecules can move in this way. The second type of motion particularly applies to molecules which are associated with only one half of the bilayer. This is motion from one half of the bilayer to the other, known as 'flip-flop'. The rate of lateral motion of a lipid molecule is thought to be great enough for it to be able to move from one end of a bacterium to another in a matter of seconds. (Lee, 1975). The rates of flip-flop must be very much slower to account for the fact that biological membranes show an asymmetric distribution of phospholipids in the two halves of the bilayer. Membrane proteins may move laterally about two orders of magnitude more slowly than lipids.

These comments about the rate of motion of membrane-bound molecules serve to illustrate another dimension in biochemical studies. We are not only interested in the structure of the systems we study but we also want to know how they vary with time. Every biological experiment has a time-scale associated with it.

14.6.1. Three-dimensional structure of membranes

It is not easy to observe the three-dimensional structure of a protein within a membrane. Physical methods such as X-ray diffraction require an ordered array, as in a crystal, in order to determine a detailed picture. Very often the energy of radiation applied by these methods is too great and the sample decomposes. The first membrane protein structure to be observed in its membrane-bound state at near atomic resolution was the protein known as bacteriorhodopsin (Henderson & Unwin, 1975). This protein is found in a membrane region of halobacteria such as *Halobacterium halobium*. Because of its colour this region is known as the purple membrane.

X-ray diffraction studies had shown that bacteriorhodopsin was an integral protein which was arranged in a hexagonal array within the membrane. Henderson and Unwin were able to take advantage of the regular structure by taking a large number of electron micrographs at extremely low beam exposures. They used a new technique of image reconstruction from these micrographs to obtain a map of the purple membrane. This map showed rod-shaped features which were aligned perpendicular to the plane of the membrane; these were assigned to α-helices of the protein. The remainder of the protein, presumably not in α-helix form, and the lipid molecules could not be discerned at the resolution of the experiment.

14.6.2. The effect of the lipid environment

A protein which can be purified from the sarcoplasmic reticulum has been used to study the effect of the membrane lipids on the enzymic activity of the protein. The sarcoplasmic reticulum is the equivalent organelle in muscle to the endoplasmic reticulum in other cells and it has the specialised function of storing and accumulating Ca^{2+} ions which are released as part of the mechanism of muscle contraction. One of the proteins involved in this process is known as the $(Ca^{2+} + Mg^{2+})$ATPase. This is the major protein of the sarcoplasmic reticulum and it can be assayed by the hydrolysis of ATP, which it catalyses in the presence of Ca^{2+} or Mg^{2+} ions.

It was possible to deplete the sarcoplasmic reticulum of lipid by treating with detergent. The ATPase is then inactive. It reassumes an active form when about 30 mole of phospholipid per mole of protein is added. The lipid requirement is relatively specific. If the phospholipid (phosphatidylcholine in this case) is replaced by cholesterol, the enzyme activity disappears. If the cholesterol is removed and replaced by phosphatidylcholine, activity is completely restored. This clearly demonstrates that the interactions between membrane protein and lipid can play a very important part in determining enzymic activity. It is thought that this could be a control mechanism for membrane proteins.

The size of the sarcoplasmic reticulum ATPase is such that it would be expected to protrude from the membrane surface so that half of it would be exposed to the aqueous medium. This allowed the workers to suggest that the enzyme requires for activity an immobilised lipid shell, one molecule thick. Since the sarcoplasmic reticulum contains lipid and protein in a molar ratio of 90 : 1, and the ATPase is the major protein, there appears to be enough lipid in total to form shells about three molecules thick around each protein molecule. Only one of these shells needs to be immobilised for enzyme activity (see Lee, 1975). This concept of a shell of lipid in a relatively immobilised state which surrounds an integral protein is perhaps the main generalisation that has been made on protein-lipid interactions in membranes. Similar results have been obtained with other proteins by different methods. Cytochrome oxidase, cytochome P 450 reductase (see Chapter 15) and rhodopsin all appear to require a lipid shell, although recent n.m.r. experiments have suggested that, on the n.m.r. time-scale at least, this shell is not present.

From what we have discussed it is clear that the molecular understanding of membrane processes is in its infancy. Much remains to be described from a structural, functional and dynamic point of view. However the basic molecular concepts of membrane structure and function appear, at least for the present, to be established and so the conceptual framework for study at a more detailed molecular level is available.

15   Drugs, agricultural and horticultural chemicals

15.1.   Introduction

Some of the greatest achievements of science have benefited mankind through the interaction of a chemical with a biological system. Medicine today is greatly dependent on chemotherapeutic agents and drugs of all kinds. Equally, agriculture and horticulture are served by fertilisers and pest-controlling agents. However, many important compounds were discovered by chance, and only recently has their mechanism been elucidated at a molecular level. Today, new drugs and phytochemicals are designed and developed with a major contribution from consideration of chemical principles. In such studies the interactions between chemistry, biochemistry, pharmacology and biology are at their strongest.

It is the task of the biologist and biochemist to elucidate the behaviour of the drug *in vivo* – its absorption into the organism, its metabolism, and its mode of action and its elimination. From such work they can guide the chemist towards his synthetic targets and the pharmacist towards the formulation of the drug or phytochemical as it is to be administered. The invention of new chemicals for specific responses in nature requires an understanding of all these features and this always means a team effort; no scientist can be sufficiently expert in all the necessary sciences. Nevertheless, a competent chemist or biochemist will appreciate how his subject and his thinking interacts with that of scientists from other backgrounds, and we show how some points of contact can be made in this chapter. Because the biochemistry of plants is less well understood, we shall not discuss them as deeply as mammalian, bacterial, or insect systems. However the general chemical principles apply to any organism and, with a knowledge of the relevant physiology and enzymology, it is easy to transfer the ideas from one system to another.

15.2.   Administration, absorption and transport of drugs

It has been appreciated for many years that drugs exert their effects at specific sites in specific tissues, for example by inhibition of enzymes or by interaction with DNA to prevent cell division. Obviously it is important that the drug should be administered to the subject in such a way that it reaches its site of action efficiently and without unwanted chemical modification. This is equally true for phytochemicals. (See Goldstein, Aronow & Kalman, 1974, p. 129.)

The most convenient route for administration of drugs to humans is oral. After oral administration, the drug is subjected to the acid and enzymes of the digestive system before being adsorbed into the bloodstream from the gastrointestinal tract. Administration via the gastrointestinal tract is known as *enteral* administration. From the digestive system the drug is transported directly to the liver, where the major drug-metabolising enzymes are located (Section 15.7). Consequently, it is pointless to administer orally a drug that is sensitive to digestion or hepatic modification, and an alternative route that bypasses the liver and digestive system must be employed. Such routes are known as *parenteral* routes; the chief are subcutaneous, intravenous and intramuscular injection, topical application and inhalation. The choice depends on the required site of action and upon the chemical properties of the drug. The absorption of a drug into an animal and its distribution throughout the various tissues is essentially a physical chemical process in which the drug is transferred from one phase to another, for example inhaled air to blood via the lungs. Once absorbed, the drug must reach its site of action and to do this it must cross cell membranes. As we saw in the last chapter, membranes can be regarded as hydrophobic envelopes through which apolar or uncharged molecules may pass. Thus a drug that is essentially hydrophobic may cross a membrane by passive diffusion down a concentration gradient. Such a process could be studied using the liposome systems described in the last chapter. General anaesthetics are essentially lipid-soluble compounds that are distributed by diffusion. In a small number of cases it is known that the drug complexes initially with a carrier molecule in the membrane and the complex can then diffuse through the membrane (facilitated diffusion) or the drug can be carried against a concentration gradient by an active transport mechanism (see Chapter 9). Active transport takes place against a concentration gradient, this process consumes ATP.

Distinguishing between these mechanisms of transport is not difficult in principle. Passive diffusion takes place down a concentration gradient, so its rate should increase as the difference in concentration between either side of the membrane increases. Active transport requires ATP, so anything inhibiting its production (for example an uncoupler of oxidative phosphorylation) should also inhibit active transport into an intact cell. Facilitated diffusion, like active transport, also depends on the presence of a carrier. Since the carrier will be present in fixed amounts in the membrane, it will only be able to carry a limited amount of substance in a given time. In fact, if the concentration of the permeating substance is increased the rate of transport will increase until all the carriers are occupied. Thus the transport system will show saturation very similar to the saturation we described for enzymes in Chapter 13.

Transport mechanisms across membranes are therefore very important, but equally important are transport mechanisms in the blood. Small ionisable molecules such as barbiturates (Figure 15.1*a*) are readily soluble, but more lipophilic molecules need assistance. The latter are often carried bound to proteins in the blood (serum proteins), as in fact are many natural lipophilic substances such as fatty acids; alternatively the pharmaceutical formulation may present them in a non-thrombogenic detergent, as in the case of the steroid general anaesthetic althesin (Figure 15.1*b*).

Figure 15.1.

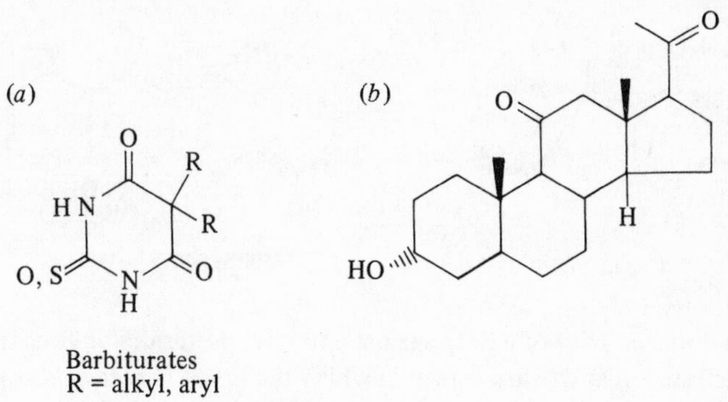

(*a*)

Barbiturates
R = alkyl, aryl

(*b*)

The chemical characteristics of the compounds thus partially determine their transport behaviour. The importance of the polarity of the molecule is especially noteworthy and is a factor that we shall return to later. Alongside the structural requirements for the drug to exert the desired biological effect, the chemist must consider these transport factors when he designs his synthetic targets. However, the structure required for activity is usually the prime consideration.

15.3. Drug design and new development

The scale of work and the expense involved in producing a new drug today are so great, of the order of tens of millions of pounds, that it is worth considerable effort to establish the biochemistry of the disease that it is desired to treat. (See Korolkovas & Burckhalter, 1976.) As we shall see shortly, with a knowledge of the biochemistry, the chemist can select compounds for synthesis. However, in contrast to the current situation, the first chemotherapeutic agents (for example penicillins) were discovered by chance, sometimes with the aid of random screening procedures. Many thousands of fungal and bacterial metabolites extracted from cultures of micro-organisms have been tested for antibacterial and antifungal activity: such drugs as penicillins and tetracyclines (broad-spectrum antibiotics), griseofulvin (for treatment of fungal diseases) arose in this way (Figure 15.2). Similar screening procedures are currently important in the discovery of fungicides, defensive compounds produced by higher plants and anticancer drugs. The reader might like to speculate on the biosyntheses of these compounds, all of which are now well known (see Pratt, 1973).

Figure 15.2. Antibiotics.

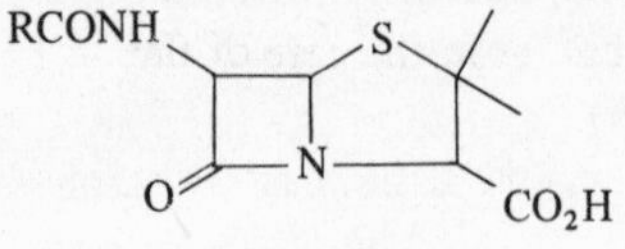

Penicillin

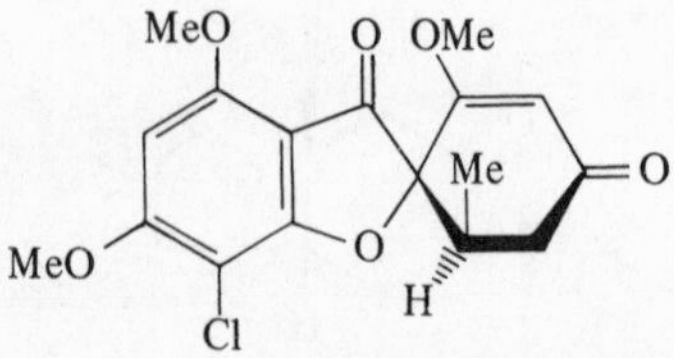

Griseofulvin

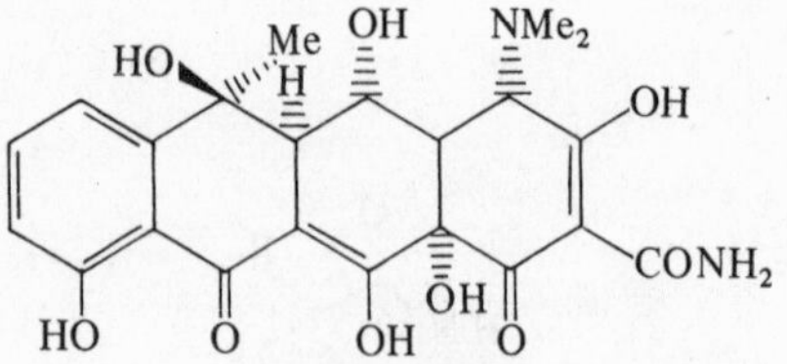

Terramycin, a tetracycline

The initial activity of a drug against a fungus, bacterium or virus is first demonstrated in an *in-vitro* screen in which the target organism is grown in culture. (Suckling, Suckling & Suckling, 1978). It is important that such screens give clear-cut and rapid information because the results determine to a very great degree the further course of research. Essentially, the screen is a model of how the drug will behave *in vivo*, and it is vital that the biochemists and biologists responsible for constructing the screen appreciate how it differs from the real-life situation. For example, it is common to use purified enzymes as primary screens for drugs that are designed to function as enzyme inhibitors. However, it is not uncommon to find that a drug is highly active against the enzyme *in vitro* but fails to affect the target enzyme *in vivo*. Usually this is because the drug fails to penetrate to the site of the target enzyme in the whole organism. Even if such tests are successful on animals, thorough clinical trials will be required before the drug can be considered for commercial production (Robinson, 1974).

Only when all of these hurdles are passed will the health authorities in most countries permit the drug to be used. It is the expense of the thorough testing and development that makes the cost of drug development so high, and any method that increases the chances of success is valuable. For these reasons so-called **rational** and **quantitative** methods of drug design have come into prominence in recent years. These methods can be divided into two classes – **lead generation** and **lead optimisation**.

In generating a lead the key is to discover compounds that are highly active against the target system: problems of toxicity and administration are secondary at this stage. Most new projects in drug design begin with at least a superficial understanding of the biochemistry of the system to be treated. With this knowledge the chemist has several ways open to design his potential drugs:

(1) Competitive inhibitors of both enzymes or receptors. These can be either transition state analogues or substrate analogues (antimetabolites).

(2) Irreversible inhibitors, including active site directed inhibitors, latent (suicide inhibitors) and more reactive alkylating agents.

Once a lead is established, the optimal compounds of that structural type need to be found. Criteria of acceptability are now widened to include pharmacological, toxicological and metabolic considerations. In the past this has usually meant a major synthetic effort in preparing analogues of the lead compound, but today physicochemical theories have been developed to help shorten the synthetic phase by directing the chemist to the most significant compounds (Hansch, 1969). It is impossible in this book to give more than an outline of the chemistry and biochemistry involved in these studies, but the following series of examples should give you a feel for the field.

15.4. Established strategies for drug design
15.4.1. *Inhibition of vitamin biosynthesis: tetrahydrofolic acid*

A vitamin is a compound required in small amounts by an organism for a healthy life but which cannot be synthesised by the organism itself. Folic acid is a vitamin for mammals; it is converted into tetrahydrofolate, which is the coenzyme responsible for the interconversion and transfer of many one carbon units (Pratt, 1973, pp. 5, 170). One especially important reaction is the conversion of deoxyuridylic acid into deoxythymidylic acid (Figure 15.3), one of the nucleotides of DNA and essential for cell division.

Figure 15.3.

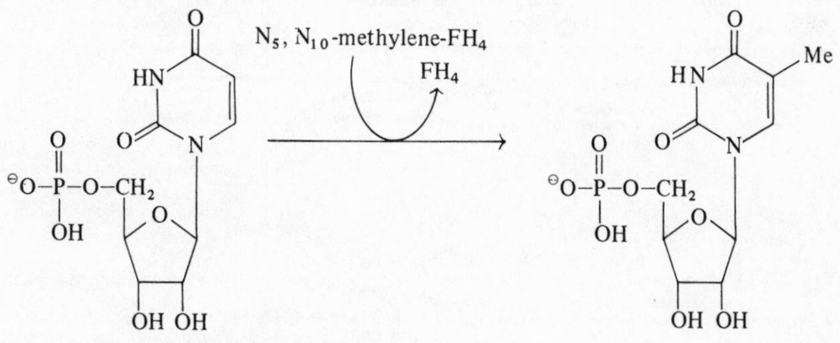

FH_4 = tetrahydrofolic acid

The final stages of folic acid biosynthesis are shown in Figure 15.4. The enzymes above the dotted line are not present in the animal but may be present in an infecting bacterium; further, the enzyme below the line is quite

Figure 15.4.

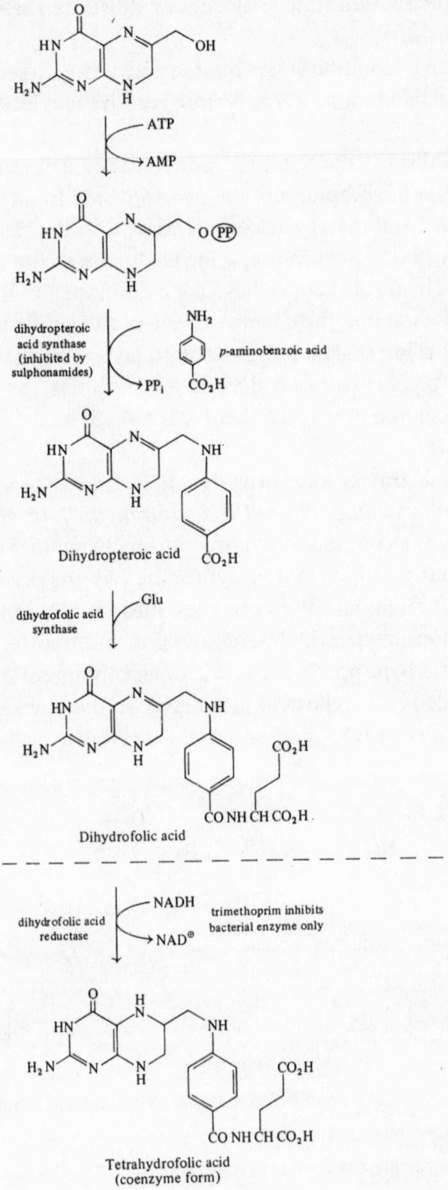
ATP
AMP
dihydropteroic acid synthase (inhibited by sulphonamides)
p-aminobenzoic acid
PP$_i$
Dihydropteroic acid
Glu
dihydrofolic acid synthase
Dihydrofolic acid
NADH
NAD$^⊕$
dihydrofolic acid reductase
trimethoprim inhibits bacterial enzyme only
Tetrahydrofolic acid (coenzyme form)

different in host and parasite. Since the host can obtain its folic acid requirements in its diet, inhibition of the folate-synthesising machinery of the parasite will prevent essential reactions such as that in Figure 15.3, and hence prevent synthesis of new DNA, cell division, thus causing the death of the parasite. This is the principle of **selective toxicity** which has become a valuable rationale for lead generation (Albert, 1973).

Sulphonamides (Figure 15.5) were first synthesised for use in dyestuffs. In 1935 it was shown that one of them, prontosil, was effective in ridding mice of streptococcal infections. It soon became clear that prontosil itself was not the active molecule but it was converted by the mouse into p-aminobenzene sulphonamide (sulphanilamide), which competitively inhibits dihydropteroate synthase.

Figure 15.5.

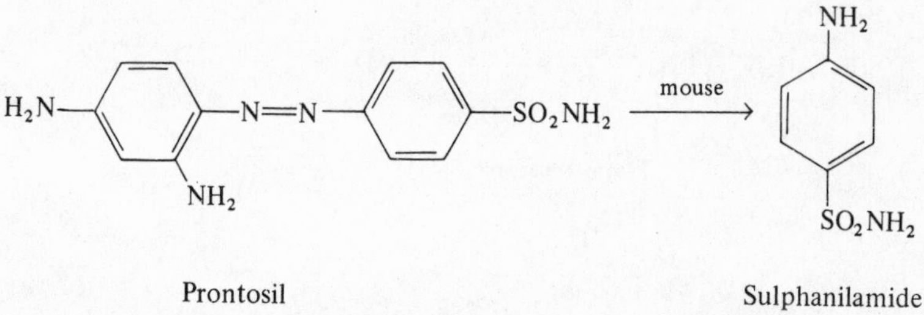

Prontosil   Sulphanilamide

The close structural relationship between sulphonamides and p-aminobenzoic acid is obvious (recall Chapter 3), and it has been found that the relationship extends further to the acidity of the sulphonamide. By choosing the appropriate amine, p-aminobenzene sulphonamides with a range of pK_a's can be obtained. The most active antibacterials are those with pK_a' closest to p-acetamidobenzoic acid (pK_a' 4.6). Thus sulphadiazine (pK_a' 6.48) is much more effective than p-aminobenzene sulphonamide (pK_a' 10.43) (Figure 15.6; Goldstein *et al.*, 1974, p. 44).

Figure 15.6.

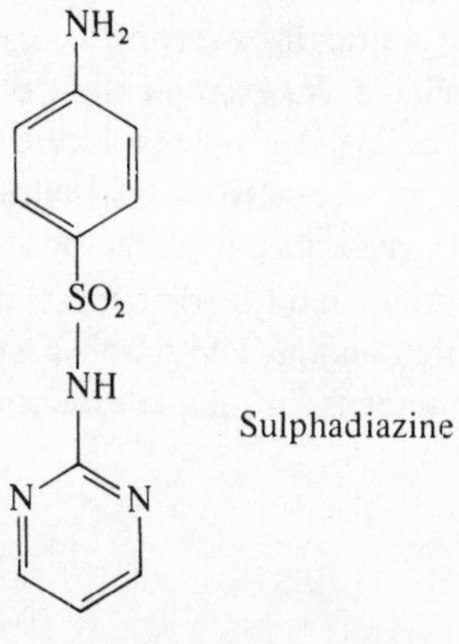

Sulphadiazine

It is not usual for the structure of the enzyme inhibitor (antimetabolite) to resemble the structure of the natural substrate so obviously. A group of very important antimalarial drugs based upon diaminopyrimidines inhibit the dihydrofolate reductase of the malaria parasite at a concentration of 10^{-9} mol l^{-1}. They do not inhibit mammalian dihydrofolate reductase until their concentration is raised 200-fold. Thus, selective toxicity is again obtained, this time presumably relying on an enzymological difference between host and parasite. It is believed that the pyrimidine portion of the drug binds to the pteridine ring binding site of the enzyme (Figure 15.7).

Figure 15.7.

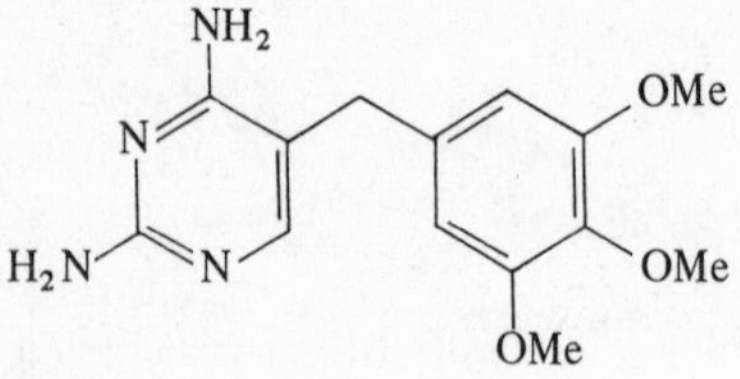

Trimethoprim

cf.

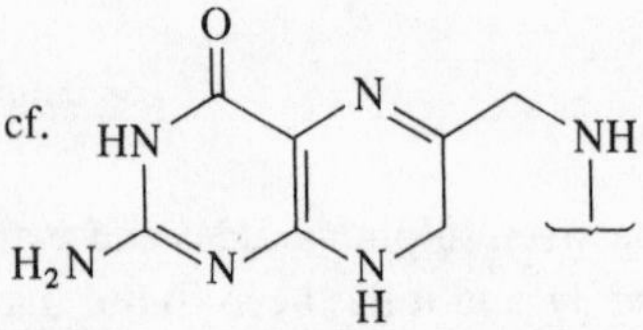

15.4.2. *Inhibition of binding to receptors: acetyl choline*

Many important physiological events are triggered by the binding of a molecule to a small region of a membrane. In Chapter 9 we met the example of a nerve cell in which a transmitter substance is emitted by the presynaptic membrane and received on the other side of the synapse by a receptor. An example of such a transmitter is acetyl choline (Figure 15.8a), a very widespread stimulator of muscle, nerve and brain cell response. Its behaviour has been especially well studied because tissues containing a very high concentration of acetyl choline receptors are readily available, for example the electric organ of the electric eel (Goldstein *et al.*, 1974, p. 73). A simple molecule like acetyl choline is a good example for discussing structure–activity relationships. Obvious features likely to have biological significance if present in the same molecule are the quaternary ammonium group and the ester function. This is readily shown to be the case by the fact that choline and simple acetate esters (Figure 15.8b) are completely inactive in screens for muscle excitation.

Figure 15.8.

(a)

$$CH_3 \overset{O}{\underset{||}{C}} OCH_2CH_2 \overset{\oplus}{N}(CH_3)_3$$

Acetyl choline

(b)

$$CH_3 \overset{O}{\underset{||}{C}} OCH_2CH_2C(CH_3)_3$$

Less immediately apparent is the need for the functional groups to be in the correct relative positions. This requirement simply arises from the fact that the receptor has a precise three-dimensional structure analogous to the relationship of an enzyme active site to its substrate. So not only must the functional groups be the correct distance apart linearly (here about 0.5 nm), they must also be in the correct conformation.

All of the biological functions of acetyl choline are not produced by the same receptor, as can be shown by pharmacological techniques using the alkaloids nicotine and muscarine. These are cyclic molecules in which the conformation is substantially dictated by the rings. Acetyl choline is more mobile conformationally, and it can be readily shown by building models how muscarine and nicotine mimic acetylcholine and are able to bind to the appropriate centre (Figure 15.9).

Figure 15.9.

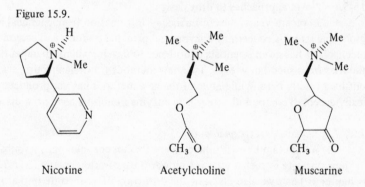

Nicotine   Acetylcholine   Muscarine

The calculation of the most probable conformation of molecules by means of molecular orbital techniques is now possible, and this allows the specification of the structural requirements for biological activity in mathematical form. Further series of drugs can then be designed based on the calculations. We can generalise that the design of a drug needs both stereochemical and functional group considerations. Some acetyl choline analogues have been used clinically as antispasmodic and antisecretory agents: their structural relationship in the protonated form to acetyl choline is obvious (Figure 15.10). The highly hydrophobic termini of these molecules help the drug to enter the central nervous system by passive diffusion through the cell membrane that separates blood and brain cells (the so-called blood–brain barrier).

Figure 15.10.

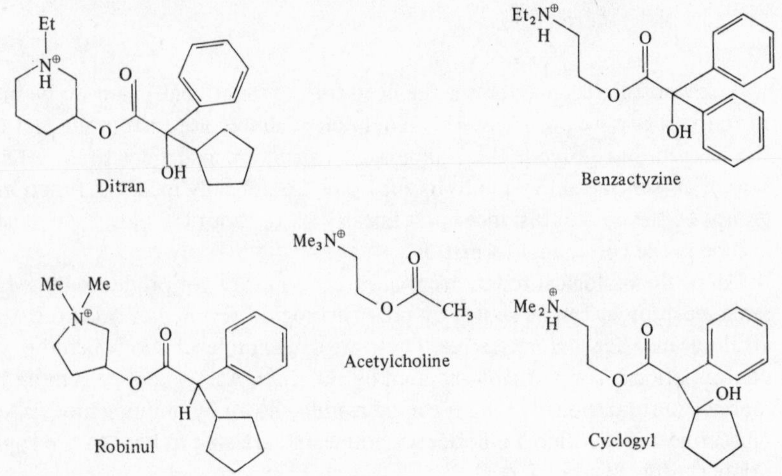

Ditran

Benzactyzine

Acetylcholine

Robinul

Cyclogyl

15.5. Novel approaches to drug design

In recent years, the confluence of information from chemical and biological streams concerning enzyme and protein structure and enzyme mechanisms has given scientists confidence to develop chemotherapeutic theories from such knowledge. Two new methods for designing drugs will be outlined briefly here. While none of the new methods has yet produced a clinically accepted compound, many new enzyme inhibitors have been discovered.

15.5.1. Transition state analogues

We are familiar with the concept that an enzyme binds its substrate at the active site in such a way as to lower the activation energy for reaction (Chapter 10) and we also saw how the distortion of a substrate into a transition state-like conformation helps to reduce the activation energy. Another way of looking at this is to consider that the enzyme binds most strongly to the transition state of the reaction. If this tight binding takes place, we would expect that a molecule with a structure resembling the transition state will be a potent inhibitor of the enzyme. Wolfenden (1972) has argued forcibly along these lines and cites several examples in support, one of which is described below.

Proline racemase catalyses the interconversion of L-proline and D-proline presumably via a carbanion (enolisation) to the carboxylate group. It is very probable that this anion, or its enzyme-bound equivalent, is close to planar geometry at the transtion state of the reaction. The α-carbon would then be sp²-hybridised, and this is the key feature that a transition state analogue should reproduce. The obvious test compound is pyrrole-2-carboxylic acid (Figure 15.11); it was found that the pyrrole analogue binds to the enzyme nearly 200 times more strongly than the substrate.

Figure 15.11.

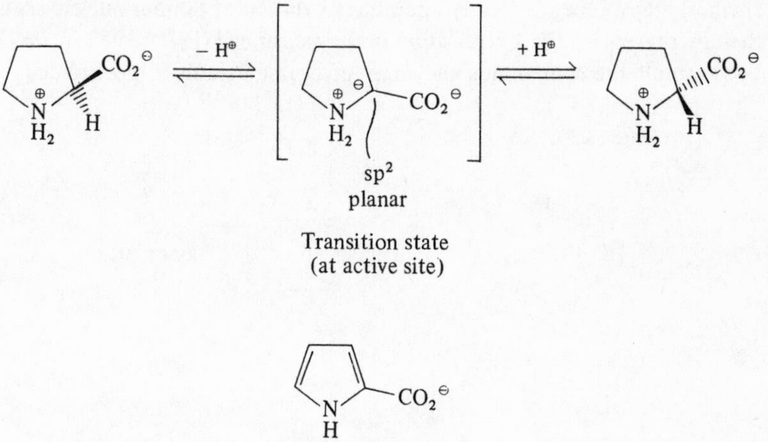

sp^2
planar

Transition state
(at active site)

Transition state
analogue

Problem 15.1.

Triose phosphate isomerase catalyses the reaction shown in Figure 15.12a. By thinking of the tautomerism of carbonyl compounds and the geometrical structure of amides, suggest:

(1) A plausible mechanism for the isomerisation.
(2) Why compound (15.12b) is a potent inhibitor of triose phosphate isomerase. (Lowe & Lewis, 1973.)

Figure 15.12.

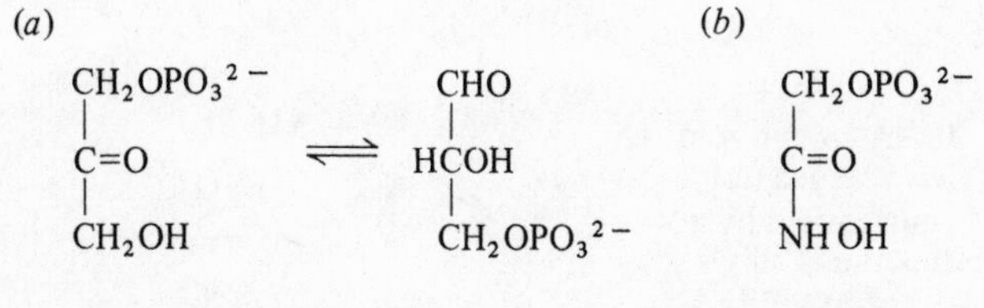

(a)

$$CH_2OPO_3^{2-}$$
$$|$$
$$C=O \qquad \rightleftharpoons \qquad$$
$$|$$
$$CH_2OH$$

$$CHO$$
$$|$$
$$HCOH$$
$$|$$
$$CH_2OPO_3^{2-}$$

(b)

$$CH_2OPO_3^{2-}$$
$$|$$
$$C=O$$
$$|$$
$$NH\,OH$$

15.5.2. *Irreversible inhibitors*

The effectiveness of the compounds described so far depends upon their competing successfully against the substrate for an active site. If a supply of substrate becomes available to the inhibited enzyme, the substrate will displace the inhibitor drug, the enzyme will function again and the chemotherapeutic effect will be lost. However, if the inhibitor can be made to react with the enzyme to form a stable covalent bond it becomes very much more difficult for the pathogen to recover: to by-pass this inhibition it must take the energetically expensive step of synthesising more enzyme. So far no enzyme inhibitors of this type have been developed for clinical use, but cancer chemotherapy depends greatly upon the alkylation of tumour nucleic acids, thereby preventing DNA replication in the cancer cell (Pratt, 1973, p. 267). For example the compounds known as nitrogen mustards (Chlorambucil,

Figure 15.13.

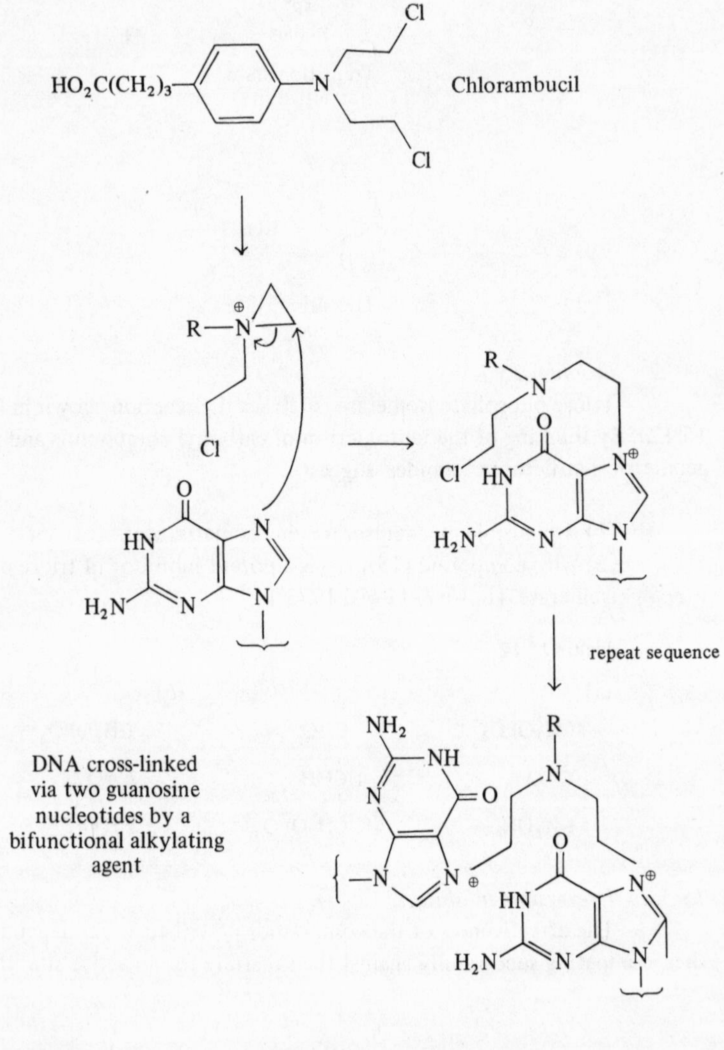

DNA cross-linked via two guanosine nucleotides by a bifunctional alkylating agent

Figure 15.13) attack guanidine residues, as shown in Figure 15.13. Nitrogen mustards are effective against several types of leukaemia, however their high chemical reactivity makes administration unpleasant in many cases.

There are obvious risks in using alkylating agents in chemotherapy; random alkylation of nucleic acids and proteins will certainly harm the patient. For this reason Baker has highlighted the concept of **active site directed irreversible inhibition** (Baker, 1969). The idea is that a weak alkylating agent or acylating functional group (such as $-CO.F$ or $-SO_2F$) will not attack any nucleophilic group that it meets; only those that are presented to it in the correct orientation will react. For the two groups just mentioned, general acid catalysis is required – usually provided by hydrogen bonding. Applying this notion to enzymes, we see that if the alkylating functional group is made part of a competitive inhibitor, binding of the inhibitor to the active site will bring nucleophiles at the active site into position to be alkylated. The weak alkylating agent is thus potentiated by binding and neighbouring group participation at the active site. The sulphonyl fluoride (Figure 15.14*a*), for example, might well be a useful inhibitor of dihydropteroate reductase (Section 15.4.1). Baker has shown that derivatives of the pyrimidine antibacterials (e.g. trimethoprim (Figure 15.14*c*) are extremely effective irreversible inhibitors of mouse leukaemia by inhibiting folate synthesis (Figure 15.14*b*).

Figure 15.14.

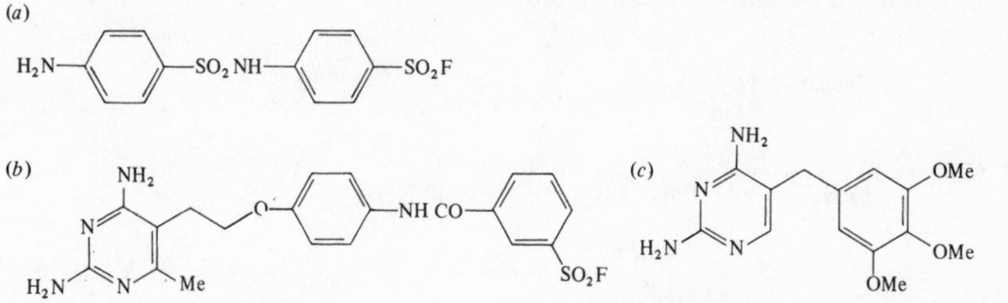

Irreversible inhibitors, as we have just seen, are usually alkylating agents such as alkyl or acyl halides, but they suffer from the disadvantage that they may react with the wrong nucleophile at the active site of the enzyme. One way round this is to feed the enzyme a pseudo-substrate with which it can react but which will be converted into a potent alkylating agent rather than an innocuous product. Such agents are known as suicide or latent inhibitors. In order to design a latent inhibitor successfully it is very useful to understand the mechanism of the action of the enzyme. In the case of reactions involving the coenzymes thiamine pyrophosphate and pyridoxal phosphate we have a very good idea of the mechanism (Chapter 4). Chemists have been active in designing inhibitors based on these schemes and the following example illustrates the argument in action. So far no latent inhibitors have found their way into clinical use.

The first step of a pyridoxal-dependent enzyme-catalysed reaction is the formation of an imine by the substrate and coenzyme. A basic group at the enzyme's active site then removes a proton to give an anion that is stabilised by delocalisation through the pyridine ring. However, if the substrate contains a suitably placed leaving group, the anion can be discharged by displacement of that group (Figure 15.15). Trifluoroalanine (Figure 15.15*a*) is an example. Loss of fluoride from (15.15*b*) affords the inhibitor proper, an α,β-unsaturated imine (15.15*c*), which is a Michael acceptor. A nucleophilic group X on the enzyme can then add to the coenzyme inhibitor complex, forming a covalently bonded species. The enzyme is now irreversibly inhibited; it has committed suicide by transforming a substrate analogue into a highly reactive inhibitor.

Problem 15.2.

The photolysis of azides, e.g. PhN_3, yeilds nitrenes and nitrogen (Figure 15.16). Nitrenes, like oxene and carbene are potent electrophiles and will insert into CH bonds.

The azide shown at the foot of Figure 15.16 was added to a preparation of electric organ containing purified acetylcholine receptor. It was found that on photolysis a product was obtained that no longer bound acetylcholine itself.

Compare the structure of the azide with that of acetylcholine and suggest how can this behaviour be understood.

Figure 15.16.

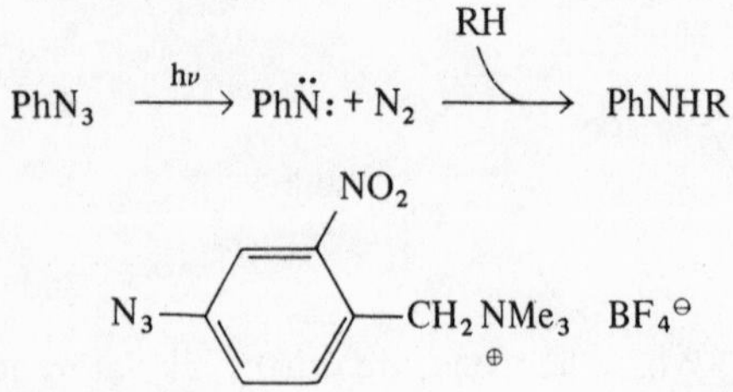

Figure 15.15.

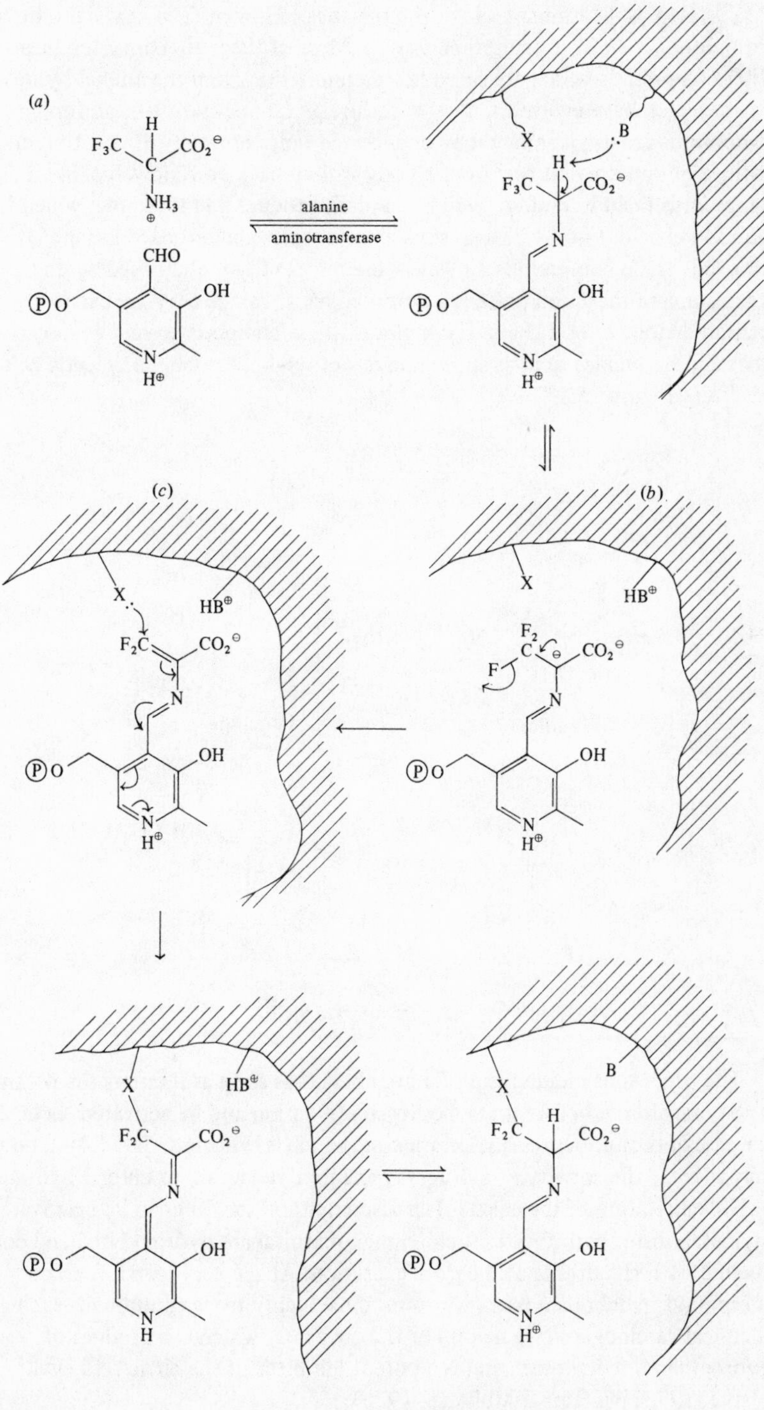

15.6. Drug metabolism

Drugs, like most biological molecules of low molecular weight, do not normally have an infinite lifetime *in vivo*. Most of them are converted into other compounds which are suitable for elimination from the animal by mechanisms which have evolved to deal with foreign compounds. Not all foreign compounds are drugs: many may be ingested quite normally as constituents of the diet and are removed by the body if they have no food value. Such compounds include vanillin, which is a plant product and tyramine, which is found in cheese. Food additives such as colourings and antioxidants are also removed. Some compounds, however, are not readily metabolised by any organism, and these can present a potential hazard since they are carried unaltered along a food chain. Examples of these compounds include insecticides and herbicides such as aldrin and 2-methyl-4-chlorophenoxyacetic acid (MCPA). (Figure 15.17).

Figure 15.17.

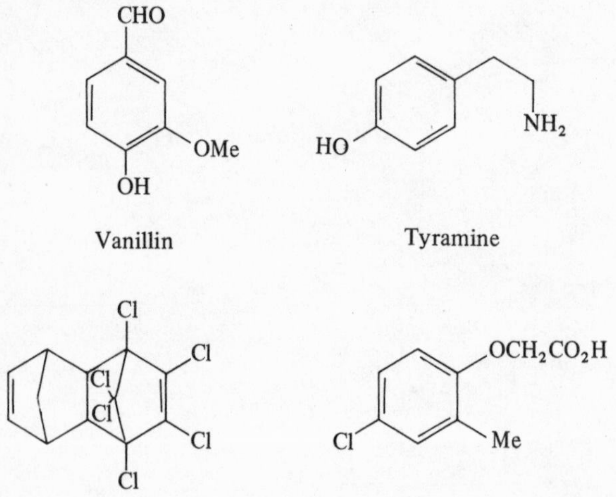

Vanillin

Tyramine

Aldrin

MCPA

The process of metabolism of a drug begins as soon as it enters the organism. It may be altered before it reaches its site of action and be activated, as in the case of the antidepressant imipramine, which is N-demethylated. Metabolism may modify the activity of a drug, for example reduction of chloral hydrate or N-dealkylation of iproniazid. It is also common for drugs to be inactivated by metabolism: barbiturates such as phenobarbital are hydroxylated and conjugated with glucuronic acid before excretion. At the very worst, foreign-compound metabolism is known to produce highly toxic compounds such as chemical carcinogens (see below) or fluorocitrate, which is a product of fluoroethanol metabolism and is a potent inhibitor of the citric acid cycle (Parke, 1973; Williams & Millburn, 1975).

Figure 15.18.

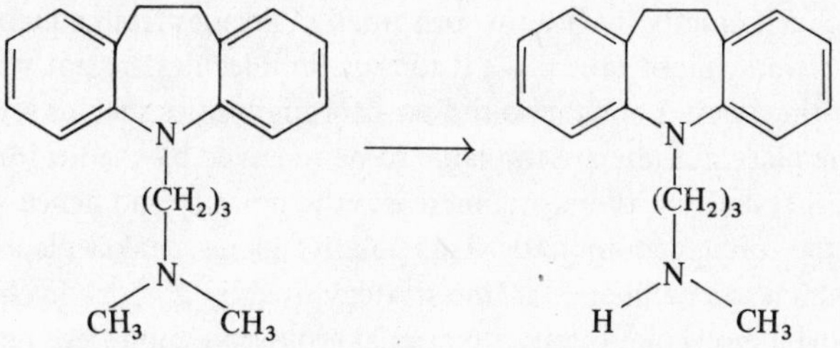

Imipramine

$$CCl_3CH(OH)_2 \longrightarrow CCl_3CH_2OH$$

Chloral hydrate

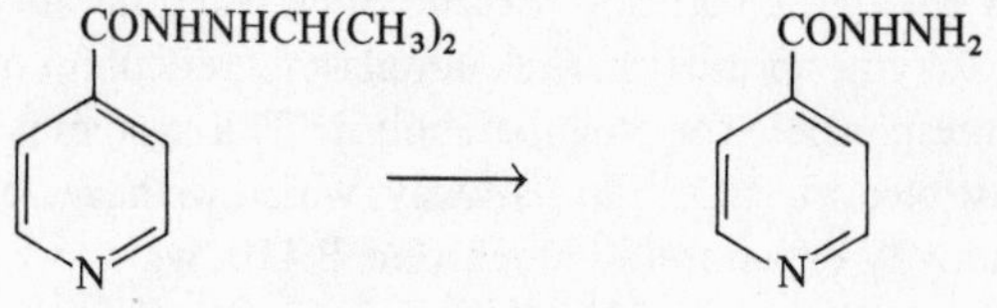

Iproniazid

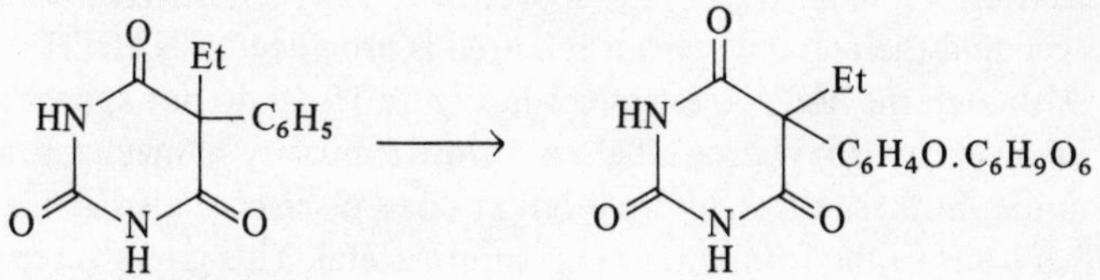

Phenobarbital

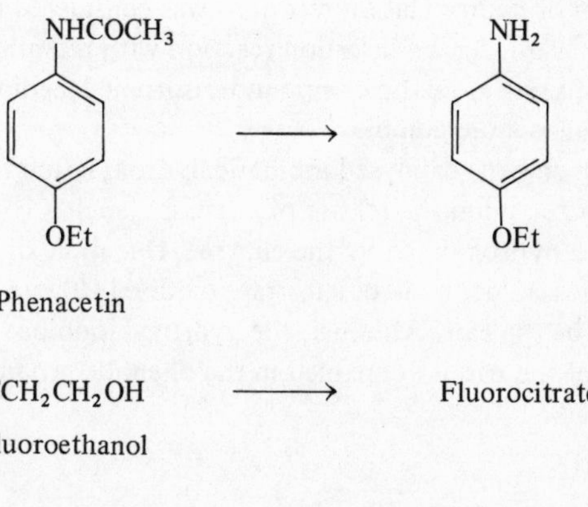

Phenacetin

$$FCH_2CH_2OH \longrightarrow \text{Fluorocitrate}$$

Fluoroethanol

The main route for excretion of modified foreign compounds is through the kidneys, which act chiefly by filtering out small molecules from aqueous solution. Filtration cannot take place if the small molecules are not water soluble, since they would remain bound to macromolecules such as serum proteins in the plasma; these are too large to be removed by the healthy kidney. Drug metabolism, therefore, increases the polarity and hence water solubility of the foreign compounds. This polarity increase takes place in two clear stages which can be likened to the strategy used by an organic chemist when degrading a molecule. Firstly functional groups are modified, removing apolar ones and introducing new polar ones. Then polar derivatives of the modified compounds are made by a process known as conjugation.

15.6.2. *Functional group modification*

One of the main ways in which a small molecule can be increased in polarity is by the insertion of a polar group like a hydroxyl group or by the unmasking of a polar group already present. Such reactions are carried out by a system of enzymes present in the endoplasmic reticulum of the liver, the main organ responsible for drug metabolism. This system is the mixed-function oxidase electron-transport pathway, which we have described earlier; the oxidative activity depends on cytochrome P 450, which we met in Chapter 9. Examples of reactions catalysed by the liver endoplasmic reticulum oxidising system (Figure 15.19) include hydroxylation, demethylation and amine oxidation. As we have seen, the source of the oxygen inserted is molecular oxygen and the reducing power required is provided by NADPH.

Although the last two examples in Figure 15.19 do not appear at first sight to involve a hydroxylation, there is a close similarity in mechanism to the first example. In these cases, hydroxylation takes place α to nitrogen, a reaction which leads to the formation of an amino acetal. This group is readily hydrolysed to a carbonyl compound - formaldehyde from morphine and phenylacetone from amphetamine.

The mechanism of hydroxylation reactions was considered in Chapters 7 and 9 and is best regarded as an insertion reaction with retention of configuration. However, in some aromatic compounds, isotopic labelling has shown that the mechanism is more complex.

An assay for an enzyme-catalysed aromatic hydroxylation reaction was being developed based upon the release of a tritium label as tritiated water from the site being hydroxylated by the enzyme. One mole of tritium was expected to be released per mole of substrate oxidised (Figure 15.20) but this was found not to be the case. Although the hydroxylation had occurred, a very large proportion of the tritium remained in the phenolic product and there

Figure 15.19. Reactions in liver endoplasmic reticulum.

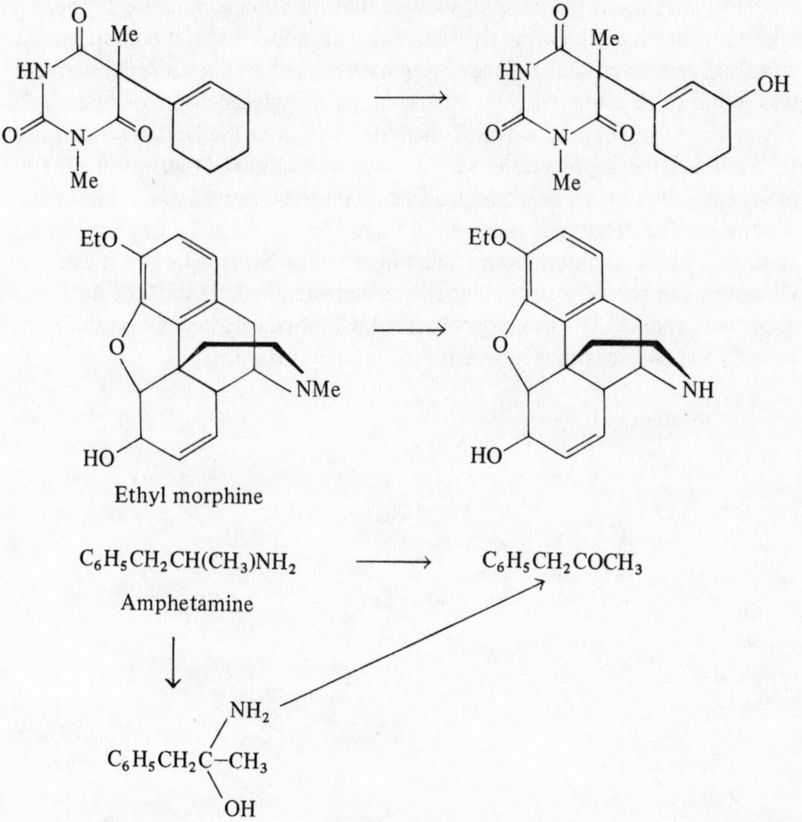

Ethyl morphine

$$C_6H_5CH_2CH(CH_3)NH_2 \longrightarrow C_6H_5CH_2COCH_3$$

Amphetamine

$$C_6H_5CH_2\underset{\underset{OH}{|}}{\overset{\overset{NH_2}{|}}{C}}-CH_3$$

Figure 15.20.
Expected reaction

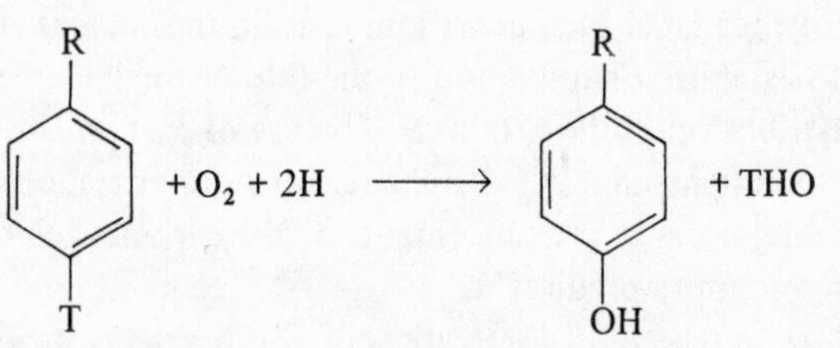

Observed reaction

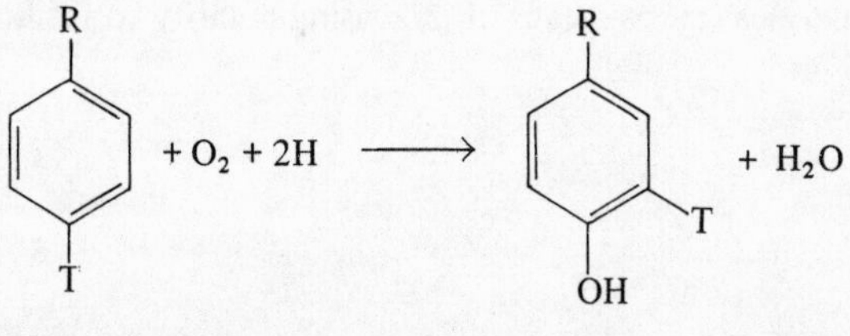

was too much to be explained by a reasonable kinetic isotope effect. Careful chemical analysis of the product showed that the label had moved to the adjacent carbon atom during the reaction. This implied that a rearrangement of substituents was taking place during hydroxylation. Such a rearrangement was found to be quite common during enzyme-catalysed aromatic hydroxylation. The mechanism was named the NIH shift, after the National Institutes of Health, Bethesda, Maryland where it was first observed (Guroff *et al.*, 1967). Many mechanisms can be envisaged to explain the observed migration of the tritium label. A likely one is shown in Figure 15.21. Here the key intermediate is an epoxide, a common intermediate in aromatic hydroxylation. It can decompose in several ways to give rearrangement, the NIH shift, or the originally expected product. The products observed will obviously depend on the relative rates of the two reactions in which the epoxide ring opens.

Figure 15.21.

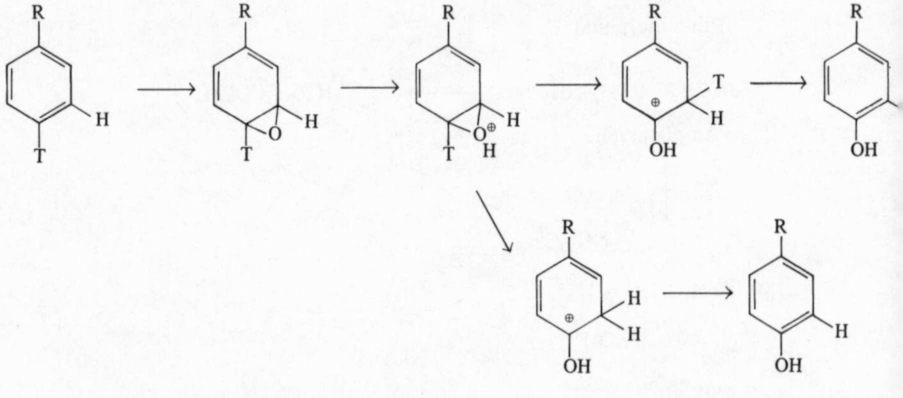

The NIH shift is a good illustration of the complexities that can arise when a relatively labile isotopic label such as tritium is used. In such experiments it is very important to establish unambiguously the fate of the label by determining its exact distribution in the products. The use of carbon-labelled compounds in tracer experiments and in isotopic methods for enzyme assay is, in contrast, more reliable, because rearrangements of the carbon skeleton are much less common and more predictable.

Apart from oxidative reactions, there are several other ways in which drugs can be metabolised in the endoplasmic reticulum of the liver and in other organs. Figure 15.22 shows some examples that illustrate firstly hydrolysis and secondly reductive cleavage as means of increasing polarity to facilitate excretion.

Figure 15.22.

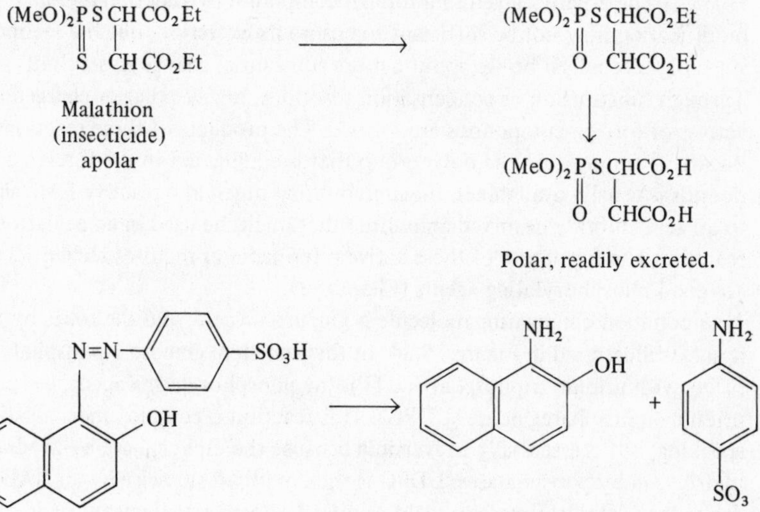

$(MeO)_2 P S CH CO_2Et$
$\quad\quad \| \quad |$
$\quad\quad S \quad CH CO_2Et$

$\longrightarrow$

$(MeO)_2 P S CHCO_2Et$
$\quad\quad \| \quad |$
$\quad\quad O \quad CHCO_2Et$

Malathion
(insecticide)
apolar

$\downarrow$

$(MeO)_2 P S CHCO_2H$
$\quad\quad \| \quad |$
$\quad\quad O \quad CHCO_2H$

Polar, readily excreted.

Orange II
(colouring)

Problem 15.3.

The structures of the drugs chlorpromazine and amphetamine are shown in Figure 15.23. Assume that you are a biochemist who wishes to study their metabolism.

(1) What possible metabolic fates can be envisaged for the two drugs? The earlier examples in this section will give you a starting point.
(2) How could your suggestions be tested experimentally? A glance back at the methods of isotopic labelling in biosynthetic studies in Chapter 5 will be helpful.

Figure 15.23.

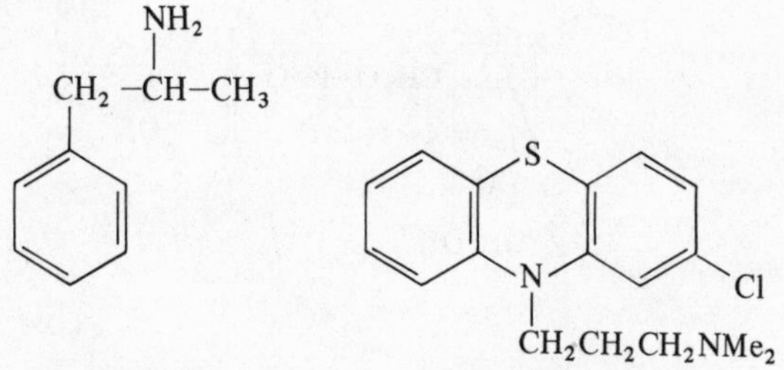

Amphetamine

Chlorpromazine

15.6.3. Conjugation

The polarity given to a foreign compound by functional group modification may not be sufficient to ensure its excretion, but the second stage of metabolism brings about a more substantial change in solubility. Through substitution or condensation reactions, highly polar or charged derivatives of foreign compounds are formed. The products of these reactions are known as conjugates. The polar group that is conjugated to the foreign compounds is usually available in the metabolising organ in a reactive form similar to an acid chloride or mixed anhydride that might be used in an acylation reaction. The formation of these active intermediates requires nucleotides that are good phosphorylating agents (Chapter 3).

A common conjugating molecule is glucuronic acid, and the route by which it acts is illustrated in Figure 15.24. In the first step, glucose 1-phosphate reacts with uridine triphosphate (UTP), the phosphorylating agent, to form uridine diphosphateglucose (UDPG). This reaction is complex mechanistically speaking, but is essentially irreversible because the diphosphate by-product is hydrolysed by phosphatases. UDPG is then oxidised stepwise by an NAD-dependent dehydrogenase to yield uridine diphosphate glucuronic acid (UDPGA), which is the conjugating reagent. By nucleophilic substitution at the glycosidic carbon of the glucuronic acid, it reacts with alcohols, amines and thiols to form new glycosides in which C–O, C–N, or C–S bonds link the aglycone to the sugar. These glucuronides are highly polar because the carboxylic acid is extensively ionised at physiological pH.

Figure 15.24.

Glucose 1-phosphate + UTP $\longrightarrow$ UDPG + PP$_i$

UDPG + 2NAD$^{\oplus}$ $\longrightarrow$ UDPGA + 2NADH

UDPG, R^1 = CH$_2$OH

UDPGA, R^1 = CO$_2$H

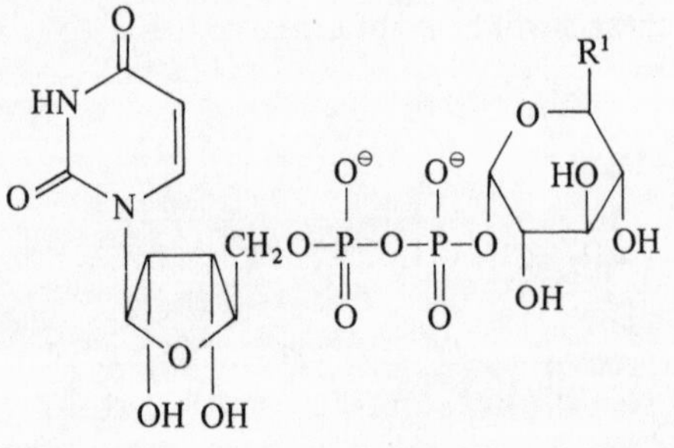

UDPGA + R^2OH $\longrightarrow$ R^2–O ... + UDP

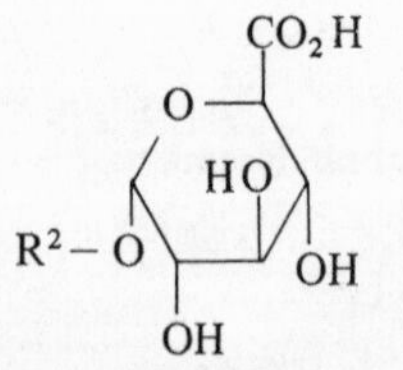

Other conjugation reagents exist, and two further examples are illustrated in Figure 15.25. We discussed the chemical reactivity of acetyl coenzyme A in Chapter 4. The sulphating agent PAPS can be considered in the context of mixed anhydride chemistry (Chapter 3).

Figure 15.25.

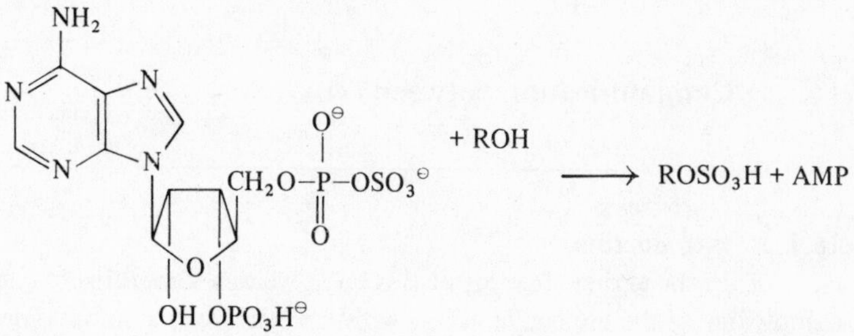

$$\text{PAPS} \longrightarrow \text{ROSO}_3\text{H} + \text{AMP}$$

+ ROH

PAPS
(3′-phosphoadenosine-5′-phosphosulphate)

$$\text{CH}_3\text{COSCoA} + \text{RNH}_2 \longrightarrow \text{RNH COCH}_3 + \text{CoASH}$$

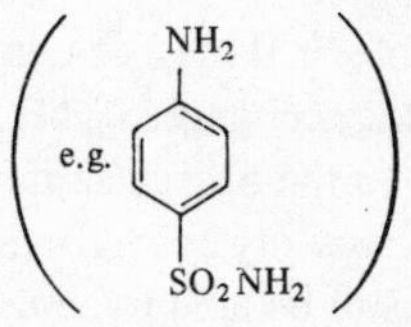

e.g.

Problem 15.4.

The data in Figure 15.26 show the products of metabolism of an anti-inflammatory drug, bucloxic acid. Try to account for the observed products using known metabolic reactions. Some have been covered in this chapter, but it will also help you to consider fatty acid degradation and the fate of acetate in the citric acid cycle (Hawkins, 1976).

Figure 15.26.

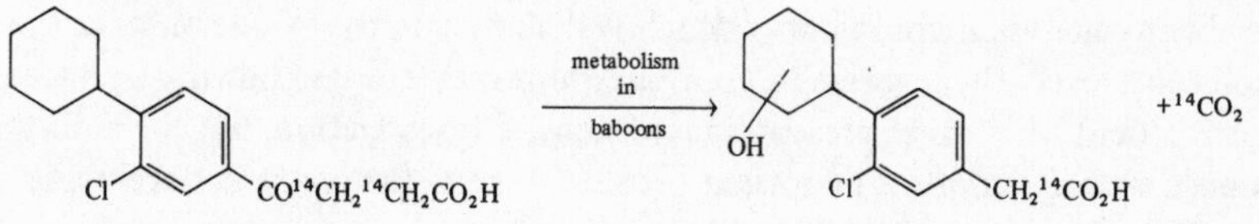

16   Communication between cells

16.1.   Introduction

In the earlier chapters of this book we saw something of the complexity of the molecular events within a cell. We examined some of the molecular mechanisms by which the various systems and pathways are thought to be associated and controlled so that the cell optimises its internal environment. However, this story is incomplete. It is equally important to understand how a cell responds to its external environment, the source of its nutrients, and the sink for its waste products.

The environment of a cell may be stable or it may be constantly changing, and the cell must be able to respond to these changes if it is to remain active, for example by adjusting its metabolism to the energy supply available. There are even further constraints on cells in higher organisms, where cells specialise in different functions. Each cell type should be able to adjust its own performance so that it is appropriate for the needs of the whole organism at any particular time. In order for this to take place it is necessary for information to be carried from one cell type which senses the environment or needs of the organism to the others from which a response is required. There must be communication between cells. Communication is also necessary in order to coordinate processes such as growth of an animal.

In this final chapter we shall look at some of the ways in which cells are thought to communicate with each other and see how far we can understand this communication in molecular terms. We hope that by now you will be able to look critically at the biological situations we present and that you will be able to analyse them in a way which will allow you to ask questions at the molecular level. The answers to your questions may not be known yet, since this is a field which is at present under intensive investigation, but the ability to pose such questions is important because it is the first stage in developing the molecular description of the biological system.

We begin our discussion of communication between cells by examining two examples of the action of hormones. Hormones, as we shall see in the next section, are chemical messengers which are carried in the bloodstream from one tissue to another. Some of the stages in the action of some mammalian hormones seem to be common to lower organisms also. For example, the cellular slime mould *Dictyostelium discoideum* is a unicellular organism which can aggregate into a multicellular system in an environment which is poor in nutrients. Aggregation is initiated by a chemical message which is transferred through space from one cell to another and so is, in a sense, hormone-like (Newell, 1977). Similarly, insect behaviour, mating, moulting and defence responses, is largely controlled by chemical messengers (Evans & Green, 1973). Finally, to illustrate a large scale biological process in which cellular communication is involved, we shall examine some aspects of the way in which animals recognise and destroy foreign cells, the immune response. This closing section of the book allows us to focus on the various levels of biological study within one functional system and in this way provides an overview of biological chemistry in relation to other biological methods, part of the perspective we have tried to present in this book.

16.2.   Hormones

In an animal many processes must be coordinated for optimum performance. Liver and adipose tissue must be able to provide glucose and fat respectively for the needs of other tissues, and must also be able to store any surplus glucose as glycogen or fat when these are available from the diet. Thus, a resting animal after a meal will be storing energy in various forms and an exercising animal will be using up these stores. The requirements for energy can change very rapidly, consequently energy-storage tissues require a mechanism by which they can respond immediately to any demand for energy. They are informed of this demand by hormones such as adrenaline.

Another example of the control of metabolic processes between different tissues is the menstrual cycle in animals. The cycle is controlled by a complex network of hormonal signals of which perhaps the best understood in molecular terms are the steroid hormones. These processes generally take place over a longer time scale than the energy-regulating ones and they have a different molecular mechanism.

However, the common factor in these processes is the hormonal control. Hormones are molecules which are synthesised in and secreted from specialised tissues known as endocrine glands. The hormone is secreted into the bloodstream and travels in the circulation to the target tissue where it exerts its specific effect. This sequence of events is common to most hormones and is summarised in Table 16.1. The table also poses appropriate molecular questions which could be asked about each stage of the overall biological process.

The area which can best be considered in molecular detail is the fifth stage, the action of the hormone at the target tissue. This is probably the most widely studied stage, since it is here that the effect of the hormone can be observed. We shall not examine any of the other steps in detail, but we can briefly mention some of the ways in which they might be approached.

Table 16.1. *The sequence of events in hormonal control*

Events during hormone action	Areas for study
(1) Synthesis of hormone	Enzymology and mechanism of synthesis; control of synthesis
(2) Stimulus for release	Mechanism of stimulation; type of signal for release; receptor for signal; transmission of signal in cell
(3) Secretion of hormone into circulation	Mechanism of secretion; events within the cell and at the cell membrane
(4) Transport to target tissue	Mechanism of transport; role of transport proteins if present
(5) Action at target cell	Specificity and recognition; mechanism of action within cell
(6) Removal	Site, mode and control of removal

Step (1) the synthesis of the hormone could be studied in a purely enzymological way to elucidate the biosynthetic route to the hormone. The rate of the synthesis could be controlled or accelerated by the stimulus (step 2) which causes the secretion of the hormone into the circulation to take place (step 3). Secretion often involves packing the hormone into membrane-surrounded vesicles which fuse with the plasma membrane of the endocrine cells and so release their contents into the circulation. So far the steps are analogous to those which take place in the transmission of a signal from nerve cell to nerve cell or from nerve cell to muscle, both important examples of communication between cells (Lester, 1977).

In step 4 the hormone is transferred via the circulation to its target tissue. If the hormone is not soluble in aqueous media, a transport system is required similar to the transport of apolar drugs which we discussed in the last chapter. Finally, the hormone reaches its target tissue where it is recognised by a specific receptor and elicits its characteristic response. After this it is destroyed (step 6), by enzyme-catalysed degradation. Table 16.2 summarises the events that take place in the action of adrenaline and of oestrogens (the latter are examples of steroid hormones, which we consider later).

16.3. The action of adrenaline in muscle

Adrenaline is the hormone which elicits the 'fight or flight' response in an animal. A suitable external stimulus which may require immediate action or a high degree of alertness from the animal causes the release of the hormone from the vesicles in which it is stored in the adrenal medulla; these vesicles are known as chromaffin granules. Adrenaline, a small charged molecule (Table 16.2), is soluble in water and is rapidly carried to the target tissue in the circulation. The concentration in the blood plasma is about 10^{-9} mol l^{-1}.

Table 16.2. *Events in the action of hormones*

	Adrenaline	Oestrogens
(1) Synthesis		
Precursor	Phenylalanine or tyrosine	Cholesterol
Site	Adrenal medulla	Ovary
Structure		
(2) Stimulus for release	External stress, leading to 'fight or flight' response	Release of specific peptide from pituitary gland
(3) Secretion	From 'chromaffin granules' formed in the cytoplasm	Diffusion from cell
(4) Transport	Soluble in plasma	Bound to plasma proteins
(5) Action at target tissue	Energy made available from glycogen stores; degradative enzyme activated by phosphorylation	Development of female sex characteristics; concentrations vary during menstrual cycle. Enzyme activity controlled at synthesis
(6) Removal	Catalysed by monoamine oxidase and catechol-O-methyl transferase.	Oxidation by mixed-function oxidases in the liver.

Problem 16.1.

Suggest a biosynthetic route to adrenaline. First consider what likely biological sources of the aromatic ring with its aminoethyl substituent would be available, and then suggest a series of reactions which would convert your starting substance into adrenaline. What enzymes and cofactors would be required? You will find some details of similar reactions and their mechanisms in Chapters 4 and 6.

There are several ways in which the effects of adrenaline can be observed. At a physiological level it causes, amongst other things, an increase in heart rate and an increase in blood pressure. These changes will cause a more rapid transfer of nutrients and of oxygen to the cells which require them. If these cells, including muscle cells, are required to perform more work they will demand a rapid increase in their total supply of nutrients for energy production. Liver responds to this demand by releasing glucose from its glycogen stores and similarly adipose tissue releases fatty acids from its store of triacyl glycerol. Muscle also has a small store of glycogen which is reserved for the use of the muscle itself: it cannot be exported to other tissues. The routes by which glycogen is mobilised in the liver and the pathways for fatty acid release from adipose tissue are shown in Appendix 1.

One way in which adrenaline might modulate the rate of energy release would be by controlling the activity of the enzymes which lead to glycogen and triacyl glycerol hydrolysis. This is indeed what happens. Let us examine the situation in muscle, which is the tissue in which the enzymology has been studied in greatest detail. Adrenaline promotes the breakdown of glycogen in muscle to form glucose 1-phosphate which is then isomerised into glucose 6-phosphate. The reason why this glucose 6-phosphate is not hydrolysed and released into the circulation (which is what happens in the liver) is that muscle does not contain the necessary hydrolytic enzyme, glucose 6-phosphatase. Muscle retains its glucose 6-phosphate for its own use for producing the ATP required for contraction. During periods of rest, the energy supply of the muscle is met by oxidation of fatty acids and any depleted glycogen stores are replenished by synthesis of glycogen from glucose, which is supplied to the muscle from the bloodstream and enters the muscle cell under the influence of the peptide hormone insulin. We thus have a linked cycle of synthesis and breakdown of glycogen in muscle which is illustrated in Figure 16.1. The action of adrenaline presumably takes place somewhere in this cycle.

We can now consider the molecular mechanisms which control the activity

Figure 16.1. Glycogen metabolism in muscle.

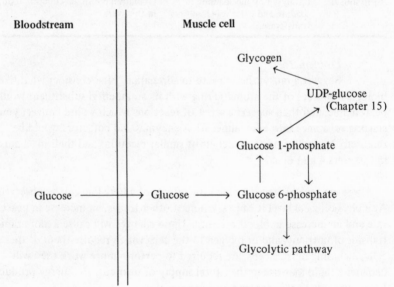

of these enzymes. Firstly, the resting muscle cell balances its own internal energy demands. This is thought to take place by an allosteric mechanism on the enzymes which synthesise and break down glycogen. Glycogen synthase, for example, is allosterically activated by glucose 1-phosphate and glycogen phosphorylase is activated by AMP. The action of the hormone is superimposed upon this internal mechanism. Adrenaline causes a great increase in the rate of glycogenolysis in muscle. This implies that the activity of the muscle glycogen phosphorylase is increased by adrenaline, and the problem of understanding the mechanism of adrenaline action resolves into one of explaining how the hormone causes this activation. Figure 16.2 illustrates the events which are thought to take place when adrenaline acts on a muscle cell. Each stage is numbered and discussed in the text which follows. Details of the experimental evidence on which this scheme is based can be found in Cohen (1976a). As we describe the scheme we would like you to consider how far the explanation goes in a molecular description of the phenomenon. You may be able to identify areas where the molecular events are rather obscure and possibly suggest ways in which one might attempt to clarify them.

(1) Adrenaline binds to a specific receptor, which is probably a protein situated in the outer surface of the plasma membrane of the cell. Receptors for other hormones such as glucagon, a peptide secreted by the endocrine pancreas, are also present in the plasma membrane. The action of glucagon is similar to that of adrenaline, and is particularly important on the liver.

(2) The binding of the hormone to the receptor causes the activation of the enzyme adenyl cyclase, which is located on the inner side of the plasma membrane. This enzyme catalyses the conversion of ATP into a cyclic phosphate diester known as cyclic AMP or cAMP. (Figure 16.3). GTP may also be involved in the activation.

(3) cAMP is the so-called 'second messenger' which diffuses in the cytoplasm to meet another enzyme, a protein kinase, which it activates by an allosteric mechanism, perhaps by causing the dissociation of a regulatory subunit as illustrated in the scheme. We met an example of a protein kinase when we discussed the control of mammalian pyruvate dehydrogenase in Chapter 13.

(4) This protein kinase phosphorylates two enzymes, causing changes in their activity. Phosphorylation of glycogen synthase causes inactivation of this enzyme, thus depressing the rate of glycogen synthesis, and

(5) phosphorylase kinase is activated by phosphorylation by the same protein kinase.

(6) The active phosphorylase kinase now converts the less active phosphorylase b to the active form a and

(7) the rate of production of glucose 1-phosphate is thereby increased.

In this way the control of glycogen synthesis and breakdown are linked. When energy is required, breakdown is stimulated and synthesis inhibited. When energy is available for storage the reverse takes place. Both are good examples of biological amplification or cascade processes (Chapter 13). The action of one molecule of adrenaline is amplified by about 3×10^6 times by the activation of enzymes which are able, because they are catalysts, to activate (or inhibit) many times the number of molecules than the single one that caused their activation.

Figure 16.2. The action of adrenaline on carbohydrate metabolism in the muscle cell.

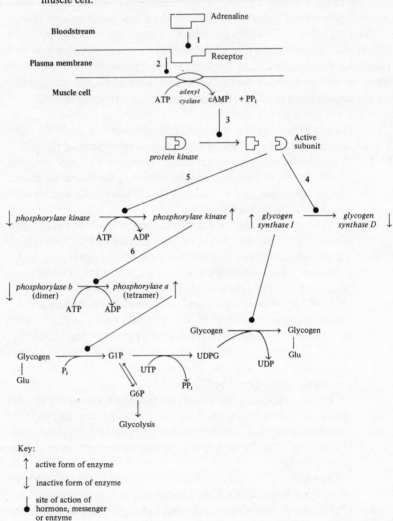

Key:

↑   active form of enzyme

↓   inactive form of enzyme

●   site of action of hormone, messenger or enzyme

We could obviously look for more molecular detail at every stage of the scheme. For example, we could consider how the adenyl cyclase is activated by the binding of the hormone to a receptor. The enzyme is in a different part of the membrane from the receptor, so the two must interact in some way. Using the concepts of the fluid mosaic model of membranes (Chapter 14), we can envisage several ways in which this coupling might take place. It will require experiments using some of the biophysical techniques described in Chapter 14 to reveal which model is appropriate (Greaves, 1977).

We could also examine how the phosphorylation of the substrates of the protein kinases causes a change in their activity. It is known that several sites are phosphorylated in glycogen phosphorylase, so that situation is certainly more complex than in the case of pyruvate dehydrogenase (Cohen, 1976*b*).

X-ray diffraction data are available for both phosphorylase a and b, and it appears that the phosphorylation of a serine residue causes a length of the amino-terminal chain to move away from the main body of the protein. This may allow access of substrate in the active form of the enzyme. More detailed analysis of the molecular events in the activation process and in the allosteric modifications of phosphorylase await X-ray data at higher resolution (Fletterick & Madsen, 1977).

Further important questions relate to how the level of activity of the enzymes is maintained and how the activation caused by adrenaline is switched off. If the cAMP concentration remains in a high steady state, the glycogen-breakdown system will clearly remain active, because a steady state implies that the rate of cAMP synthesis and breakdown are equal and that the degree of activation of the protein kinase will remain at a constant level. If the activation due to the adrenaline is removed, by removal of the adrenaline from its receptor or by inactivation of the adrenaline, the activation of the adenyl cyclase will be removed and the concentraton of cAMP will therefore drop. The break-

Figure 16.3.

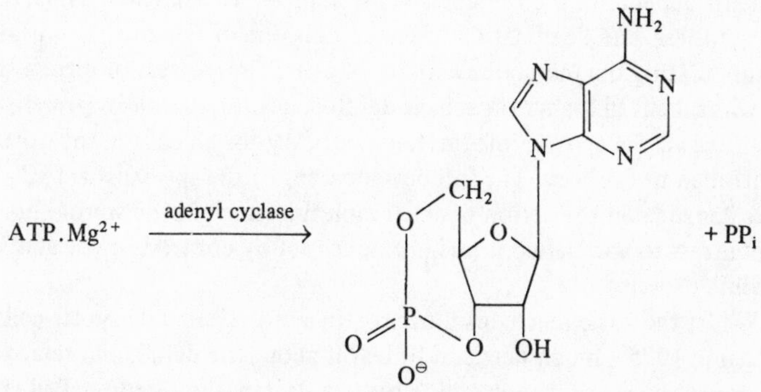

Cyclic AMP

down of the cAMP is catalysed by a hydrolytic enzyme known as a phosphodiesterase, which hydrolyses cAMP to AMP. Removal of the activation of the
rest of the system depends on the action of protein phosphatases.

Problem 16.2.

Methyl xanthines such as caffeine and theophylline (Figure 16.4) are
inhibitors of phosphodiesterases. Drinks such as tea and coffee contain caffeine.
Can you suggest how they may act as stimulants?

Figure 16.4.

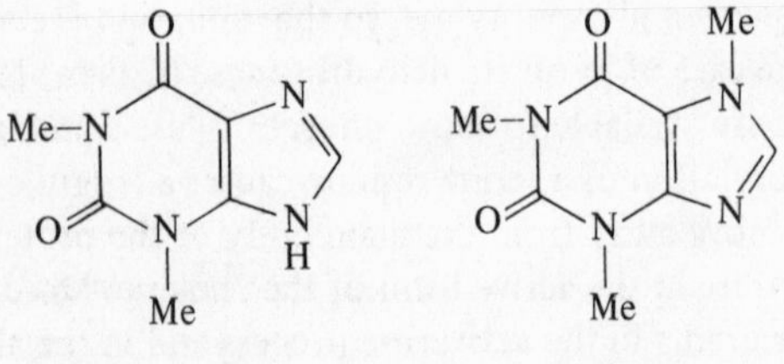

Theophylline   Caffeine

The mechanism we have just described, involving the mediation of cAMP between the hormone and its intracellular response, seems to be quite common.
The action of the peptide hormone glucagon on liver and adipose tissue seems
to be dependent on cAMP, and so does the action of some peptide hormones
produced by the pituitary gland on steroid hormone producing tissue. Although
the synthesis of steroid hormones may be in part controlled by cAMP, the
mechanism of their action at their target tissue is very different, as we shall
see in the next section.

16.4.   Steroid hormones

Practically all cellular functions and the synthesis of all cellular components depend on the presence of the appropriate enzymes. Thus, controlling
the synthesis and degradation of enzymes is one of the most important ways
of modulating the metabolic activity of cells. The process of differentiation
by which cells in higher organisms develop specific functions (usually whilst
retaining all the genetic information necessary for all cells in the organism) is
controlled in this way. The full development of the specialist activity of some
cells depends on the action of small molecules such as the steroid hormones;
in contrast to adrenaline, these hormones act by controlling the synthesis of
specific proteins.

Whilst the basic mechanism of protein synthesis in eukaryotic cells is clear
(Watson, 1975), much needs to be learnt about the detail, and relatively little
is known about the way in which protein degradation is controlled (Katanuma,
et al., 1976). The overall concentration of the required enzymes, which is

determined by the rates of synthesis and degradation, will define the extent
to which the hormonal effect is observed.

In order to synthesise a protein, the ribosomes, the protein-synthesising
systems of the cell, require a messenger RNA (mRNA), which carries the coded
sequence of amino acids for the protein. The mRNA is synthesised by tran-
scribing (copying) the appropriate section of DNA in the nucleus of the cell.
This process is catalysed by an enzyme known as RNA polymerase. Protein
synthesis can then be precisely controlled by determining what part of the
total DNA present in the cell is transcribed. DNA in eukaryotic cells occurs in
a complexed form known as chromatin. Apart from DNA, chromatin contains
proteins of two main types. One type is generally basic and is called histone
protein, but other non-histone proteins are found and are often acidic in
character. It seems that all of these proteins are involved in the mechanism of
determining which regions of DNA are transcribed and at what time.

The steroid hormones are derived from cholesterol by oxidation by a cytochrome P 450-dependent mixed-function oxygenase. The basic reaction is
shown in Figure 16.5.

Figure 16.5. The oxidation of cholesterol to pregnenolone and (*bottom*) some
steroid hormones.

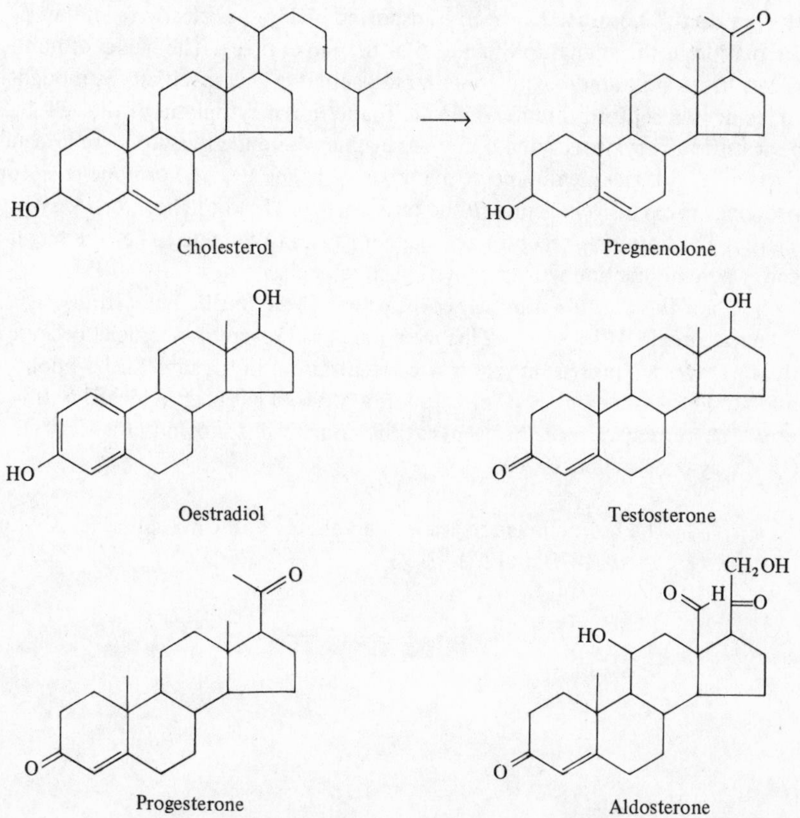

Cholesterol

Pregnenolone

Oestradiol

Testosterone

Progesterone

Aldosterone

Problem 16.3.

It is not immediately apparent from Figure 16.5 that a hydroxylation reaction is involved. With the products shown, can you suggest a mechanism and predict how many moles of oxygen and NADPH would be required to oxidise one mole of cholesterol to pregnenolone? (Hint: consider first how cholesterol could be oxidised in the side-chain to give a product which could be cleaved to pregnenolone.)

Following the formation of pregnenolone, a further series of reactions depending on the tissue concerned leads to the various steroid hormones (Figure 16.5). In the remainder of this section we shall consider oestrogens and progesterone, which are the steroid hormones responsible for the development of female sex characteristics, and, as with adrenaline, we shall only consider the mechanism of their action at the target tissue.

The key piece of evidence for the mode of action of the steroid hormones comes from the following experiment. If one injects an animal with a sample of a tritiated hormone, for example progesterone, and after a few minutes kills the animal and examines the target tissue (in this case the uterus) by auto-radiography, one finds fogging of the photographic plate in areas which correspond almost entirely to the nuclei of the cells. This suggests that in some way the radioactive hormone has been transported highly specifically from the circulation of the animal to the nuclei of the target tissue. The nuclei of non-target tissue are found not to contain radioactivity. This specificity is thought to reside in receptor proteins which are found in the cytoplasm of the cell, in contrast to the systems acted on by adrenaline where the receptors are integral parts of the plasma membrane of the target cell. The steroid hormone receptor proteins are exclusively found *in* the target cell and bind the hormone very tightly. The hormone–receptor aggregate migrates to the nucleus of the target cell, where interaction with chromatin can take place.

Some of the steroid hormone receptors have been purified by affinity chromatography (Chapter 7). This technique has proved to be crucial because the receptors are present in very low concentrations in the target cells, about 6000 molecules per cell. A purificaton of 4000-fold has been achieved in this way. The current view of the events at the target cell is shown in Figure 16.6 (O'Malley & Schrader, 1976).

(1) The hormone enters the cell and binds to the cytoplasmic receptor (K_b about 10^{-10} mol l^{-1}).

Figure 16.6. Events at the target cell following the arrival of a steroid hormone.

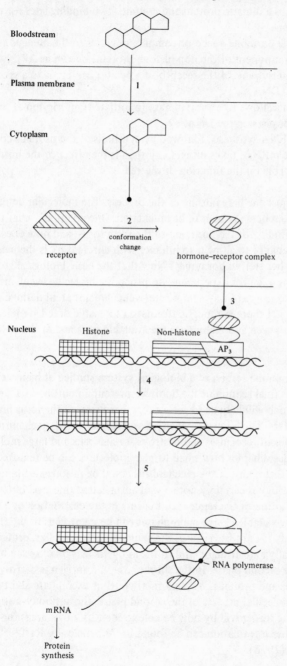

(2) The binding causes a conformational change in the receptor, which is a dimeric protein and contains two binding sites for the hormone.

(3) The hormone–receptor complex migrates to the nucleus and binds to a component of the non-histone proteins known as AP_3.

(4) One subunit of the receptor dissociates and binds to a specific site on the DNA.

(5) This allows RNA polymerase to initiate transcription at the correct place in the DNA.

(6) mRNA synthesis, followed by synthesis of the proteins coded for by the mRNA, takes place. Only then is the effect of the hormone detectable on the function of the cell.

Clearly, this is just the bare outline of the process. The molecular details of each step in the scheme remain to be elucidated. One might ask what the nature of the binding of the hormone to the receptor is, and how does its conformation change to allow it to interact with chromatin? Is the mechanism the same for other steroid hormones? Now that the basic biological strategy is clear, the way is open for experiments to probe the molecule events in detail. This is a good example of a system where events are not at all defined in a chemical sense yet there is enough informaton to enable detailed molecular questions to be posed and techniques are available for some of the answers to be found.

16.5. The immune response: a biological system studied at many levels

In this final section of the book we present an outline of a large biological field which will help us to gather together many of the ideas presented in earlier chapters. The immune system of vertebrates is the mechanism by which the organism recognises and destroys foreign cells and large molecules. We saw in the last chapter how small foreign molecules can be removed by oxidation and that many of the mechanisms could be rationalised in molecular terms. Immunology is certainly not so well advanced at this level but a biochemical description of the sequence of events in the destruction of a foreign cell is partially available. Three main phases can be identified. In the first phase, the foreign body (or antigen) causes the synthesis of antibodies, proteins which specifically and strongly bind to the antigen. This binding can be regarded as the second phase. In the third phase, the antigen is destroyed by specialised cells and proteins. We shall firstly outline each phase and then focus on the molecular aspects of the second phase, the antibody–antigen interaction. The reader will by now be able to identify other areas for experiment and further information can be found in the articles by Raff (1976) and Porter & Reid (1978).

Phase 1

When an antigen, for example a foreign cell or large molecule like a protein, enters the circulation of an animal, the animal reacts by synthesising a series of specific proteins which interact very strongly with the antigen. These proteins are known as antibodies or immunoglobulins, and the initiation of their synthesis is another example of regulation of protein synthesis, which may involve mechanisms similar to those we discussed in the action of the steroid hormones. Specifically sensitised cells are also produced which bind to the antigen. Many different antibodies are synthesised and they bind to different parts of the antigen, each with a high degree of specificity. Frequently these antigenic or binding sites are found on the surface of the foreign cells. The cell surface proteins which characterise human blood groups are such surface antigens. It is thought that specific antibodies can be made for a very large number of antigens, perhaps several million, and the mechanism by which this diversity is created is of great interest (Rabbits, 1976).

The cells which are responsible for the reaction to the antigen are the white blood cells, which stem from bone marrow and the thymus gland and are known as lymphocytes. The cells derived from the bone marrow are known as B cells and those from the thymus gland as T cells. The function of the B cell appears to be to synthesise and secrete antibodies. In contrast, T cells are stimulated directly by the antigen to produce immune responses which do not involve antibody. Each B cell appears to specialise in the production of one specific antibody (Raff, 1976).

Figure 16.7.

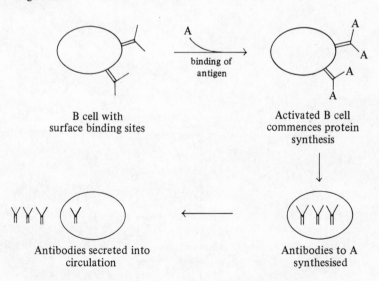

B cell with
surface binding sites

binding of
antigen

Activated B cell
commences protein
synthesis

Antibodies secreted into
circulation

Antibodies to A
synthesised

When the antigen enters the animal, the appropriate B and T cells are activated by binding of the antigen to the surface of the cell. It is at this stage that the B cells rapidly divide and secrete antibody after its rapid synthesis. As we mentioned before, the way in which the B cells are programmed to produce a specific antibody to a given antigen is not known. The scheme of events is illustrated in Figure 16.7.

Phase 2

The antibodies produced by the activated B cells are of several types. Most of them are variations on the simplest type, the immunoglobulin-γ or IgG. IgG molecules are proteins which consist of two pairs of chains held together by disulphide bonds and they can be schematically represented as a Y-shaped molecule. The binding site for the antigen is located at the two ends of the arms of the Y, and thus each IgG molecule can bind to two molecules of antigen. A schematic diagram of an IgG molecule is presented in Figure 16.8.

Phase 3

Having been labelled with antibody, the antigen is now marked for destruction, and the removal of the antigen–antibody complex can take place in several

Figure 16.8. An IgG molecule.

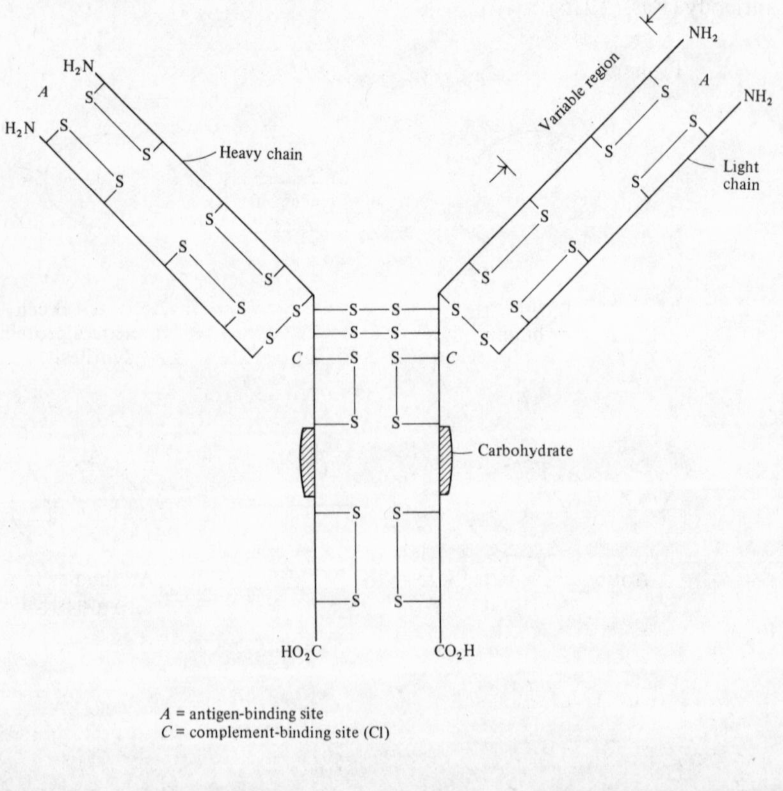

A = antigen-binding site
C = complement-binding site (CI)

ways. Protein antigens, for example, may be engulfed by white blood cells known as macrophages and digested within these cells. Foreign cells, however, have a cell wall which may require more drastic treatment. In order to deal with this barrier a system of plasma proteins known as the complement system comes into play. We mentioned this system of proteins in Chapter 13 as an example of activation of proteins by limited proteolysis and of amplification by a cascade mechanism.

The complement system is a complex series of proteins which together cooperate in punching a hole in the wall of the foreign cell which has been labelled with antibody. The punching of a hole in the cell wall causes the foreign cell to break up and the resulting fragments are easily digested by the scavenging cells, the macrophages (Porter & Reid, 1978).

In this long and complex series of biological events there are obviously many points where we could ask for clarification at a molecular level. We can identify some of them by considering what important molecular interactions take place. These include the binding of the antigen to the B cell, the formation of the antigen–antibody complex itself and the interactions of the various proteins of the complement system which lead to the breaking open of the foreign cell. In all of these questions perhaps the most important initial problem is what is the structure of these interacting molecules? Let us consider the example of the antibody–antigen complex. A knowledge of the structure of the antibody and its corresponding antigen should allow us to see how the specificity of binding arises, what sorts of bonds are present and what strength they give to the overall interaction. We might also be able to see if any change takes place in the conformation of the antibody when the antigen binds to it. Such a change may be necessary in order to trigger the complement system. Two of the methods which can give us detailed information of the three-dimensional structure of the antibody binding site are X-ray diffraction and n.m.r. spectroscopy. (Chapter 11, see also Chapter 12). Both approaches have been used with success in this field. As with all structural studies an important problem was the obtaining of sufficient pure material. Fortunately, relatively large amounts of pure proteins which are indistinguishable from normal antibodies can be obtained from tumours induced, for example, in mice.

One of the most striking results to emerge from the X-ray diffraction studies was the observation that immunoglobulin molecules are constructed in a common way. Each domain of the molecule (Figure 16.8) consists of two layers of anti-parallel β-pleated sheet joined together by a disulphide bridge. This structure is known as the immunoglobulin fold, and is illustrated in Figure 16.9.

Within this basic structure it has been possible to define the antigen binding site for several immunoglobulins. The example shown in Figure 16.10 is of a protein which binds to phosphoryl choline, and is the result of an X-ray study (Capra & Edmundson, 1977). We can see from the figure how the amino acid residues of the protein complement the groupings of the phosphoryl choline, thus conferring specificity on the binding site.

Figure 16.9. (*Top*) The structure of one chain of immunoglobulin. (*Bottom*) The mode of association of two chains to form a dimer. (From Dwek *et al.* (1977), *NMR in Biology,* Academic Press.)

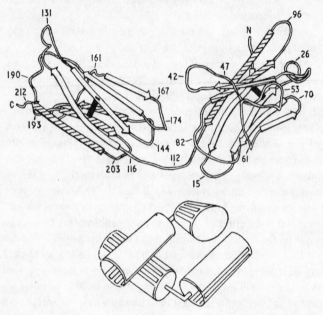

The main interactions are hydrophobic, ionic and hydrogen bonding. Combinations of such bonds are, of course, common in binding sites in proteins, for example in enzyme active sites, and they provide a great range of specificity and of binding strength because of the additive effects of multiple specific non-covalent bonds. In contrast, a covalent bond would take a large amount of energy to form (not just the thermodynamic energy of the compound once formed but also the activation energy for the reaction in which the bond is made) and a correspondingly large amount of energy to break.

The X-ray study shows that the positively charged nitrogen of the phosphoryl choline is electrostatically bonded to a glutamate residue and that the negatively charged phosphate is hydrogen bonded to an arginine and a tyrosine residue. The two methylene groups of the phosphoryl choline are supported by hydrophobic interaction with the indole ring of a tryptophan residue.

Another immunoglobulin whose binding site has been studied in detail, this time by magnetic resonance, binds dinitrophenol derivatives. This work shows the power of combining a wide range of biochemical and biophysical techniques in the solution of one problem. The first step in the sequence was to predict a three-dimensional structure of the immunoglobulin from a knowledge of the

Figure 16.10. The antigen binding site of an immunoglobulin which binds to phosphoryl choline. (From Capra & Edmundson, 1977.)

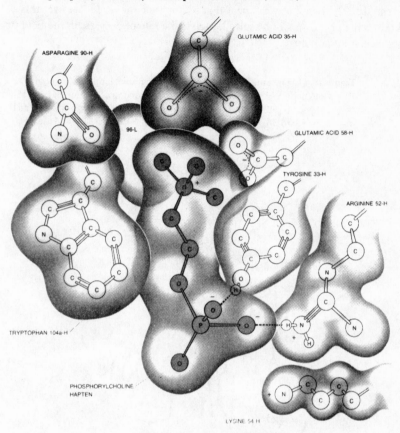

three-dimensional structure of the phosphoryl choline binding just described. This was possible because the amino acid sequences of both proteins were very similar (40–50 % homology) and the basic structure was defined by the immunoglobulin fold. The three-dimensional structure which resulted showed a potential binding site which involved several aromatic amino acid residues. A series of experiments which used chemical modification, e.s.r. and n.m.r. allowed the dimensions and nature of the site to be defined more closely and the final result is shown in Figure 16.11. The rather hydrophobic dinitrophenol appears to be bound in a hydrophobic box formed of the aromatic amino acids' side chains. Some hydrogen bonding also takes place, as the diagram indicates (Gettins & Dwek, 1977).

These studies re-emphasise how important a knowledge of structure is in a molecular study. Once structures are defined we can build on them by introducing the element of time which allows us to describe the changes which take place during the operation of the complex biological sequence of events. In the case of the immune system the molecular details of the subsequent steps are unknown, but we expect that by the time the reader receives this book some problems will be solved, such is the rate of progress in the field of molecular immunology.

Figure 16.11. The antigen binding site of an immunoglobulin which binds to dinitrophenol (DNP) derivatives. The DNP is bound in a position parallel to tryptophane (Trp) 93 and is able to hydrogen bond to tyrosine (Tyr) 34 and asparagine (Asn) 36. (From Dwek *et al.* (1977), *NMR in Biology*, Academic Press.)

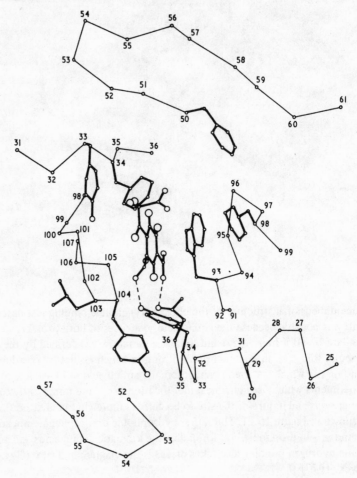

16.6. Conclusion

In this last chapter we have reached frontiers in the molecular understanding of biology at the time of writing. The topics we have discussed are typical of modern biology, which is a patchwork of some areas in which we have a detailed knowledge of molecular structures and their relation to biological function and others where only the barest outline of the biological process itself can be discerned. Many other important topics could have been selected for inclusion in the second part of this book and the interested reader will find further details and suggestions for further study in the references given in each chapter. The systems which we have described give a representative survey of the range of problems which a biochemist with an interest in molecular function may be called upon to tackle, and if a molecular approach to biology is a valid and viable one, the molecular ideas described in these chapters should be rapidly superseded by a deeper understanding, made available, perhaps, by developments of techniques and approaches which we have described. Much remains for biologists and chemists of all specialities to tackle and their efforts will be all the more effective if each understands the concepts and approaches of his colleagues. We hope that this book has helped in achieving this goal.

Appendix 1. Some pathways discussed in the text

(a) Glucose oxidation and the citric acid cycle

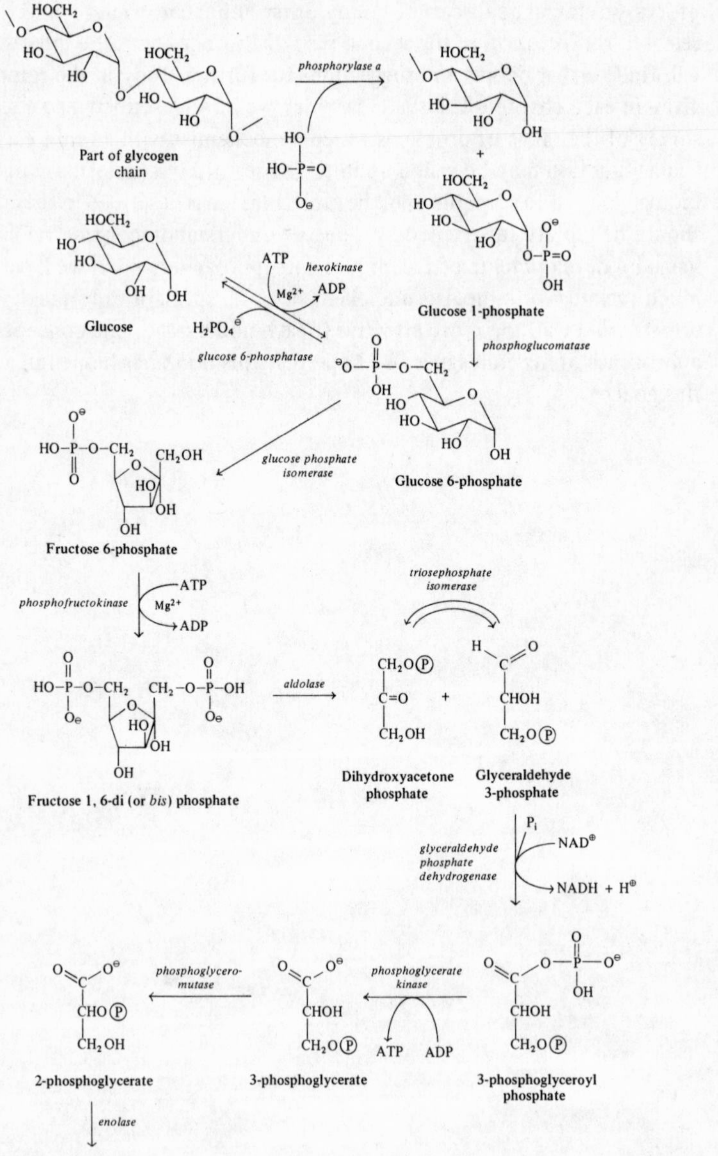

Part of glycogen chain

phosphorylase a

HO—P=O

+

Glucose 1-phosphate

Glucose

ATP   *hexokinase*

Mg²⁺   ADP

$H_2PO_4^{\ominus}$

glucose 6-phosphatase

phosphoglucomatase

Glucose 6-phosphate

glucose phosphate isomerase

Fructose 6-phosphate

phosphofructokinase

ATP

Mg²⁺

ADP

triosephosphate isomerase

Fructose 1, 6-di (or *bis*) phosphate

aldolase

Dihydroxyacetone phosphate

Glyceraldehyde 3-phosphate

glyceraldehyde phosphate dehydrogenase

P_i

NAD$^{\oplus}$

NADH + H$^{\oplus}$

2-phosphoglycerate

phosphoglycero-mutase

3-phosphoglycerate

phosphoglycerate kinase

ATP   ADP

3-phosphoglyceroyl phosphate

enolase

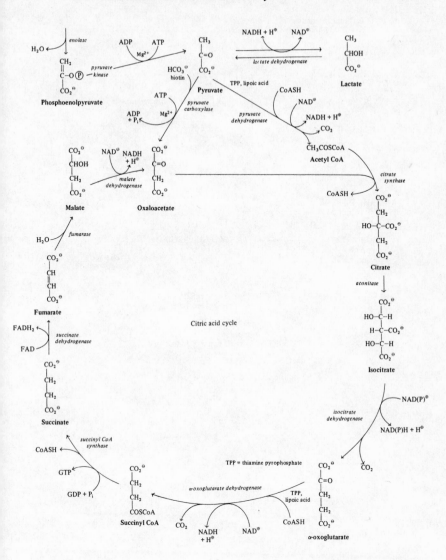

(*b*) Fatty acid oxidation
(see Fig. 13.18 for fatty acid synthesis)

$$
\begin{array}{c}
\text{H}_2\text{C-O-C-R} \\
\quad\quad\quad \overset{\text{O}}{\|} \\
\text{HC-O-C-R} \\
\quad\quad\quad \overset{\text{O}}{\|} \\
\text{H}_2\text{C-O-C-R} \\
\quad\quad\quad \overset{\|}{\text{O}}
\end{array}
\quad
\xrightarrow[\substack{\text{(adipose}\\\text{tissue)}}]{\textit{lipase}}
\quad
\begin{array}{c}
\text{CH}_2\text{OH} \\
| \\
\text{CHOH} \\
| \\
\text{CH}_2\text{OH} \\
\textbf{Glycerol}
\end{array}
\quad + \quad \text{RCO}_2{}^{\ominus} \quad \text{(transported to oxidising}
$$
tissue)

$R = \text{CH}_3(\text{CH}_2)_n -$
$n = 12\text{--}18$

Triacyl glycerol

$$\text{RCH}_2\text{CH}_2\text{CO}_2{}^{\ominus} \xrightarrow[\substack{\textit{fatty acyl CoA}\\\textit{synthetase}}]{\overset{\text{CoASH} \quad \text{ATP} \quad \text{AMP} + \text{PP}_i}{\curvearrowright}}$$
$(n = 11\text{--}17)$

$$\text{RCH}_2\text{CH}_2\overset{\text{O}}{\overset{\|}{\text{C}}}\text{-SCoA}$$

⟶ FAD
↘ FADH$_2$ *acyl CoA dehydrogenase*

$$\text{RCH=CH-}\overset{\text{O}}{\overset{\|}{\text{C}}}\text{-SC}\rho\text{A}$$

⟶ H$_2$O   *enoyl CoA hydratase*

$$\text{RCH-CH}_2\text{-}\overset{\text{O}}{\overset{\|}{\text{C}}}\text{-SCoA}$$
$\quad\;|$
$\quad\text{OH}$

⟶ NAD$^{\oplus}$
↘ NADH + H$^{\oplus}$ *3-hydroxy acyl CoA dehydrogenase*

$$\text{RC-CH}_2\text{-}\overset{\text{O}}{\overset{\|}{\text{C}}}\text{-SCoA}$$
$\;\;\overset{\|}{\text{O}}$

⟶ CoASH   *acetyl CoA acetyl transferase*

$$\text{RC-SCoA} + \text{CH}_3\text{C-SCoA}$$
$\;\;\overset{\|}{\text{O}} \qquad\qquad\quad \overset{\|}{\text{O}}$

(returns to   **Acetyl CoA** (to citric acid cycle)
acyl CoA
dehydrogenase
step)

Appendix 2. A note on curly arrows

Curly arrows are widely used in organic chemistry to provide a dynamic aspect to structural formulae. They offer a way to visualise a possible mechanism for a reaction. Because of the great success of curly arrows it is tempting and easy to over-value mechanisms that look nice on paper, and we suggest that the following points be borne in mind when drawing curly arrows.

(1) A curly arrow represents the formal movement of an electron **pair** from the tail of the arrow to the atom or bond contacted by the arrowhead. In this way, the making or breaking of a covalent bond can be illustrated:

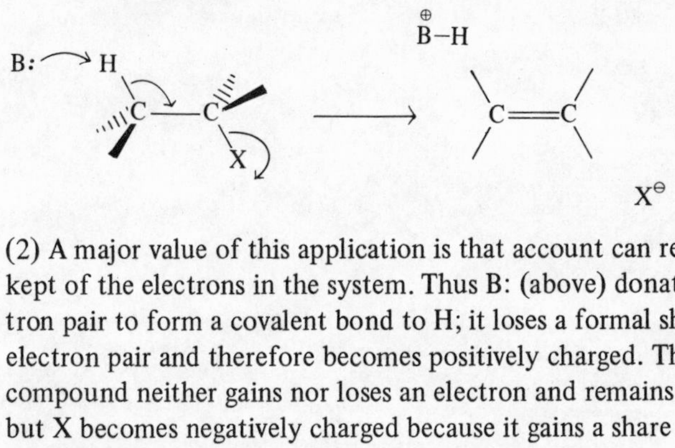

(2) A major value of this application is that account can readily be kept of the electrons in the system. Thus B: (above) donates an electron pair to form a covalent bond to H; it loses a formal share in one electron pair and therefore becomes positively charged. The carbon compound neither gains nor loses an electron and remains uncharged, but X becomes negatively charged because it gains a share in an electron pair.

(3) Curly arrows should never be used to illustrate the transfer of protons:

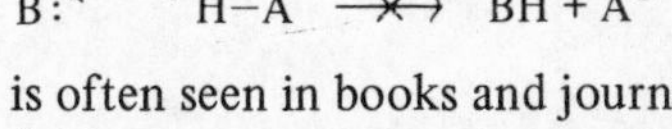

$$B:\overset{\frown}{}H-A \quad \xrightarrow{\;\;\times\;\;} \quad \overset{\oplus}{B}H + A^{\ominus}$$

is often seen in books and journals but is misleading and wrong.

(4) Curly arrows are often used to generate canonical forms for a molecule. Although the procedure works well, it is strictly incorrect

because canonical forms (linked by ⟷ represent stationary electronic states in which the transfer of an electron pair symbolised by the arrow is inappropriate.

(5) It is not reasonable to represent the direction of attack of one reactant upon another by means of curly arrows, except in the crudest possible terms.

(6) The drawing of more than one two-electron curly arrow on a single part of a structure is confusing and is to be discouraged.

(7) One-electron transfer in radical reactions can be represented by fish hooks:

$$
-\underset{/}{\overset{\backslash}{C}}-H \quad \cdot X \longrightarrow \quad -\underset{/}{\overset{\backslash}{C}}\cdot + HX
$$

Again this procedure helps to keep track of electrons.

(8) Particular care must be exercised in the application of curly arrows in reactions involving redox active inorganic metal ions (e.g. Fe^{II}, Fe^{III}). Such species can undergo important changes in electronic structure (e.g. spin state) which cannot be accommodated easily in curly arrow mechanisms.

Answers to problems

Answers are outlined here for those problems that are either without literature references or have not been answered in the main text.

Problem 3.2.
(1) Zn^{2+} is a Lewis acid.
(2) It can coordinate to a carbonyl oxygen and polarise the group increasing the positive charge on carbon. This facilitates attack by a nucleophile at carbon, and hence catalyses hydrolysis.

Problem 3.3.
(1) At pH 8 phosphate esters are chiefly monoionic. Zn^{2+} will readily complex with a phosphate anion.
(2) P—O bond cleavage occurs.
(3) A phosphorylated enzyme intermediate could react with R^2OH instead of water to cause transesterification. The hydroxyl group of a serine probably acts as the phosphate carrier.
(4) No. Phosphorylation of the enzyme must precede formation of inorganic phosphate. Attack of nucleophiles on phosphorus is aided by Zn^{2+} which shields the anion and polarises the P=O bond.

Problem 3.4.

The withdrawal of electrons by the electronegative substituents ($-NO_2$, $-Cl$) enhances the rate of nucleophilic attack at the ester carbonyl group. Also the dissociation of the tetrahedral intermediate is favoured because of charge delocalisation in the product anions.

Problem 3.5.

Acid-catalysed nucleophilic addition to an iminium ion. Protonation of the adduct to form a good leaving group - acid catalysis again. Nucleophilic addition affording a tetrahedral intermediate. Decomposition of the intermediate yielding products.

Problem 3.6.

(1) Mg^{2+} probably chelates with ATP.

(2, 3) Acetyl adenylate is a possible intermediate because it can be converted by the enzyme into reactants or products according to the reaction conditions.

(4) C—O bond cleavage of acetyl adenylate occurs.

Problem 4.1.

Problem 4.2.

(1) Hydrolysis of the amide of glutamine followed by condensation of the ammonia produced with CO_2. Phosphorylation by ATP affords the mixed anhydride, carbamyl phosphate (a very complex enzyme-catalysed reaction sequence).

(2) Acylation of aspartate by carbamyl phosphate.

(3) Condensation of amine and amide to form a cyclic dihydropyrimidine.

(4) Dehydrogenation – see Chapter 8.

(5) Alkylation of nitrogen by a diphosphate – a substitution reaction at a glycosidic carbon atom – see Chapter 6.

Problem 4.3.

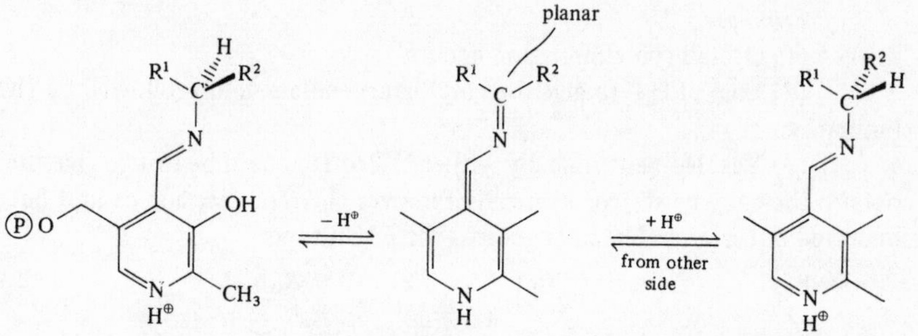

Problem 4.4.

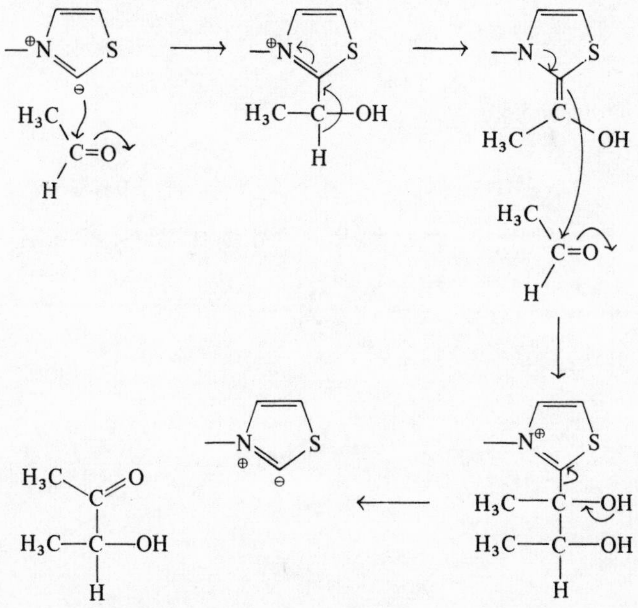

Problem 4.5.

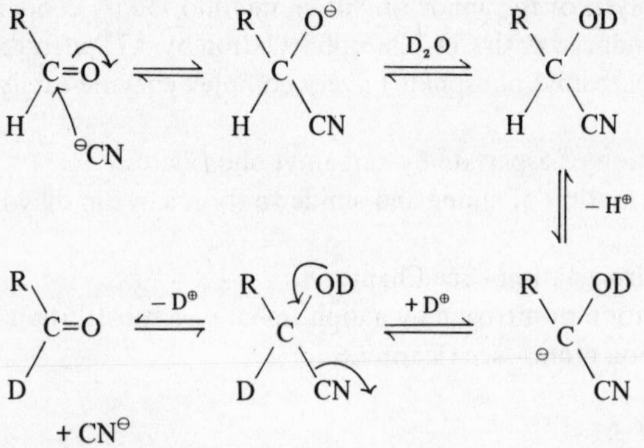

Problem 5.2.

(1) Overall *syn* elimination occurs.

(2) Loss of H_S^+ to give the vinylogous enolate anion followed by HO^- elimination.

(3) Yes. Non-enzymically, either H_R or H_S could be lost to give the enolate: the enzyme stereospecifically removes H_R. The reaction cannot be concerted but is probably analogous to the E1CB type.

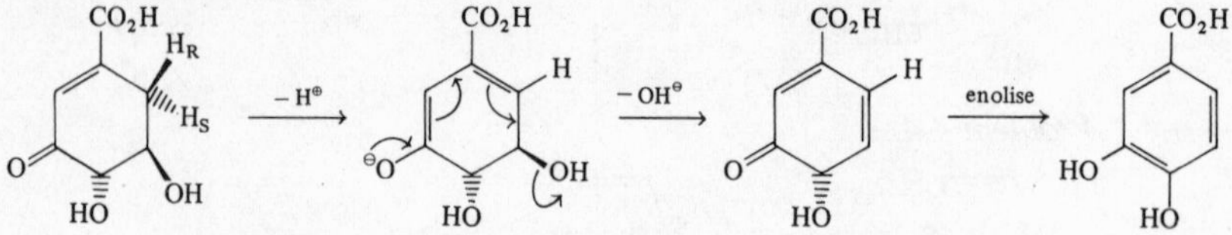

Problem 5.3.

(a)

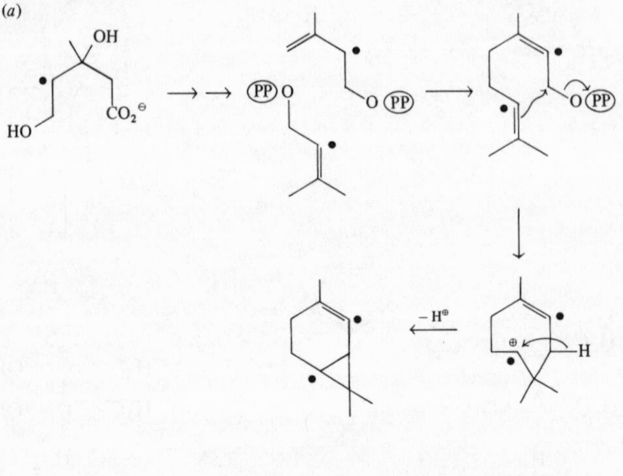

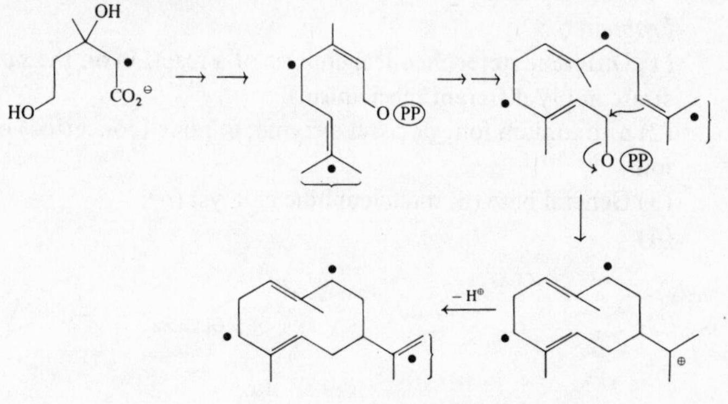

OH
$CO_2^{\ominus}$
HO
O PP
O PP
$-H^{\oplus}$

(*b*)

T
OH
(1) $SOCl_2$
(2) – HCl
T
trans elimination
defines stereo-
chemistry of T
CrO_3/acetone
T
O
aq. $OH^{\ominus}$
exchange
via enol
H
O
locates label
but does not
define stereo-
chemistry

Problem 6.1.

A two-step mechanism is probable.

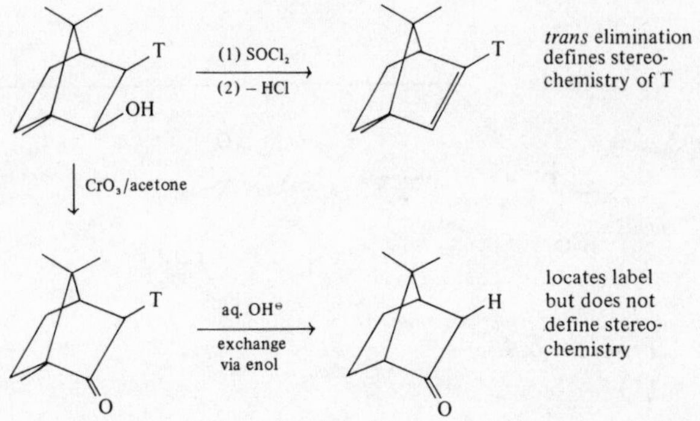

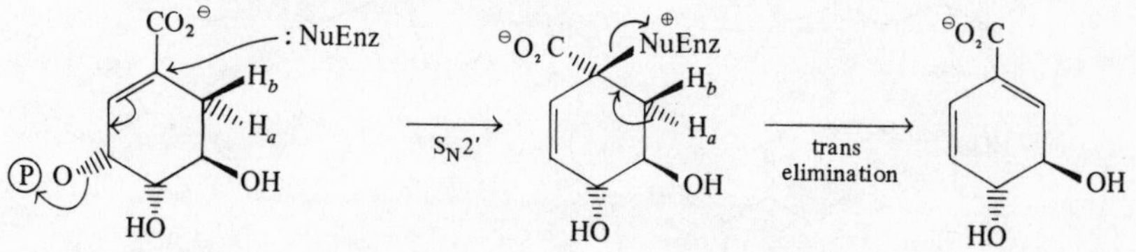

$CO_2^{\ominus}$
: NuEnz
H_b
H_a
P O
OH
HO
S_N2'
$^{\ominus}O_2C$
NuEnz
H_b
H_a
OH
HO
trans
elimination
$^{\ominus}O_2C$
OH
HO

Problem 6.2.

If reaction is concerted, β-anomer will form. If a cationic intermediate is involved, it could be stabilised by the C-2 OH and attack by nitrogen must again be β.

HO
O
O PP
HO OH
HO
O
HO OH
HO
O
HO
OH
HO
O
NH_2
HO OH

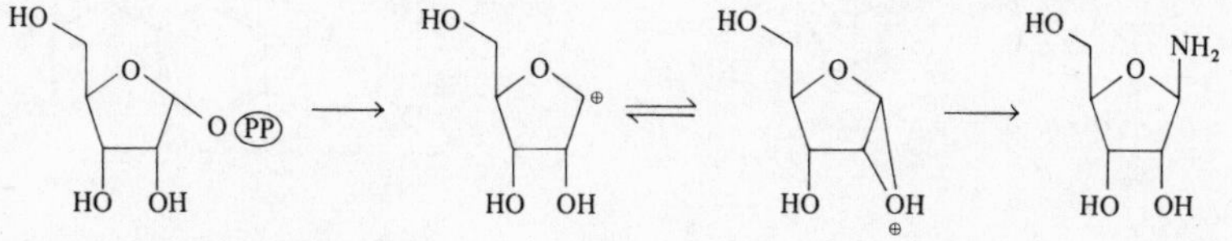

Problem 6.3.

(1) Different stereochemical courses of a reaction on the *same* substrate imply different mechanisms.

(2) α: oxonium ion, glycosyl enzyme; β: none (concerted) or oxonium ion.

(3) General base (β) o-nucleophilic catalyst (α).

(4)

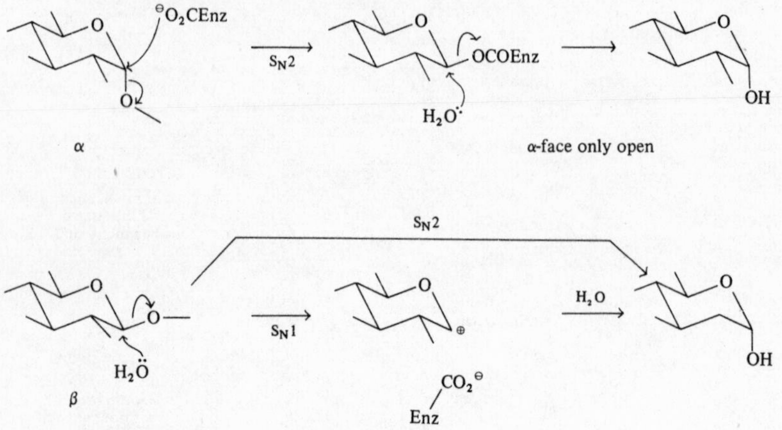

Problem 6.4.

(1)

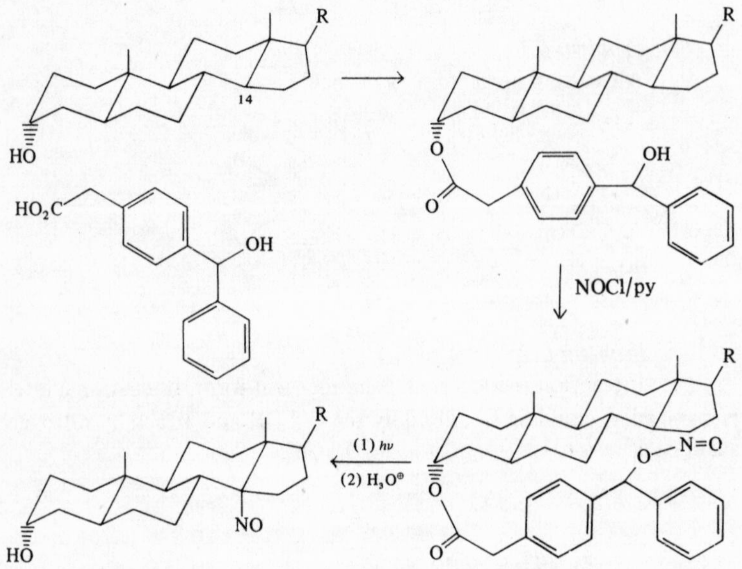

(2) Selective esterification of the less hindered alcohol. NO· is generated in just the right place for the residual alkoxy radical to abstract the 14-α-H as in Breslow's reactions.

Problem 7.1.

(*a*) Involves a primary carbocation, whereas (*b*) goes through a tertiary cation; (*b*) is therefore the more probable. Simple labelling of the starting material will not help because all atoms end up in the same relative positions by each mechanism. Therefore the first intermediates must be synthesised and fed. The correct mechanism will be shown by which intermediate is incorporated into the final product.

Problem 8.1.

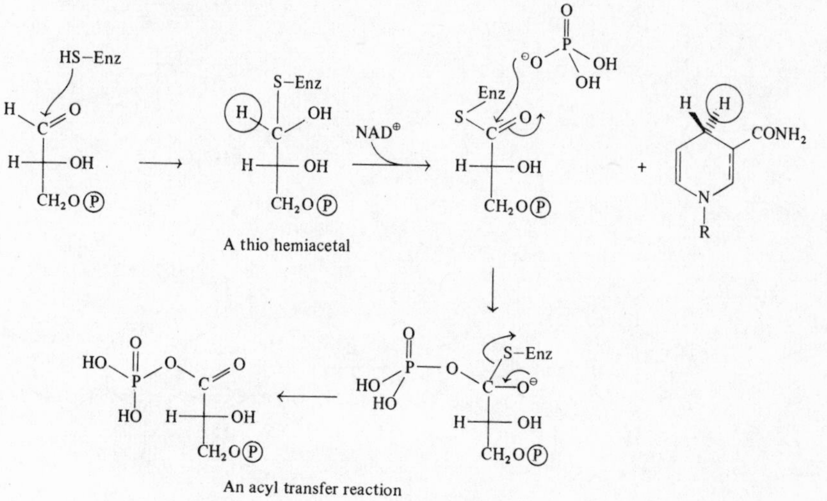

A thio hemiacetal

An acyl transfer reaction

Problem 8.2.

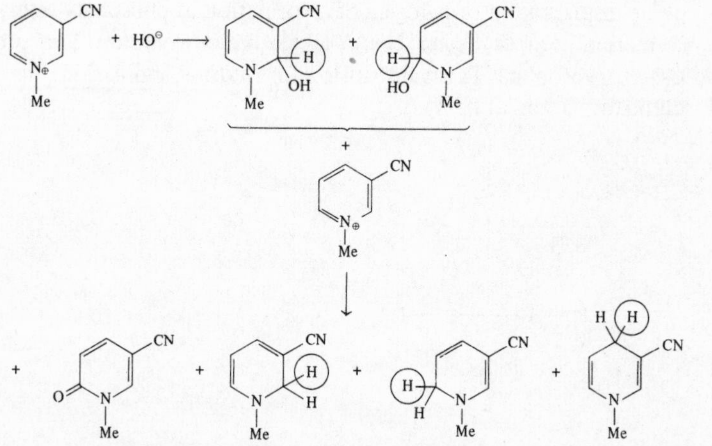

Problem 8.4.

(1) A carbanion α to the carbonyl group.

(2) By nucleophilic addition of the carbanion to N-5 of the flavin.

(3)

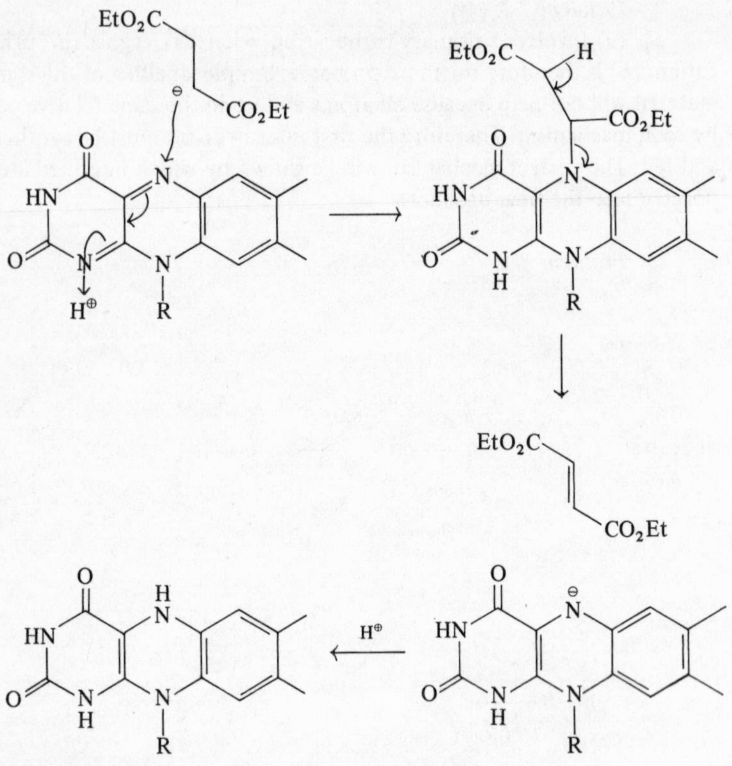

Problem 12.1.

Lys: amide formation, diazotisation, addition to C-2; Cys: nucleophilic displacement of halogen; Ser: formation of phosphate ester; Asp, Glu: formation of methyl ester; His: *N*-alkylation or acylation; Trp: substitution at C-3 of tryptophan; Tyr: substitution for fluorine, iodination; Met: nucleophilic displacement of X; Arg: e.g.

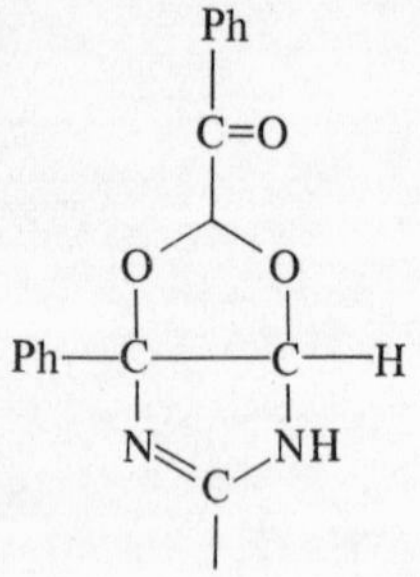

Problem 13.1.

A possible mechanism is shown in Figure 13.16. One oxygen atom is transferred to the phosphate and the other two remain in the carboxy group to appear in the product. Since if $C^{18}O_2$ were the substrate no ^{18}O would be found in the phosphate, this experiment supports the idea that bicarbonate is the substrate for carboxylation.

Figure 13.16.

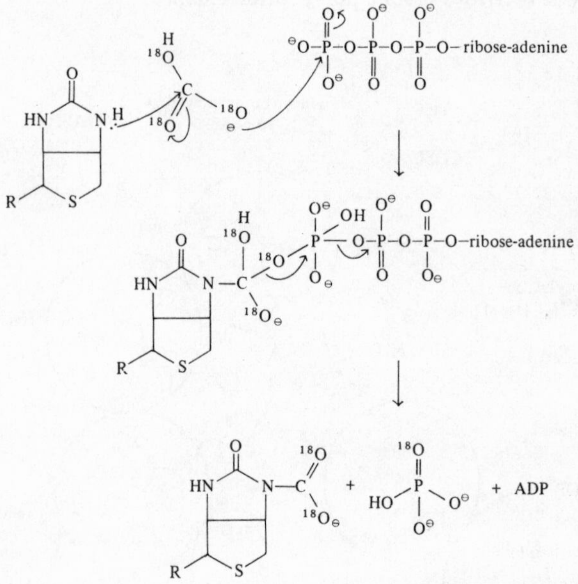

Problem 13.2.

The primary product is acetaldehyde, which is formed by the decomposition of acetyl thiamine pyrophosphate (Figure 13.17). Reduction of the aldehyde by a NAD-dependent alcohol dehydrogenase then affords the 'desirable' product, ethanol. (Compare Figure 4.26)

Figure 13.17.

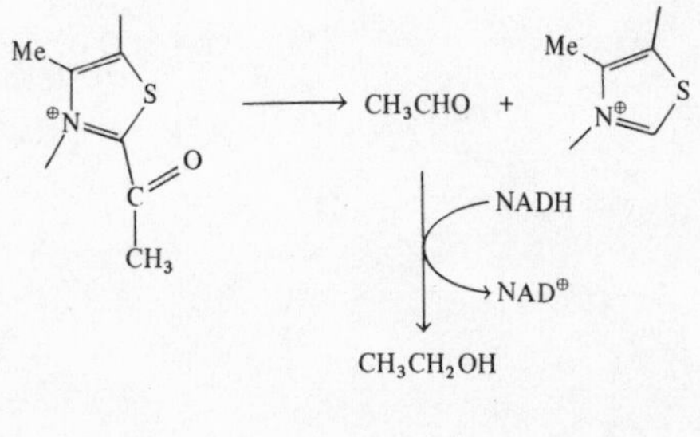

Problem 13.3.
Figure 13.18 shows a scheme of the fatty acid synthase complex.

(1) It consists of six catalytic subunits, the growing fatty acyl chain is transferred from one subunit to the next as a thiol ester by the acyl carrier protein. This has a similar function to the biotin in acetyl CoA carboxylase. In *E. coli* these subunits can be dissociated from each other, but the synthases from mammalian systems carry several catalytic activities on one polypeptide chain.

Figure 13.18.

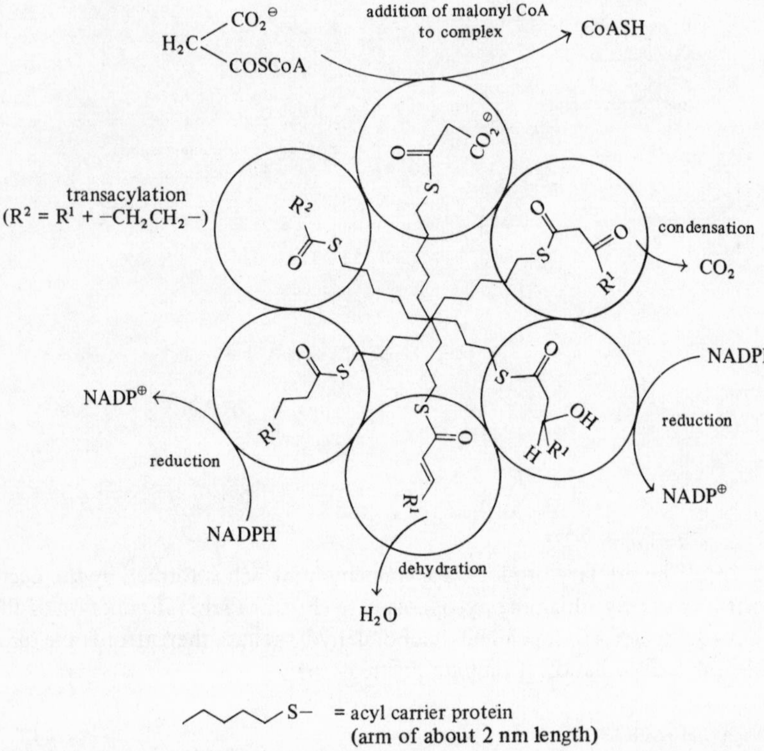

$S-$ = acyl carrier protein (arm of about 2 nm length)

(2) Two major possibilities are thought to be important in regulation: (*a*) The supply of acetyl CoA and malonyl CoA, the latter being determined by the activity of acetyl CoA carboxylase. Note that in order to make sensible suggestions on possible control mechanisms we must bear in mind the overall metabolic context of the system being examined. (*b*) There is some evidence for feed-back inhibition by the major product of the synthase, palmitoyl CoA.

Problem 13.2.
See Figure 4.25.

Problem 14.1.
The graph indicates that the ordering of the fatty acid chains decreases steadily as the bilayer is penetrated. We can interpret this as indicating that the fatty acyl chains of the phospholipids increase in their motion as the distance from the polar end of the bilayer increases. This change in mobility has been called the 'flexibility gradient'.

The steady increase in motion shown in Figure 14.12 is not shown by other magnetic resonance methods such as deuterium n.m.r., which suggests that the change is more sudden and not gradual. The reasons for this apparent discrepancy are not fully agreed and need not be discussed here, but the difference in the detailed result obtained by two very similar physical methods serves to illustrate how careful one must be when interpreting the results obtained by using relatively large extrinsic probes such as spin labels. No single technique can be relied on to give a complete answer to a biochemical problem.

Problem 15.2.
The drug binds to the N^+Me_3 binding site; on photolysis the azide reacts with a neighbouring bond, covalently linking the drug to the protein and blocking the binding site.

Problem 15.3.
Chlorpromazine: *N*-dealkylation in side-chain, *N*-oxidation in side-chain, oxidation of S to sulphoxide or sulphone, aromatic hydroxylation, scission of side-chain, conjugation of phenols with glucuronic acid or sulphate. Amphetamine: deamination, *N*-oxidation, *N*-dealkylation, aromatic hydroxylation, hydroxylation on β-carbon, conjugation with glucuronic acid.

Problem 15.4.

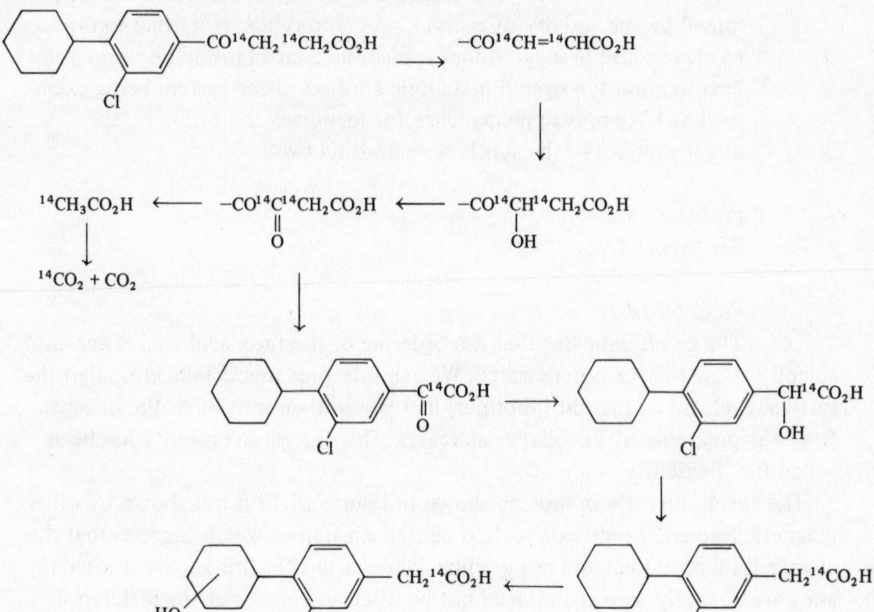

Problem 16.1.

Tyrosine is hydroxylated on C-3 to give dihydroxyphenylalanine (DOPA). This reaction is catalysed by a mixed-function oxidase, and thus requires reducing power and molecular oxygen. DOPA is then decarboxylated (pyridoxal phosphate) and dopamine, the resulting phenylethylamine hydroxylated on the benzylic carbon giving noradrenaline (another mixed-function oxidase). N-methylation (S-adenosylmethionine) leads to adrenaline.

Problem 16.2.

By inhibiting phosphodiesterase the methyl xanthines will cause the cAMP concentration to remain elevated once adenyl cyclase has been activated. This also takes place in nerve cells which are stimulated by adrenaline. The activity of glycogen phosphorylase in muscle and liver will remain high, allowing the abundant supply of glucose to tissues such as nervous tissue and brain which use glucose as the preferred energy source. The increased glucose supply and the activation of nerve cells may relate to the stimulative effect experienced from drinking tea or coffee.

Problem 16.3.

The pathway is thought to be that shown in Figure 16.12.

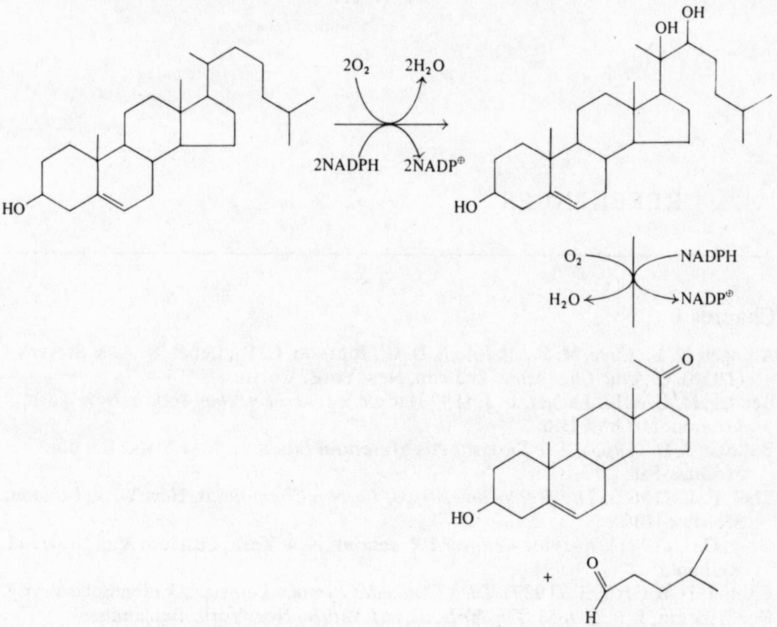

REFERENCES

Chapter 1

Allinger, N. L., Cava, M. P., de Jongh, D. C., Johnson, C. R., Lebel, N. A. & Stevens, C. L. (1976). *Organic Chemistry*, 2nd edn, New York: Worth.

Bender, M. L. & Brubacher, L. J. (1973). *Catalysis and Enzyme Action*, New York, London: McGraw Hill.

Bu'lock, J. D. (1965). *The Biosynthesis of Natural Products*. New York, London: McGraw-Hill.

Eliel, E. L. (1962). *The Stereochemistry of Carbon Compounds*. New York, London: McGraw-Hill.

Gray, C. J. (1971). *Enzyme Catalysed Reactions*. New York, London: Van Nostrand Reinhold.

Fischer, H. & Orth, H. (1937). *Die Chemie des Pyrrols*. Leipzig: Akademische Verlag.

Hendrickson, J. B. (1965). *The Molecules of Nature*. New York: Benjamin.

Metzler, D. E. (1977). *Biochemistry: Chemical Reactions of Living Cells*. New York, London, San Fransciso: Academic Press.

Mislow, K. (1966). *Introduction to Stereochemistry*, New York: Benjamin.

Morrison, R. T. & Boyd, R. N. (1973) *Organic Chemistry*, 3rd end. Boston: Allyn and Bacon.

O'Connor, C. (1970). *Quarterly Reviews*, 24, 533.

Powers, J. C., Baker, B. L., Brown, J. & Chalm, B. K. (1974). *Journal of The American Chemical Society*, 96, 238.

Pracejus, H., Kehlen, J., Kehlen, H. & Matschiner, H. (1965). *Tetrahedron*, 21, 2257.

Robinson, R. (1917a). *Journal of the Chemical Society*, 876.

Robinson, R. (1917b). *Journal of the Chemical Society*, 762.

Robinson, R. (1974). *Tetrahedron*, 30, 1477.

Scrimgeour, K. G. (1977). *Chemistry and Control of Enzyme Reactions*. New York, London, San Francisco: Academic Press.

Suckling, C. J., Suckling, K. E. & Suckling, C. W. (1978). *Chemistry Through Models*. Cambridge University Press.

Chapter 2

Hinkle, P. C. & McCarty, R. E. (1978). *Scientific American*, 238, 104–21.

Metzler, D. E. (1977). *Biochemistry: Chemical Reactions of Living Cells*. New York, London, San Francisco: Academic Press.

Suckling, C. J., Suckling, K. E. & Suckling, C. W. (1978). *Chemistry Through Models*. Cambridge University Press.

Chapter 3

Bender, M. L. & Brubacher, L. J. (1973). *Catalysis and Enzyme Action*, New York, London: McGraw-Hill.
Cooperman, B. S. & Hsu, C. M. (1976). *Journal of the American Chemical Society*, 98, 5652, 5657.
Gray, C. J. (1971). *Enzyme Catalysed Reactions*, New York, London: Van Nostrand Reinhold.
Jencks, W. P. (1969). *Catalysis in Chemistry and Enzymology*, New York, London: McGraw-Hill.
Suckling, C. J., Suckling, C. W., & Suckling, K. E. (1978). *Chemistry Through Models*. Cambridge University Press.
Tedder, J. M., Nechvatal, A., Murray, A. W. & Carnduff, J. (1972). *Basic Organic Chemistry*, Part 4. New York, London: Wiley.

Chapter 4

Arigoni, D., Rétey, J. & Lüthy, J. (1969, 1970), Nature, *London*, 221, 1213: 226, 5245.
Battersby, A. R. & Staunton, J. S. (1974). *Tetrahedron*, 30, 1707.
Bender, M. L. & Brubacher, L. J. (1973). *Catalysis and Enzyme Action*. New York, London: McGraw-Hill.
Bruice, T. C. & Benkovic, S. J. (1966). *Biorganic Mechanisms*, Vol. 1. New York: Benjamin.
Cornforth, J. W. (1976), *Science*, 193, 121.
Cornforth, J. W., Redmond, J. W. Eggerer, H., Buckel, W. & Gutschow, C. (1969). *Nature, London*, 211, 1212.
Eliel, E. L. (1962). *The Stereochemistry of Carbon Compounds*, Chapter 3. New York, London: McGraw-Hill.
Jencks, W. P. (1969). *Catalysis in Chemistry and Enzymology*, p. 125. New York, London; McGraw-Hill.
Joule, J. A. & Smith, G. F. (1978). *Heterocyclic Chemistry*, 2nd edn, New York, London: Van Nostrand Reinhold.
Lynen, F. (1975). *European Journal of Biochemistry*. 55, 561.
Mahler, H. R. & Cordes, E. H. (1971). *Biological Chemistry*, 2nd edn, New York, London: Harper and Row.
Pearson, R. G. (1967). *Chemistry in Britain*, 3, 103.

Chapter 5

Banthorpe, D. V. & Baxendale, D. (1970). *Journal of the Chemical Society*, C, 2694.
Barton, D. H. R., Mellows, G., & Widdowson, D. A. (1971). *Journal of the Chemical Society*, C, 110.
Battersby, A. R. Wightman, R. H. Staunton, J. & Hanson, K. R. (1972). *Journal of the Chemical Society, Perkin Transactions 1*, 2355.
Battersby, A. R. & Staunton, J. (1974), *Tetrahedron*, 30, 1707.
Battersby, A. R., McDonald, E. Wurziger, H. K. W. & James K. J. (1975). *Chemical Communications*, 493.
Brown, S. A. (1972). *Biosynthesis Specialist Periodical Reports*, 1, 19.
Bruice, T. C. & Benkovic, S. (1966). *Biorganic Mechanisms*, vol. 1, pp. 300, 319. New York: Benjamin.
Eliel, E. L. (1962). *The Stereochemistry of Carbon Compounds*, New York, London: McGraw-Hill.
Hill, R. L. & Teipel, J. W. (1971). In *The Enzymes*, ed. P. D. Boyers, 3rd edn, vol. 5, p. 556. London, New York, San Francisco: Academic Press.
Johnson, W. S., Gravestock, M. B. & McCarry, B. E. (1971). *Journal of the American Chemical Society*, 93, 4332.
Knowles, P. F., Marsh, D. & Rattle, H. W. E. (1976). *Magnetic Resonance of Biomolecules*. New York, London: Wiley.

Mislow, K. (1966). *An Introduction to Stereochemistry.* New York: Benjamin.
Mulheirn, L. J. & Ramm, P. J. (1972). *Chemical Society Reviews,* 1, 259.
Nigh, W. G. & Richards, J. H. (1969). *Journal of the American Chemical Society,* 91, 5847.
Olah, G. A. (1973). *Chemistry in Britain,* 9, 281.
Robinson, R. (1974). *Tetrahedron,* 30, 1477.
Ruzicka, L., Eschenmoser, A., Jeger, O. & Arigoni, D. (1955). *Helvetica Chimica Acta,* 38, 1890.
Sammes, P. G. (1971). *Quarterly Reviews,* 25, 135.
Scharf, K. H., Zenk, M. H., Onderka, D. K., Carroll, M. & Floss, H. G. (1971). *Chemical Communications,* 765.
Scott, A. I. (1976). In *Techniques of Organic Chemistry,* vol. X, part II, ed. J. B. Jones, C. J. Sih & D. Perlman, New York, London: Wiley-Interscience.
Simpson, T. J. (1975). *Chemical Society Reviews,* 4, 497.
Stork, G. & Burgstrahler A. W. (1955). *Journal of the American Chemical Society,* 77, 5068.
Tedder, J. M., Nechvatal, A., Murray, A. W. & Carnduff, J. (1972). *Basic Organic Chemistry,* Part 4. New York, London: Wiley.
Willadsen, P. & Eggerer, H. (1975). *European Journal of Biochemistry,* 54, 247.

Chapter 6

Allinger, N. L., Cava, M. P., De Jongh, D. C. Johnson, C. R., Lebel, N. A. & Stevens, C. L. (1976). *Organic Chemistry,* 2nd edn. p. 372. New York; Worth.
Battersby, A. R., Kelsey, J. E., Staunton, J. & Suckling, K. E. (1973). *Journal of the Chemical Society, Perkin Transactions 1,* 1609.
Breslow, R. (1972). *Chemical Society Reviews,* 1. 553.
Breslow, R. Wife, R. L. & Prezant, D. (1976). *Tetrahedron Letters,* 1925.
Cornforth, J. W. (1969). *Quarterly Reviews,* 23, 125.
Gray, C. J. (1971). *Enzyme Catalysed Reactions,* p. 246, New York, London: Van Nostrand Reinhold.
Gunsalus, I. C., Pedersen, T. C. & Sligar, S. G. (1975). *Annual Reviews of Biochemistry,* 44, 377.
Hesse, R. (1969). *Advances in Free Radical Chemistry,* 83.
House, H. O. (1972). *Modern Synthetic Reactions,* 2nd edn, chapters 7 and 8. New York: Benjamin.
Kaufman, S., Bachan, L., Storm, C. B. & Wheeler, J. W. (1974). *Journal of the American Chemical Society,* 94, 6799.
Morrison, R. T. & Boyd, R. N. (1973). *Organic Chemistry,* 3rd edn, p. 459. Boston: Allyn & Bacon.
Nonhebel, D. C. & Walton, J. C. (1974). *Free-Radical Chemistry,* Cambridge University Press.
Nonhebel, D. C., Tedder, J. M. & Walton, J. C. (1979). *Radicals.* Cambridge University Press.
Tedder, J. M., Nechvatal, A., Murray, A. W. & Carnduff, J. (1972). Basic Organic Chemistry, Part 4, pp. 224, 253. New York, London: Wiley.

Chapter 7

Akhtar, M. & Wilton, D. C. (1973). *Annual Reports on the Progress of Chemistry,* B, 69, 146.
Barton, D. H. R. & Kirby, G. W. (1962). *Journal of the Chemical Society,* 806.
Battersby, A. R., McDonald, E., Williams, D. C. & Wurziger, H. K. W., (1977). *Journal of the Chemical Society, Chemical Communications,* 113.
Joule, J. A. & Smith, G. S. (1978). *Hetereocyclic Chemistry.* New York, London: Van Nostrand Reinhold.

Lang, T., Suckling, C. J. & Wood, H. C. S. (1977). *Journal of the Chemical Society,
Perkin Transactions 1*, 2189.
Nonhebel, D. C. & Walton, J. C. (1974). *Free-Radical Chemistry*, pp. 341, 417. Cambridge
University Press.
Simpson, T. J. (1975). *Chemical Society Reviews*, 4, 497.
Schwartz, M. A. & Holton, R. A. (1970). *Journal of the American Chemical Society*, 92,
1090.
Suckling, C. J. (1977). *Chemical Society Reviews*, 6, 215.
Tedder, J. M., Nechvatal, A., Murray, A. W. & Carnduff, J. (1972). *Basic Organic Chemistry*, part 4. New York, London: Wiley.

Chapter 8

Bruice, T. C. (1976). *Progress in Biorganic Chemistry*, 4, 1.
Dunn, M. F. (1975). *Structure and Bonding*, 23, 89.
Hamilton, G. A. (1971). *Progress in Biorganic Chemistry*, 1, 84.
Hemmerich, P. (1976). *Fortschmitte der Chemie Organische Naturstoffe*, 33, 451.
Joule, J. A. & Smith, G. S. (1978). *Heterocyclic Chemistry*. New York, London: Van
Nostrand Reinhold.
Staunton, J. (1978). *Primary Metabolism*, pp. 66–70. Oxford University Press.

Chapter 9

Breslow, R. & Schmir, M. (1971). *Journal of the American Chemical Society*, 93, 4960.
Breslow, R. & Khanna, P. L. (1976). *Journal of the American Chemical Society*, 98, 1297.
Cohen, C. (1975). *Scientific American*, 233, 36.
Coleman, J. E. (1974). In *MTP International Review of Science. Biochemistry*, vol. 1,
p. 185. ed. H. Gutfreund, Oxford, London: Butterworth.
Cotton, F. A. & Wilkinson, G. (1972). *Advanced Inorganic Chemistry*. New York: Wiley
Interscience.
Dunn, M. F. (1975). *Structure and Bonding*, 23, 103.
Fenton, D. E. (1977). *Chemical Society Reviews*, 6, 325.
Furhop, J.-H. (1974). *Structure and Bonding*, 18, 30.
Holm, R. H., Koch, S., Tang, S. C., Frankel, R. B. & Ibers, J. A. (1975). *Journal of the
American Chemical Society*, 97, 916.
Hughes, M. N. (1972). *The Inorganic Chemistry of Biological Processes*. New York,
London: Wiley-Interscience.
Kadish, K. M., Morrison, M. M., Constant, L. A., Dickens, L. & Davies, D. G. (1976).
Journal of the American Chemical Society, 98, 8387.
Kassner, R. J. (1972). *Proceedings of the National Academy of Sciences, USA*, 69, 2263.
Lindskog, S. (1970). *Structure and Bonding*, 8, 153.
Lipscomb, W. N. & Quiocho, F. A. (1971). *Advances in Protein Chemistry*, 25, 1.
Perutz, M. F. (1978). *Scientific American*, 239, 68.
Perutz, M. F., Muirhead, H., Cox, J. M. & Goaman, L. C. G. (1968). *Nature, London*,
219, 131.
Pratt, J. M. (1975). In *Techniques and Topics in Bioinorganic Chemistry*, p. 113, ed. C. A.
McAuliffe. London: Macmillan.
Schrauzer, G. N. (1977). *Angewandte Chemie International Edition*, 15, 417.
Williams, D. R. (1971). *The Metals of Life*. New York, London: Van Nostrand Reinhold.
Williams, R. J. P. (1970). *Quarterly Reviews*, 24, 331.
Williams, R. J. P. (1978). *Chemistry in Britain*, 14, 25.
Williams, R. J. P. & Moore, G. R. (1977). *FEBS Letters*, 79, 229.
Williams, R. J. P. & Wacker, W. E. (1967). *Journal of the American Medical Association*,
201, 18.

Chapter 10

Bender, M. L. & Brubacher, L. J. (1973). *Catalysis and Enzyme Action*. New York,
London: McGraw-Hill.

References 374

Brown, J. M. & Bunton, C. A. (1974). *Chemical Communications*, 969.
Bürgi, H. R., Dunitz, J. D., Lehn, J. M. & Wipff, G. (1974). *Tetrahedron*, 30, 1563.
Jencks, W. P. (1969). *Catalysis in Chemistry and Enzymology*. New York, London: McGraw-Hill.
Klotz, I. M., Royer, G. P. & Scarpa, I. S. (1971). *Proceedings of the National Academy of Sciences, USA*, 68, 263.
Knowles, J. R. & Alberry, W. J. (1977). *Accounts of Chemical Research*, 10, 105.

Chapter 11

Butler, P. J. G. & Klug, A. (1978). *Scientific American*, 239, 40.
Craigmyle, M. B. L. (1975). *A Colour Atlas of Histology*. London: Wolfe.
d'Albis, A. & Gratzer, W. B. (1974). In *Companion to Biochemistry*, ed. A. Bull *et al*. London: Longman.
Fiddes, J. C. (1977). *Scientific American*, 237, 54.
Gould, H. & Mathews, H. R. (1976). In *Laboratory Techniques in Biochemistry and Molecular Biology*, vol. 4, ed. T. S. Work & E. Work, Amsterdam: North-Holland
Griffith, O. M. (1976). *Techniques of Preparative, Zonal and Continuous Flow Ultracentrifugation*, Palo Alto: Beckman Instruments Inc.
Hall, D. O. (1977). *Trends in Biochemical Sciences*, 2, 99.
Heil, A. Müller, G., Noda, L., Pinder, T., Schirmer, H., Schirmer, I. & von Zabern, I. (1974). *European Journal of Biochemistry*, 43, 131.
Herzenberg, L. A., Sweet, R. G. & Herzenberg, L. A. (1977). *Scientific American*, 234, 108.
Huber, R. (1979). *Nature, London*, 230, 538.
Knowles, P. F., Marsh, D. & Rattle, H. W. E. (1976). *Magnetic Resonance of Biomolecules*, New York, London: Wiley.
McCammon, J. A., Gelin, B. R. & Karplus, M. (1977). *Nature, London*, 267, 585.
Raff, M. C. (1976). *Scientific American*, 234, 30.
Spiro, T. G. & Gaber, B. P. (1977). *Annual Review of Biochemistry*, 46, 583.
Thomas, J. O. (1974). In *Companion to Biochemistry*, ed. A. Bull *et al.*, London: Longman.
Unwin, P. N. T. (1977). *Journal of Molecular Biology*, 114, 491.
Van Holde, K. E. (1971). *Physical Biochemistry*. Englewood Cliffs, N.J.: Prentice-Hall.
Weakley, B. S. (1972). *A Beginner's Handbook in Biological Electron Microscopy*. Edinburgh: Churchill Livingstone.
Wu, R. (1978). *Annual Review of Biochemistry*, 47, 607.

Chapter 12

Cleland, W. W. (1975). *Accounts of Chemical Research*, 8, 145.
Ferdinand, W. (1976). *The Enzyme Molecule*. New York, London: Wiley.
Halford, S. E. (1974). In *Companion to Biochemistry*, ed. A. Bull *et al.* London: Longman.
Jones, S. R., Kindman, L. A. & Knowles, J. R. (1978). *Nature, London*, 275, 564.
Lowe, G. & Sproat, B. S. (1978). *Journal of the Chemical Society, Chemical Communications*, 783.
Mildvan, A. S. (1977). *Accounts of Chemical Research*, 10, 246.
Milner-White, E. J. & Watts, D. C. (1971). *Biochemical Journal*, 122, 727.
Price, N. C., Cohn, M. & Schirmer, R. A. (1975). *Journal of Biological Chemistry*, 250, 644.
Roberts, D. V. (1977). *Enzyme Kinetics*. Cambridge University Press.
Rose, I. A. (1970). *Journal of Biological Chemistry*, 247, 1096.
Schulz, G. E., Elzinga, M., Marx, F. & Schirmer, R. H. (1974). *Nature, London*, 250, 120.
Steitz, T. A., Fletterick, R. J., Anderson, W. F. & Anderson, C. M. (1976). *Journal of Molecular Biology*, 104, 197.
Sternberg, M. J. E. & Thornton, J. M. (1978). *Nature, London*, 271, 15.

Thomas, J. O. (1974). In *Companion to Biochemistry*, ed. A. Bull *et al.*, p. 87. London: Longman.
Wimmer, M. J. & Rose, I. A. (1978). *Annual Review of Biochemistry*, 47, 1072.

Chapter 13

Bates, D. L., Dawson, M. J., Hale, G., Hooper, E. A. & Perham, R. N. (1977). *Nature, London*, 268, 313.
Ferdinand, W. (1976). *The Enzyme Molecule*. New York, London: Wiley.
Huber, R. & Bode, W. (1977). In *NMR in Biology*, ed. R. A. Dwek *et al.*, London, New York, San Francisco: Academic Press.
Lehninger, A. L. (1975). *Biochemistry*, 2nd edn, New York: Worth.
Obermayer, M. & Lynen, F. (1976). *Trends in Biochemical Sciences*, 1, 169.
Perutz, M. F. (1978). *Scientific American*, 239, 68.
Reed, L. J. (1974). *Accounts of Chemical Research*, 7, 40.
Roberts, D. V. (1977). *Enzyme Kinetics*, Cambridge University Press.
Srere, P. A., Mattiason, B. & Mosbach, K. (1973). *Proceedings of the National Academy of Sciences, USA*, 70, 2534.
Wood, H. G. (1976). *Trends in Biochemical Sciences*, 1, 4.

Chapter 14

Colley, C. M. & Ryman, B. E. (1976). *Trends in Biochemical Sciences*, 1, 203.
Cullis, P. R. & McLaughlin, A. C. (1977). *Trends in Biochemical Sciences*, 2, 196.
Henderson, R. & Unwin, P. N. T. (1975). *Nature, London*, 245, 249.
Korn, E. D. (1966). *Science*, 153, 1491.
Lee, A. G. (1975). *Progress in Biophysics and Molecular Biology*, 29, 3.
Singer, S. J. & Nicolson, G. L. (1972). *Science*, 175, 720.
Zwaal, R. F. A., Demel, R. A., Roelofsen, B. & van Deenen, L. L. M. (1976). *Trends in Biochemical Sciences*, 1, 112.

Chapter 15

Albert, A. (1973). *Selective Toxicity*, 5th edn, London: Chapman & Hall.
Baker, B. R. (1969). *Accounts of Chemical Research*, 2, 127.
Berkoff, C. E., Cramer, R. D. & Redl, G. (1974). *Chemical Society Reviews*, 3, 273.
Evans, R. M. (1965). *The Chemistry of the Antibiotics Used in Medicine*. Oxford: Pergamon.
Goldstein, A., Aronow, L. & Kalman, S. M. (1974). *Principles of Drug Action.* New York, London: Wiley.
Guroff, G., Daly, J. W., Jerina, D. M., Renson, J., Witkop, B. & Udenfriend, J. (1967). *Science*, 157, 1524.
Hansch, C. (1969). *Accounts of Chemical Research*, 2, 252.
Hawkins, D. R. (1976). *Chemistry in Britain*, 12, 379.
Korolkovas, A. & Burckhalter, J. H. (1976). *Essentials of Medicinal Chemistry*. New York, London: Wiley.
Lowe, G. & Lewis, D. J. (1973). *Chemical Communications*, 713.
Parke, D. V. (1973). *Chemistry in Britain*, 9, 102.
Pratt, W. B. (1973). *Fundamentals of Chemotherapy*. Oxford University Press.
Robinson, F. A. (1974). *Chemistry in Britain*, 12, 129.
Suckling, C. J., Suckling, C. W. & Suckling, K. E. (1978). *Chemistry through Models*. Cambridge University Press.
Williams, R. T. & Millburn, P. (1975). *MTP International Review of Science*, Biochemistry, Series 1, vol. 12, p. 211, Oxford, London: Butterworth.
Wolfenden, R. (1972) *Accounts of Chemical Resarech*, 5, 10.

Chapter 16

Adler, J. (1976). *Scientific American*, **234**, 40.
Capra, J. D. & Edmundson, A. B. (1977). *Scientific American*, **236**, 50.
Cohen, P. (1976*a*). *Control of Enzyme Activity*. London: Chapman & Hall.
Cohen, P. (1976*b*). *Trends in Biochemical Science*, **1**, 38.
Evans, D. A. & Green, C. L. (1973). *Chemical Society Reviews*, **2**, 75.
Fletterick, R. J. & Madsen, N. B. (1977). *Trends in Biochemical Sciences*, **2**, 145.
Gettins, P. & Dwek, R. A. (1977). In *NMR in Biology*, ed. R. A. Dwek *et al.* New York, London, San Francisco: Academic Press.
Greaves, M. F. (1977). *Nature, London*, **265**, 681.
Katanuma, N., Kominami, E., Banno, Y., Kito, K., Aoki, Y. & Urata, G. (1976). *Advances in Enzymology*, **14**, 325.
Lester, H. A. (1977). *Scientific American*, **236**, 106.
O'Malley, B. W. & Schrader, W. T. (1976). *Scientific American*, **234**, 32.
Newell, P. C. (1977). *Endeavour* (New Series), **1**, 63.
Porter, R. R. & Reid, K. B. M. (1978). *Nature, London*, **275**, 699.
Rabbitts, T. H. (1976). *Trends in Biochemical Sciences*, **1**, 86
Raff, M. C. (1976). *Scientific American*, **234**, 30.
Watson, J. D. (1975). *The Molecular Biology of the Gene*, 3rd edn, New York: Benjamin.

INDEX